A Rainbow Book

THE COMPLETE GUIDE TO FLIGHT INSTRUCTION

GREGORY M. PENGLIS

RAINBOWS BOOKS, INC.

The Complete Guide to Flight Instruction
by Gregory M. Penglis
Cover and Interior Design by Marilyn Ratzlaff
Cover Photo: Ken Cowan
$29.95
ISBN # 1-56825-012-6

Library of Congress Cataloging-in Publication Data

Penglis, Gregory M., 1959-
    The complete guide to flight instruction / Gregory M. Penglis.
        p.   cm.
    Includes index.
    ISBN 1-56825-012-6   : $29.95
    1. Flight training.
TL712.P46   1994
629.132'52--dc20   93-38698
                                                              CIP

DEDICATION

Dedicated to anyone
who has ever given
or taken a flight lesson.

ACKNOWLEDGMENTS

To my dear wife Leslie,
    without whose tireless energy, help, work,
    inspiration, and support this book would have
    remained simply an idea.

To my dear friend Doug,
    who generously let me borrow his computer and
    taught me how to use it, without which I would
    still be typing on a 1940 Smith Corona manual.

CONTENTS

1 • DREAMING OF FLIGHT

No one knows how the bug bites; where the desire to break the earthly bond comes from. All we know is that sometime in a quiet moment of ecstasy, those of you destined to be pilots will gaze skyward at a shimmering silver flash, your heart and mind will bond in one unique vision as you softly whisper, "I can do that."

Maybe you will be at your desk, behind a keyboard performing accounting, when your eyes take you out a small window to a small airplane winging into a local field. You may think back to your childhood, when life was simple and choices were easy. You remember a time when life was filled with possibilities, never limitations. Can you recall the purity of a child's dream of flight? Can you challenge that vision, make the dream come alive, and risk becoming a pilot? If you dare to try and attain a privilege few will achieve, you have taken the first step in a long and fascinating journey.

Along the way this journey contains pitfalls and successes, frustrations and exaltations. You will love it and you will hate it. There will be many times when nothing else in the world matters, and times when you could easily throw in the towel. There will be times when you will doubt your ability to do anything as your self-worth crumbles in humiliation. There will be times when you will rise and soar with such pride that immortality itself seems tepid by comparison. You will also be drenched in the cold sweat of sheer terror. Your boundaries are the full range of human emotions.

None of this will be apparent as you gaze out the

window vicariously content with the grace of that little airplane. Your dream is steeped in utter fantasy. You do not know, what you do not know. You cannot know until you take the journey. The world is filled with dreamers who nourish ignorant shreds of vision their whole life. How often have you heard people say "I have always wanted to learn to fly?" They never will. They are prisoners of their ignorant vision, captivated by the dream with no desire to act on it. If you dare to act on your dream; if you dare to learn to fly; if you want to take all or any part of this journey; then come with me. Just beyond the long path lies a lifetime of flight.

Welcome to a flight training perspective unavailable from any other source. On the dark side you won't find any sugar coating. You will be exposed to the raw essence of the world of flight training. Armed with the knowledge and experience of those who have gone before you, you can choose the best path for your journey. My goal is that you will never have to say "if I only knew then what I know now." A potential student with money, whose defenses consist solely of blind optimism, is a lost lamb in a meadow of hungry wolves. The world of flight training you dream of in such idyllic splendor does not exist. So if you are going to have any success in your endeavors, you have to have your eyes wide open. My job is to open those eyes.

Okay, so you have this dream, now what? Well, you have to take that dream in your head and turn it into a possibility. Before you can learn to fly, you have to believe that you can learn to fly. To do that you need some time. You must eat, sleep and breathe this dream. You must try it out, see how it fits. This is a normal process. How much time this process takes depends entirely on you. No one, however, wakes up one morning, looks out the window, and says, "What shall I do today? I know, I'll learn how to fly."

Certain symptoms of this process will allow you to monitor your progress. For example, do thoughts of flight progressively invade your daydreams? Does your mind wander through past air shows when you should be concentrating on work? Can you envision yourself tossing your job for a career in aviation? These are all great indications. However, there is one dead giveaway that you will be flying soon. Do you race your car down the highway faster and faster, pulling harder and harder on the steering wheel, hoping to take off? Better get in an airplane before you have a serious problem on the ground.

The more you convince yourself that you have the ability to fly, the less the dreams will suffice. You are going to have to act on your dream. Soon you will not be able to imagine not following through. As thoughts of flight dominate every aspect of your life, you move from believing flight is something you merely want to try, into something you absolutely have to do. Some would label this an obsession. To me, this is just the drive necessary to complete the journey. By convincing yourself that you can't imagine not flying, through the back door you will believe that you have the ability. You have now taken flight from a dream to a possibility. The process is a bit roundabout, but it works.

A word to parents of budding pilots. Young people take a faster route through this process because it would never occur to them that they wouldn't have the ability, and therefore the confidence, to do anything. Shortly after they learn how to read, kids will memorize all the statistics on every airplane listed in those big, colorful books you have. This is a ploy — what they really want are lessons. Since they can't fly, this is the next best thing. I started my attack at the age of five and drove my parents nuts until they finally relented and gave me my first flight lesson at twelve. Desperately hoping to scare away my adamant desire to fly, it only confirmed what I already knew. I had to fly. The dreams of children are as valid as of any adult. As parents, you are the custodians of those dreams. How you handle that responsibility is up to you.

As adults, you can do it on your own without even telling a soul, but you won't. Like any good dreamer, the first thing you must do is share your dream. This brings about your first pitfall — criticism. You know you want to fly. However human nature requires that you receive recognition and validation of your dream from those you care about . . . "You want to do what? That's dangerous! You wouldn't catch me up in one of those things." You asked for it. All you want is a little reinforcement. What you get is your dream as seen through eyes other than yours. People always look at things from their own perspectives, not yours. Only a pilot can understand your dream. The best you can hope for is someone close who respects your dream because they respect you. Eventually you will find the person who says, "I think it's a wonderful idea for you to learn to fly." With that out of your system, you may move on.

Sometimes, though, parents, spouses, friends, relatives, people you sleep with for a while, bosses, business partners, in fact anyone with a vested interest in you can set up road blocks. Ostensibly for safety reasons, these blocks are usually rooted in petty and selfish jealousies. For those who must convince significant persons of the safety of flight, I have provided some statistical ammunition. In 1988, according to the Department of Transportation, 805 people died in general aviation accidents. In that same year, 47,093 people died in automobile accidents. The American Cancer Society tells me that 390,000 people die every year because of smoking. The public perception that aviation is dangerous is because of Hollywood, and the media's incredible sensationalism of any accident. All of life carries some risk. Flying carries some of the least risk of any activity. If you have to fly, you have already accepted the risk. Those close to you may not be so willing. If flying is to be your passion, use whatever information it takes to remove your road blocks.

To the professional pilot, flying is like breathing; they would die without it; if not physically, certainly spiritually. All of us who have made flying our profession started out as dreamers. We took that dream and made it a reality. If you are ready to live your wildest dreams, you are ready to begin flight training.

2 • SELECTING A FLIGHT SCHOOL

➤

You have set yourself up. Everyone you know is sick of hearing that someday you intend to get a pilot certificate. Once again we move through the back door. When you have doubt your dream of flight by itself is not enough to get you into the air. This is why you put yourself in the position of having to live up to the expectations you created in those you care about. First-time skydivers have this down to an art form. After working up the courage to make the appointment to jump, they close off the possibility of backing out by telling anyone who will listen, about the big day. Although they can live with themselves if they don't jump, they cannot live with the disappointment of their friends; so they jump. Interesting motivation. A lot of people start flying pretty much the same way.

The incredible ignorance with which people plunge into flying is staggering. Folks who would normally shop around, willingly put their lives and fortunes in the hands of the first nice instructor they see. Flight instructors are more than eager to take the money. Woefully underpaid, working upwards of seven days a week and available sixteen hours a day, they have to take any business they can get just to scrape out a living. Flight instruction is nothing more than a business. The owners and upper management of the flight schools count on your ignorance and romantic delusions to separate you from your money. The more they can get you to pay for your training, the greater their windfall. The profit comes from the difference between the actual operating cost of the aircraft, compared to the rental rate; and the money that goes to the instructor, compared to how much they charge for instruction. The worst example I ever saw of gouging was one school where the instruction rate was $45 per hour, of which the instructors received between $6-8 depending on their status.

Most instructors are aspiring commercial pilots who are really just apprentices building flight time to fulfill Federal Aviation Administration (FAA) requirements for future jobs. To qualify for those jobs takes far more flight hours than it does to get the ratings. The only way to get the hours is to become an instructor and attract enough students to pay for them. Those students the instructor brings in who desire a career in aviation will become instructors themselves and do the same thing. Flight instruction is a pyramid scheme.

You have an incredible variety of options from which to choose your training. You may hitch up with a freelance instructor. You could go with a full four-year university degree and flight program. You could end up anywhere in between; it all depends on your needs. There are advantages and disadvantages to every type of training. As you investigate various facilities, there is one thing that you must keep in mind. The FAA requires that no matter where or how you train, every pilot must cover the same material, learn the same procedures and fly to the same standards. Since the FAA issues all pilot certificates, there is absolutely no difference in the value of a certificate earned from training at a major academy or from a freelance instructor operating off a dirt strip. If you meet the standard of the certificate, you may exercise the privileges of the certificate. Anything about a flight school that is not directly related to the training requirements spelled out in the FAA regulations, is nothing but sales fluff.

You must identify the training environment that most closely meets your needs, separate out the sales fluff, and find the best facility in that environment. A potential student pilot's perception of flight training is usually so distorted that his major decisions are based totally on the fluff rather than on an accurate evaluation of the quality of training available. To identify the best environment to train, ask yourself some basic questions:

Are you going to be a recreational pilot or do you desire a career?
Will you want to train full-time or are you content to fly leisurely on weekends?
Will the bulk of your flying keep you in rural areas, congested urban airspace, or a healthy mixture of both?
What environment will best suit your needs?

Let's debunk some myths about flight training. You can get full-time intensive flight instruction from small flying clubs. You can take a leisurely pace through a professional flight academy. That old flight instructor perched in the corner wearing a ragged flannel shirt, may also be an airline pilot with 20,000 hours experience. That eager young face in the shiny new uniform complete with glittering epaulets may have only yesterday earned his certificate to teach. You don't know. Appearances in this business mean nothing. Despite the clear difference in experience, the eager young face may belong to the better instructor. As you wade through the sales propaganda and ask yourself never-ending questions, please remember this truth: The relationship you have with your instructor and the ability of that instructor to teach you are the greatest factors in the successful outcome of your training. Your challenge, therefore, is to find the perfect instructor for you, in your chosen training environment.
Let's explore some options.

INDEPENDENT FREELANCE INSTRUCTORS
There are independent freelance instructors who operate from a phone answering machine. With virtually no overhead, they can offer some of the best rates around. You are best off with a freelancer if you personally know the instructor or have a really good recommendation, you are

an advanced student, or you own your own airplane. When an instructor is affiliated with a club or school, there is a peer review and support system that does not exist for the independent operator. There is no quality control for the freelancer and the latest information and techniques can take a while to filter down.

FLYING CLUBS

The next step up is the flying club. With little in the way of facilities (my club operates from a trailer), the overhead is low, so the costs are kept to reasonable levels. You have to be a member of the club to receive instruction. Membership more closely controls who flies the club aircraft which contributes to lower insurance rates. The money you pay for training goes directly for the aircraft, the instructor, and a little to keep the club going. Clubs are relaxed places to learn to fly. However, with a willing instructor, you can make your training as intensive as you want. The variety, quality, and consistency of instruction available at flying clubs varies in the extreme. You can easily find WWII retired bomber pilots teaching alongside 9-to-5 yuppies. Some of the greatest characters in aviation reside in flying clubs.

Smaller clubs sometimes suffer from limited aircraft availability and flight delays due to maintenance. With lower fixed costs than larger facilities, they don't have the incentive to "keep them flying" 24 hours a day. Sometimes clubs have leaseback aircraft where negotiating with the owner is required before any maintenance can be initiated. Flying clubs should be researched thoroughly, for they offer some of the best and worst training to be found.

FIXED BASE OPERATORS (F.B.O.)

The classic way to learn to fly is from the Fixed Base Operator, or F.B.O. This is your traditional flying school. They usually do other things as well like fueling, maintenance, pilot shops, or whatever. With superior facilities when compared with a club, you will have aids to your training like simulators, video libraries, classrooms, and flight planning areas. To support the extra fluff, you are going to pay more money. On demand instruction is the rule, which means anyone can walk in off the street and hop into an airplane; you don't have to join anything, like the club. Consequently, the insurance cost to the school goes up, and so does the overall cost to you.

Flying schools are filled with young, ambitious, career-minded instructors doing their time. The teaching is more structured and consistent than a club. Many schools operate from a syllabus and standardize their procedures. Most clubs are a collection of freelance instructors who have banded together for economic advantage. The structured environment of the flight school makes it easier for you to change instructors as everyone is teaching more or less the same thing. Structure, however, limits freedom. Your instructor may not be able to personally tailor your instruction or give you as wide a range of experience. I used to teach at a school that prohibited landings on anything but hard-surfaced runways. The FAA requires pilots to be proficient in a technique for landing on soft fields: mud, grass, gravel, turf, stuff like that. Nobody ever graduated who had made a soft-field landing on a real soft field. This is very common. When I moved to the club next door, I obtained permission to operate off a nearby gravel strip. I could then offer more realistic training.

FLIGHT ACADEMIES

On a distinct and specialized level is the flight academy. They operate under a different set of training rules, Part 141 of the FAA regulations, which gives them a sales advantage they exploit to the fullest. By operating from a strict and regimented FAA-approved curriculum; by detailing each lesson in writing; by carefully monitoring student progress with phase checks; these schools are allowed to recommend pilots for various certificates with fewer total flight hours than are mandated for all other schools operating under Part 61 of the regulations. The logic of this from the FAA point of view is that such a closely supervised program should be able to turn out a qualified pilot in fewer hours than a regular school.

The catch, however, is that you still have to know the same information and fly to the same standard as any other candidate for a certificate. Even though you can legally be recommended with fewer hours, you won't be unless you meet the standards. Since large flight academies are by far your most expensive option, the only way to come out ahead is to qualify for your certificates on or about the minimums. Only superior training from your particular instructor and resourceful independent drive from you will make this possible. A candidate for the private certificate from a Part 61 school must have 40 hours before being eligible for the flight test. The Part 141 student needs only 35 hours of flight time. This difference is meaningless because according to the FAA, the national average for attaining a private certificate is after a pilot has over 70 hours of flight time.

The eligibility requirements haven't changed for decades, but everything else about aviation has changed. Today's airspace is far more congested and complex than when the standards were established. A private pilot today has to know as much as someone who was half way to an instrument rating 20 years ago. Flight academies base all their training cost estimates in their sales material on minimum times that virtually no one can meet. This is deceptive advertising and should be investigated by the FAA.

The major academies have the most fluff. You will pay for fancy carpeting, nice offices, inflated salaries of management, colorful brochures, worldwide advertising, salespeople euphemistically renamed counselors, impressive simulators, and in one case I know, a corporate jet. The training is carefully controlled and amazingly consistent; everybody learns exactly the same thing. However, the range and depth of experience is the most restrictive because everybody learns exactly the same thing. Without flexibility, there is no innovation, no breakthroughs, and no improvement. I attended a Part 141 academy where the instructors taught exactly the same lessons, went to the same airports, sent students on the same cross-countries, and beat to death the same rote procedures. All of this was officially approved. Most of the instructors were former students. They were very well versed in the standard procedures; unfortunately, they had never been exposed to anything else.

What you can get from a major academy is contacts. They are a great place to meet people moving up in the industry. People are attracted to the large academies because they think it will be good for their career. Big schools also dangle the prospect of an instructor job at the successful completion of your training. You are led to believe that by attending the school you can buy your way into a job. Some of you will, and some of you won't. This is a cold and ruthless business. Employment is dependent totally on the need and availability of pilots. You will only get a job if your school can make more money by hiring you. Besides, you may find that a great place to train is a lousy place to work. Think carefully why you are willing to pay the inflated prices of an academy. You can only save money if you earn your certificates in minimum time, and the odds are stacked against you. You will only get a better deal if you receive superior training, and that depends on your instructor — nothing else.

INSTRUCTORS

When you begin visiting flight schools keep in mind that you are searching for an instructor, not a school. The school will have very little impact on you. The school is there so that the airplanes have a place to park, your instructor has a desk from which to teach and store his books, and you have a nice place to go between flights. In exchange for that the school takes all your money. Your goal is to find the most compatible instructor in the most suitable training environment. You should always remember that anytime you communicate with anyone associated with flight instruction, you are being sold. That friendly voice on the phone who requests your name and number is selling you. The instructor who chats with you and wants to make an appointment to meet you is selling you. The manager who offers you a tour of the facility is selling you. Every word you hear is tainted and slanted towards sales, which creates quite a pile to wade through.

Appearances are designed to be deceiving. Many schools do whatever it takes to look like what you think a flying school should look like. Flying schools are a lot like restaurants. Sometimes the best food can be found in some dull, dreary dive, at really great prices. You must have the inside scoop however, to find the dive. Other restaurants have dress codes, reservations, elaborate decor, a la carte hidden charges and factory-assembled food — and everyone goes there to be seen. It's all fluff! You have to decide how much fluff you are willing to pay for. In your search remember: You can never ask too many questions, talk to too many instructors, visit too many flight schools, or be too careful in making your final decisions.

What makes a good instructor? This is impossible to generalize because every student has individual needs and every instructor has individual talents. You have to analyze the learning qualities which allow you to excel, and match them with the instructor who possesses the qualities you most desire, admire, and respect in a teacher. Your success depends on your ability to learn from one particular instructor. You are going to spend many hours together cooped up in a fast and noisy classroom. You will experience many moments of stress and frustration, so you have to be able to work together and laugh a bit when the lesson is over. It cannot be a problem for you that the individual in the other seat is going to occasionally see you at your worst. You must be able to take criticism not as a personal attack, but as the only way to become a good pilot.

Instructors are as unique as fingerprints, complete with their own style and technique. They do not vary their methods to suit the student. You have to find the style you wish to emulate. There are an infinite variety of correct ways to fly, yet each instructor is convinced that the wisdom of Solomon exists only in his way of doing things. The looser the program at any given facility, the greater the range of correct methods you will find.

GUIDELINES

Here are some guidelines for your search. For you aspiring career pilots, look for an instructor with similar goals. Your training will be direct, intense, and geared toward acquiring your ratings as quickly and efficiently as possible. Pick someone between 25 and 35 years old. Young instructors are hotheads and have little patience. Older instructors still trying to get into the business are racing a biological clock, where flight time might be a higher priority than flight instruction. Your instructor should be full of zest for teaching. They should demand your best and be able to get it. The ideal instructor should have one year of full-time instructing experience and around 1000 hours flight time.

The logic is that a brand new instructor has all the legal qualifications to teach, but none of the practical experience. They can't give you a quick solution to your difficulties because all your problems are new to them. After a year, instructors realize that most problems fall into particular categories, which makes for easy diagnosis. It can take two years or more of instructing to accumulate the hours to qualify for the next job. This is when instructors are burning out. Their minds aren't focused on teaching. Their patience and enthusiasm are waning. Every student seems to make the same stupid mistakes. You can't beat the wisdom and experience if you can take the temperament. They will tell you what you will do incorrectly long before you ever get to the airplane. High time instructors are best suited to advanced students.

When you find an instructor you like, you will want him around long enough to complete your training. An instructor actively seeking other employment could be gone in an instant. If you want to learn to fly just for fun, you will do just as well with a full-time or part-time instructor with the equivalent one year/1000 hours experience. Training part-time your lessons will be more relaxed, but it will take longer to get your certificate.

Some instructors have been in the business just too long. These crusty, old geezers have wisdom to be revered. However, they may believe what was good enough in 1932 is still good enough today. Style is another critical factor to watch. An ex-military instructor may be suitable for a youth desiring boot camp treatment, but totally inappropriate for the middleaged business executive. Easygoing instructors may be fun to fly with, but may be a waste of your time. Your life depends on your training. An instructor who is too nice may skip the hard and valuable lessons. Be wary of becoming close friends with your instructor, lest you lose the student/instructor relationship. It is emotionally more challenging to accept criticism from a friend. You are paying for a service. Take responsibility for your training and get the most for your money.

You must be a good observer and listener when you visit flight schools, for there is much more to see than they may want to show you. Find out how the instructors are treated. Generally, happy instructors make for good instruction. Ethical instructors who stay at their jobs out of economic necessity, even though mistreated by the boss, will usually tell you if and how you might be cheated. The knowledge you gain could put you in a pretty good negotiating position. Be skeptical when the sales pitch gets overbearing or claims of speedy training times seem unreasonably short.

You may have friends in flight programs. Be careful with their recommendations. People would rather lie to you than admit they are getting anything less than the best training. All students have to believe that they have the best instructor. Their true feelings may be revealed only after a transfer of instructors or flight schools, where they will once again have the best of everything.

Take note of how much time an instructor is willing to spend with you answering your questions. If they have time for you now at no charge, then you aren't as likely to feel like a sausage link in a grinding machine during your training.

A high turnover rate of instructors is a red flag to be investigated. Only a high rate of airline hiring should cause much change.

Take a look at the condition of the airplanes and meet the mechanics who work on them. Do the mechanics have the resources and management cooperation to do a good job? Most employees freely complain to perfect strangers about bad management, so it never hurts to ask. Maintenance is critical when our national training fleet is on average over 20 years old. If you want to have some real fun, ask the mechanics about the instructors and the instructors about the mechanics.

You could even go to the FAA Flight Standards District Office (FSDO) and see what they have on your prospective school. Take a look at the layout of the school. Are the instructors clumped together like rats in a maze of office dividers, in a room so noisy you can't learn anything?

How is the ground training handled? This is a great indicator. Proper ground training can save many hours of mistakes in the air. Some schools charge you less for ground training because they perceive that students don't value the time as much as flight instruction. All this does is force the instructor to keep you in the air longer to maintain his income or to stretch out your ground sessions.

You can save money if small classes are offered. The school gets the economy of scale and can charge you less than the hourly instructor rate while paying the instructor the usual amount and keeping the difference. Large classes, however, are worthless. Concepts may pass you by and the class won't wait for you to catch up, so you have to learn it all again from your instructor. Large classes are extremely profitable for the school because they involve virtually no overhead, just the cost of a room and one employee.

As you tour various schools, remember that the closer you look, the more you will see. The selection of your training is one of the most critical decisions you will make, so take your time.

When you understand what you are looking for, it is time to go out visiting. On a cool, crisp, early Saturday morning, you may find yourself winding down a dusty, gravel road. Passing cows that pay no notice and aviation relics in various states of disrepair, you observe a faded, hand-painted sign of your prospective flight school weather-beaten by sun and rain. Parking so as to avoid the biggest potholes and kicking up dust which would settle on immaculate, 30-year-old autos, you have arrived. Between a couple of overhanging willows, a shack with peeling paint lies nestled. You walk up a broken rock path, past a ragged half a windsock draping from a rusted pole. The screen door creaks behind you as you timidly advance back into aviation history. Welcome to my old flying club from college.

As the assault of espresso-strength coffee hits you, your eyes leisurely sample the yellowing pictures of classic aircraft adorning fake, wood-paneled walls. "May I help you?" drawls the friendly voice at the far end of a pair of old cowboy boots. With the obligatory bent propeller in the corner and the ratty T-shirt tails from generations of first solos, you realize that this is aviation in its essence. All those old war film characters are alive and snoozing before you. Behind magazines that slowly lower with curiosity, lie the bloodshot eyes of gurus who question whether you are worthy of flight training. Don't expect sensitivity training if one of the old codgers takes you on as a student.

This is the old school, most likely the military old school. Flying clubs are like treasure chests — you never know what you will find until you take a look. Instructors of all ages, levels of experience, style, method, and technique can be found. You just have to go out and find yourself a match.

Your country airport will most likely not have a control tower. The single runway might be only slightly wider than your wingspan. The asphalt and numbers will have faded into the natural colors making the field clearly visible only from directly above. Here you learn to fly by instinct; what we pilots call "the seat of the pants." It is so named because your butt is the part of you directly connected to the airframe, and through it you can feel everything you need to know to fly the airplane. Decorum precluded naming it "butt flying." When flying by instinct, you will learn how to read the sky, predict weather from watching the clouds, and analyze the wind by watching the trees being blown around. Try to get some training in this environment. You will miss a great privilege if you pass up the chance to gain a flying instinct from those who invented it.

There is another end of the spectrum. Shrouds of mist curl around fortress pillars as you stand before — the academy. Up a long stairway and through huge oak doors, the freshly polished floor squeaks under your shoes. You walk into a beehive of bustle. A friendly minion appears to lead you through a maze of streaming pilots and instructors, darting from all directions at once. You pass the dispatch office, where sweat-soaked students trade airplane keys with fresh victims. You walk under a cloud of cigarette smoke while foreign nationals read magazines in Arabic and Japanese. While figuratively led by the hand, you pass a room burning with the smell of electronic machinery and the buzz of fake airplane noises. Up the stairs, around the corner and away from the din, you curl up in a comfy chair across from the Cheshire Cat grin of your aviation professional counselor.

The modern training environment of the academy is geared toward turning out pilots equipped for the rigors of the airlines and corporate aviation. Strict discipline with faithful adherence to procedures is the rule. You fly precisely by the numbers. Once this method is ingrained, you can confidently handle congested airspace anywhere in the country. If you like the idea of preening yourself in a sparkling uniform, this is the place for you.

Having trained and worked in both worlds, I have no problem letting my romantic preference for the flying club predominate, while my obvious disdain for the academy pours out in unbridled sarcasm. You should experience both worlds because each has its advantages and limitations, and you need the widest possible range of experience. Many pilots segregate themselves to either environment. A good pilot is comfortable in all situations.

You may feel now that rural flying is all you will ever do, right up to the day you have to make an emergency landing at a big terminal. The same goes for the big airport jock who has to glide into a small field without the usual aid of folks on the ground. Be flexible. You can always attend a major academy and hop to a small flying club for a weekend lesson or two. It is your training — go wherever you want. There is no reason for exclusive loyalty to any one flight school. If I had to pick the ideal place to train, it would be a good flying club or small school, moderately priced, at a busy tower-controlled field, with both country airports and busy terminals within easy reach, where the instructors cover a wide range of age and experience, and they have a variety of aircraft makes and models in which to train.

So much for the environment, now you have to find an instructor. Who are these instructors anyway? Who would willingly sit in an airplane with their hands folded, talking a neophyte through their early lessons. Who are these steel-nerved, calm, cool, rocks of stability who allow you to risk both of your lives in pursuit of aviation excellence? Well, there are two reasons people become instructors: an absolute love of aviation and a desperate need for someone else to pay for it. The exceptions are people with real jobs and retired airline or military pilots who actually enjoy instructing on the side.

It pays to know what to look for in an instructor because there are a few types that must be avoided. Take "the screamer" for example.

"Jesus Christ! What the hell are you doing? Did you leave your brains on the ground today? You will never become a pilot."

This is not what I would characterize as constructive criticism. Sad but true, there are always some students who are subject to this abuse. Fortunately, you pay the bill so you get to go elsewhere. Screaming accomplishes nothing, except to build a mental block with the student preventing all learning. With such general statements, you never know what you did wrong, so it is impossible to take corrective action, and this only creates frustration. The screamer shatters dreams, destroys confidence, and ruins self-esteem. The only reason to ever raise your voice in an airplane is when the safety of the aircraft requires immediate action.

Take some young kids in their early 20s. Give them experience in fast multi-engined aircraft. Throw in some wealth and privilege. Put them in a fancy uniform. Mix it all together with genuine flying ability and the accompanying arrogance, and you have an "egomaniac."

"You can't fly, only I can fly. Watch me. I can walk on water."

The higher you rise, the further you have to fall. When these instructors make their inevitable mistakes, they will deny it or blame something unconnected. Even the beginning students can tell when something is wrong, even if they don't know exactly what it is. You won't trust an instructor who constantly criticizes you while ducking his own failings. Mistakes are a natural part of learning. You never learn anything from correctly doing something you already know how to do.

The dangerous thing about egomaniacs is that they take unsafe chances, believing the physical laws of nature do not apply to them. Ego in proportion is good. No one would strap on an airplane without it. No student would fly with an instructor who didn't exude confidence. Ego untempered by judgment and responsibility can be fatal. So how do these people become instructors? It's all too easy.

Anyone who has the qualifications and demonstrates the ability can get the certificate. The examiner has no control over what happens after that.

While the egomaniac goes looking for trouble, "the coward" just kind of backs into it. As students, they were so cautious they never took up a challenge. Because of this, you will never see a strong crosswind, experience marginal weather, land on a short runway surrounded by real trees, or learn any maximum capability of the airplane. It is just as important to learn your limitations as it is to learn your capabilities. You have to have an instructor who will allow you to safely make mistakes, develop judgment, and learn resourcefulness to safely handle the unexpected.

The number one cause of accidents for private pilots, according to the FAA, is continued flight into deteriorating weather. If more folks could see adverse weather as students, they might come up with alternate plans sooner because they are able to recognize the folly of exceeding their capabilities. Unfortunately, knowledge is no guarantee of exercising good judgment. Be careful of labeling an instructor a coward. You may have been protected from a situation where you had no knowledge of the danger.

The screamer, egomaniac, and coward are hazards and must be avoided. There are also more benign types that are just plain useless. For example, the "Gary Cooper." Strong, silent types are great in classic westerns, but they have no place teaching flying where feedback is essential for training.

You ask, "How was my landing?"

"Fine."

"Okay, do you have any critique?"

"Nope."

"Anything really bad?"

"Nope."

"Shall I try again?"

"Yup."

"Will you have anything to say then?"

"Nope."

"Does your mind often wander to your future airline career?"

"Yup."

Feedback is critical to developing the fine touch of a smooth pilot. How will you know what that is if there is no one to tell you.

Everyone knows about the "time-builder." This is the instructor who has you start the engine and then sits around for the next hour discussing philosophy before you fly. They want as much loggable flight time, as quickly as possible, regardless of how much abuse is laid on the student in the process. Be careful with this label. Because it is universally recognized in aviation, it carries weight. Students may brand their instructor when in reality they are using them as a scapegoat for their own failings.

With our current system, all instructors who desire careers in aviation are time-builders. You can acquire all the ratings you need to work for hire in about 300 hours. However, the FAA requires you to have 1200 hours before you can carry persons or property for hire beyond the strict limits of the VFR. Unless you are fabulously wealthy, you will have to have someone else pay for the difference in hours. One of the few exceptions and viable options to build that time is to become an instructor. Eminently qualified pilots, who have neither the aptitude nor desire to teach, must nonetheless go through this apprenticeship building the flight time before moving on to commercial jobs. Most instructors make every possible effort to teach as well as they can while trying to meet the requirements of future jobs. I know some really great instructors who absolutely hate teaching.

Some types of instructors are just plain fun. There is "good old uncle Bob." This is the guy who treats you like one of the family. There is no pressure, you have a good time, and you are treated with fatherly responsibility. They are ideal for teaching young students because of their patience and role model value. They are usually found on weekends teaching at the local flying club.

How about "the jock." Broad-shouldered with big hands

that engulf the controls, they are the epitome of self-confidence, with just too much starch in the old uniform. Please give them a sense of humor. No one should take himself that seriously. However, if you can learn to project that same confidence, your future passengers will love you.

If you want to learn the really neat stuff, look for "the professor." Perfectly content discussing the theoretical pathway of a disturbed molecule traversing the upper cambered paradigm complete with vector forces therein, you are in for the greatest collection of noodle-brained factoids ever assembled. It's fascinating information; just sit around and listen.

This is what makes aviation so much fun. The characters you meet will be a never-ending delight. I must confess that I suffer from one of my own caricatures. I am the "Gomer Pyle." Filled with such bubbly enthusiasm for flight, my students were frequently nauseated. "Gooooooooly, let's go flying! I don't see any hurricanes out there. Oh boy!" Well, we all have our idiosyncrasies.

After all the phone calls and visitations, you sit at your kitchen table buried under a pile of brochures, business cards, posters, official paperwork, handbooks on flying, and stuff of which you have no idea. After carefully considering the options, you haven't a clue what to do. You may be considering the academy, not because the training is any better but because the school has convinced you that it will look good on your resume. That's fluff. Your future employers care only how much money you will make them; whether you have any violations, incidents, or accidents on your record; how proficient are your instrument skills; and the most important quality, who you know in the company.

While you sit there befuddled by the propaganda, think about what you really want out of aviation. Are you prepared to commit at least three years of your life to aviation before getting your first commercial job? This includes one year of training and then instructing until you get the job. If you are going to fly just for fun, are you willing to invest the time it takes to be a good pilot and then to keep working on your flying to maintain that proficiency? Time is a precious commodity.

The worst place to be is waffling between the recreational or professional pilot. You begin training half-heartedly with the intention of deciding somewhere along the way if you want to make a career out of this. Half-hearted efforts bring half-hearted results. Without a clear objective there is no focus, no drive, no determination, and ultimately no career. You will only succeed if failure is not an option. The hard times will be cushioned by having a goal to keep you going. This is a far cry from the vision of flight you held while gazing at aircraft out your office window. Welcome to your first rule of aviation: Every hour of flight requires some 6 to 10 hours of preparation, negotiation, study, and frustration. And you haven't even had your first flight yet.

You may need outside help. If you are considering a university flight program, your local library should have information. I have no experience in this area. Since all airlines require a college degree, this option may make sense. Like many professions, the type of degree is irrelevant. You could major in Russian literature for all the airlines care. If you intend to get a degree in aviation management because you think it is the best path to a job, you are basing your decision on fluff. You are no more guaranteed a job in your field than a marketing major is guaranteed a job in advertising. Rather than wallow in fluff, why not ask an airline pilot the path he took to get the job. You may want to check your school with the Better Business Bureau. We all have this romantic vision of flight. Flight is beautiful; flight instruction is a business.

What makes an airplane fly? Money! How will you support your flying habit? Money! What if you want to buy an airplane? Big money! Flight schools with the largest overhead will require the most money. Each additional rating and certificate will take even more money. Maintenance costs lots of money because every product associated with aviation contains an automatic markup.

When you have spent all your money becoming an instructor, you will spend your time going after other people's money. People and goods move by air because the saving in time is worth the money. The commercial pilot who can save the most time will make the boss the most money. For those who fly purely for personal reasons, you will need an endless source of money. Money drives the system. Money provides the instructors and airplanes. Money keeps the flight schools going.

Flight training costs so much money because it is an incredibly inefficient way to utilize an airplane. Aircraft, when engaged in commerce, maximize their utility by loading passengers and cargo up to the highest gross weight and fly at altitudes and power settings that result in the lowest operating costs. However, when aircraft are engaged in flight training, they go nowhere, move nothing, operate at inefficient and constantly changing altitudes and power settings, all of which is paid for by one person.

Have you figured out how you are going to pay for this dream? Money is your greatest impediment. This is why military training at taxpayer expense is so attractive. I wear glasses so that was not an option, and it took me years to raise the money for training. To give you an idea of cost, private training should run you about $3500, that is unless you exceed the national flight hour average, go to a big, expensive school, or both. For the amateur pilot, this is the only certificate you will ever need.

If you are planning on a career, figure on $25,000 in training costs before you can earn a living as an instructor. The average income for the full-time instructors I know is about $10,000 a year. This is barely enough to cover living expenses today, not to mention money for future ratings. If you have to borrow money for training, don't figure on paying it back anytime soon. By the time you are a working flight instructor, you will most likely have drained all your savings, maxed out your credit cards, loaned yourself out well into the next decade, and be in debt to family and friends who you hope will turn any loan into a gift. You will also discover how long a human can live on macaroni because every dollar you earn will go right back into paying your debts. Aviation is unique in that you cannot earn back the money it costs to get into the business in your first job.

One of the main attractions of the professional academies and university programs is that they are sometimes the only options that accept student loans. You are forced by your lack of money to take the most expensive choices. What a system. Take the time to plan your finances so that you may complete your training. There is nothing worse than falling short of your goal, running out of money. When you know you could have made it, you have to take a job outside aviation, and you are still burdened with student loans.

Money carries its own psychology. Many students claim a lack of money when they stop training when the real reason may be fear. Fear is natural. Using money to cover for fear is denial. Flight training will appear expensive if you are wishy-washy in your approach. If you absolutely have to fly, no sum is unreasonable; you will find the money somehow.

Analyze why you think flight training is too expensive. If one option is too expensive, see if you can find a better deal. If you can plunk down a large sum, you can usually get a discount for a block of hours. You can save money by concentrating your lessons because the time to catch up from the previous lesson is reduced. Lessons spread over time, however, will cost less per week. It all depends on your situation. Generally the commercial pilot will do better with the lump sum, and the private pilot can more easily pay as they go.

I have saved the best option in civilian flight training for last. Within our shores lies the ideal path to the airlines. Unfortunately, the citizens of this country are not eligible. Since our tax dollars support all our airports, the national airspace system, and the national weather system, we are directly subsidizing our competition. I refer to Lufthansa and Japan Airlines which have training facilities for their pilots in the United States. I had a really nice chat with the chief instructor at the Japan Airlines (JAL) facility in Napa, California. This is how they do it. College graduates with no flight experience are recruited and after careful selection sent to Napa for training. It costs them nothing. They are already under contract to JAL receiving a salary and full benefits throughout their training. They have an airline seat waiting for them at the successful completion of their instruction. The primary trainer where they spend their first 160 hours is a brand new, high-performance A36 Bonanza. Multi-engine training for the next 100 hours is done in a turbine King Air. After an additional 25 hours of line operations flight training (LOFT), they go to Moses Lake in Washington to train in Boeing 747's. They have no training cost, no student loans, no need to instruct for years, and their job security is guaranteed.

You may wonder why we make our pilots go through hell before they get to the airlines when a system like this is available to the rest of the world. I do. The reason they come, of course, is money. Fuel prices around the world average twice what we pay here because of additional government taxes. Some countries, especially in Europe, charge for tower services, flight plans, and every landing regardless of aircraft type. Also, our weather in California, Arizona, and Florida is ideal for flight training. We have a gold mine here which the rest of the world is happy to exploit, except, of course, our own airlines.

The academy I attended had more foreign nationals than Americans. The big schools actively recruit foreign students for the large lump sums of money they bring and the potential to recruit their friends back home. Foreign students subsidized by rich parents can exist on substandard wages, taking instructor jobs from qualified Americans, and artificially suppressing the salaries. Foreign carriers training their pilots in the U.S. benefit from reduced training costs, our airlines benefit because we pay for all our own training, and the schools benefit because they take money from everybody. My solution to this imbalance is to charge all foreign students and foreign carriers operating schools in this country the same price for fuel that they would pay in their own country. The extra money raised would then be used to provide scholarships for American pilots who cannot afford the prohibitive costs of training.

You now know that the best you can hope for in flight training is inferior to what is possible. The journey to become a commercial pilot is a long and costly struggle. Darwin's rules apply. Despite all the adversity, it is still the best career you can have, which is why we do it. Be careful when you select your training. Identify your needs and find the best match. Weigh all the evidence. Read everything and talk to everyone you can. Bring a good healthy dose of skepticism to your visits. When you find the right instructor, go for your goal with all the gusto you have. The only losers in our current system are students and instructors. To you this book is dedicated. Good luck.

3 • THE DEMO RIDE

————————————————————————————————➤

As we go through this journey, we will lose people along the way. Some of you have discovered that flight instruction involves more than you expected. You realize that too much hard work, sweat, and study will be required to live your dream. Flight schools want all your money; instructors want all your time. You begin to believe that getting a pilot certificate is too complicated an endeavor for you to handle. Too much information must continually whorl in your brain, poised for instant recall. How can anyone grasp all that knowledge, maneuver such complicated machinery, and have all the right stuff? Maybe you are content to just sit back, watch the air shows, and tell folks that you almost learned to fly. You take the safe choice; the easy way out. You go back to your accounting desk and spend the rest of your wretched life looking at airplanes out the window. Wimp . . .

What a cop out! You haven't even been up in an airplane yet. Well, maybe you have sat in the back of a big jet sipping martinis. That is riding — not flying. Our society reeks of spectators. This is why television is so popular. You get all of the thrills, you think, with none of the risk or hard work. Well, are you just going to sit there and watch your whole life pass you by, or are you going to go fly airplanes?

Stout-of-heart individuals, undaunted by the challenge ahead, will persevere toward their goal. Such characters have already signed on the dotted line, mortgaged their future, bought the books, and begun their training. They have also skipped this chapter and gone on to the more

interesting stuff that lies ahead. But what about you inde-
cisive types? There you sit wallowing in doubt, anxiety, and
nervous excitement. You really want to fly. You want to tell
your friends just as much that you are learning to fly. Pilots
are special, not everyone can pass the rigors. You want to
be special, too. The sky beckons. Slick videos fill you with
adrenaline. Poetry drives your heart to fly while trepida-
tion rules your brain. What is the ingredient in the mix
that will allow you to try this new challenge? To see the
whole picture, you need just one more piece of the puzzle.

For people like you, the flight schools have created the
demo ride. This is exactly what you do not need, which is
why it is exactly what you will get. Euphemistically camou-
flaged under the title "introductory lesson" or "discovery
flight," the demo ride is a baited hook so irresistible, so
calculated in its diabolical nature, so guaranteed to snare
any receptive victim, that only those who are just trying to
prove they don't like flying anyway, can escape. Once you
have bitten the apple; once you have felt the speed and
power; once you have lifted your body above the familiar
earth into the sky; once you have seen these things under
the careful orchestration of your most sincere instructor,
who is shamelessly stealing your wallet, your family life
and your free time; once you have been through this experience; unless flight absolutely terrorizes you, you will be a
student pilot.

A side note to parents. Sometimes young people are
pushed into flying by parents who are pilots or frustrated
pilots. Sometimes they actually get through a demo ride
and a few painful lessons. No one can fly to please someone
else. The nature of flight training will weed them out.
Encourage a child to fly and you will have an insight into a
wonderful world. Force a child to fly and you will earn
everlasting resentment.

The demo ride is nothing more than a quick jaunt
around the patch. You will be back on the ground in about
20 minutes. For a bargain introductory price, you get just a
tiny sample. After just enough time for you to enjoy the

experience, you are back on the ground wanting more. Drug dealers work the same way; after a taste you want more.

The only reason the demo ride exists is to get you to sign up for lessons. The school will lose money on the flight. However, the probability of you signing up should you make the flight justifies the artificially low price. Once they deceptively show you how much fun and easy flying is, you can't help but commit yourself to training. Flying is certainly one of the most fun things life has to offer; however, it is anything but easy. A good demo ride will show you only the fun without requiring any effort on your part.

Attracted by the low price and freedom from any obligation, you are led to believe that the demo ride will give you the information you need in order to make the decision to fly. It won't. You cannot learn anything about flight training by watching an instructor fly, except for those brief moments when you get to wiggle the controls. Flight schools all know this. The demo ride is the most effective way to sell you specifically for what it does not tell you. Because it works so well, virtually every school has some variation.

Most students have some form of a demo ride before they begin training. This makes sense on the surface. Why would anyone want to invest in something they haven't even tried? That's what they want you to believe. You don't need it. If you have already decided to learn to fly, this is just a waste of time and money. You won't learn anything so why bother. If you are in doubt about flight training but enjoy riding in airplanes, this won't help you either because you will still be just riding in an airplane. If you want to honestly evaluate flight training and your potential as a pilot, you will be helped least of all. Unfortunately, you are the target group for the demo to sell you on lessons.

If you really want to know what training is all about, plan to spend some time at an airport one day. When you have found an instructor you like, take about 1.5 hours on the ground learning how an airplane works. When you think the flying part will make some sense to you, spend another 1.5 hours in the air. Your instructor can show you the whole private flight course in that time. You can also have ample opportunity to get the feel of flight and have some idea what you are doing. After this experience you will have far more insight into flight training than is possible from a sales gimmick. For the aspiring professional, isn't it worth a few hours of your time to find out if aviation is where you want to spend your life?

For those of you who don't know how the psychology works, this is how many of you end up in flight training. It starts at the best place for daydreaming, your desk at work. As you gaze at the small craft passing by, you try to imagine yourself at the controls. Could this be a real possibility? You have visited the local field a few times and spent a while pouring over the same sales brochures. Can you imagine yourself in those stunning pictures? You now have an emotional stake in this, and a real risk to consider.

There is always the possibility of failure, that you will not be able to live your dreams, or that you may actually hate flying. You may secretly hope for this as it would be the easy way out. The real risk is that you will absolutely love flying. It will be every ecstasy you ever imagined. You will have to follow through now and get the certificate. The fear of success is always greater than the fear of failure. Failure is easy. You just go on with your dreary life dwelling on what might have been. The demo ride is the last chance to duck out before you have to decide to succeed.

So the real risk begins after the demo ride. It is a rule of aviation that everything works backwards. You take the demo ride hoping it will convince you to stay out of aviation. The flight school, however, orchestrates the whole experience to remove any negatives, leading you to the conclusion that only positives exist in flight training. To your horror, you find you love everything about flying. That is how they get you. Motivations and desires are tricky things; manipulating them is just good business.

You will now pace up and down, weighing your decision to take a demo ride. You will pace by the phone deciding when to call. Because you don't know how the system works, and you haven't done your background investigations, flight training is a mystery and you are at the mercy of the flight school. You have been flirting with the idea of flight until this point. It is as if you were on a diving platform: Do you dive in head first or walk back down the ladder? If you do not call, you have only regret with which to look back.

You will make the call because you don't want to have to work yourself up like this again. You book your introductory lesson. The instructor writes on their schedule — demo ride. You have no idea that this demo can't possibly answer your questions. Several days ahead you begin to watch the sky. Listening to the weather reports takes on a new urgency. What if it is too cloudy or windy on flight day? What if something goes wrong with the airplane? What if something comes up and you have to cancel? What will they think of you at the school? What if the instructor gets sick? What if you run out of what-if questions? Waiting for that first flight is a healthy mix of wanting to go and hoping for an excuse not to. You might as well relax because you have no control over the above events. If you really want to fly, someday you will.

You are already starting to think like a pilot. Anything can come up to cancel a flight. That's aviation. Just remember: Take-offs are optional; you never have to go. Landings, however, are mandatory. Most pilots mentally get in gear for a flight one day ahead, so relax until then. You will not be allowed to fly if the conditions are unsafe. You won't go unless the instructor thinks you will enjoy it. If there is a glitch, you simply reschedule.

This is your first real insight into flight training although you won't see it. If your first flight is canceled, you will initially feel relieved because you don't have to go through with it. You will then feel disappointed and cheated because you didn't get to go and you have to pump all that adrenaline through your system next week. Don't cancel on your own if you see a couple of white puffs of cloud out your window. The weather always looks worse than it is to the first-timer. Call the flight school before drawing any rash conclusions.

The big day arrives — oh boy! After a night of fitful sleep, you awake to coffee you do not need. The morning breeze feels like a hurricane, giving you the pretense to call the airport and confirm that the airplanes can still fly, the airport is open, the instructor hasn't fled to the airlines, and the cirrus clouds at 30,000 feet won't be a factor. You will be assured by a calm, sleepy voice muffled slightly by doughnuts. With wings on your car, you hurtle down to the airfield. There is something magical about the drive to an airport the first time you are to fly. The airplanes look somehow different. You never really noticed the color and detail until now. Whether one engine or two, sitting on their tails or not, each airplane radiates its own personality, each airplane carries its own soul. Each airplane better be flying soon or the school won't make any money. This is why you are there.

After making sure that none of the wall photos are missing, you timidly step up to the desk and announce with slight apprehension that you are there for your introductory lesson. When a suitable amount of time twiddling your thumbs has passed, the instructor will stroll out. This may be your first meeting. Funny, the instructor looks just like a regular person. This is not what you expected. You aren't sure what you expected, but this wasn't it. Despite the first impression though, he may still walk on water. You hope so since your life will soon be in this stranger's hands. You will then engage in some small talk to convince yourself that this instructor can be trusted and that you will enjoy the experience. What a sales job.

You will be led by the hand to the airplane. Take in its wonder. Look at all those instruments. Play with the controls. This is fun because there is no pressure on you, unlike real lessons. You are not expected to know, or be able to do, anything. You are in a sense, like a baby. The weather is checked for you. The route of the flight, standardized by the school for economic reasons, will follow the same path taken by all the students before you. Any instructor worth his salt will have preflight checked the airplane before you arrive. The tanks will have fuel and the tires will have air. Any obstacle to your blissful and smooth experience will have already been dealt with.

First impressions are critical. You can't see yet how the system really works. You hop into the airplane, fire it up, call the tower if there is one, and head for the active runway. How much of this you get to do depends on how much you ask to do and what the instructor lets you do. They are controlling what you experience, so to make it appear in the best light, they will try to do as much as they can get away with. If this is how you are evaluating flight training, try to do as much as you can. Depending on your ability to negotiate, you may actually sneak in some flight training. If you are like most people, you will just sit there in awe. You don't want to do anything because you refuse to do something wrong and appear stupid. Since you can't possibly know what to do, how would you know what was right or wrong?

But that is not how the psychology works. Instructors love demo rides because they have a captive paying passenger who doesn't expect to fly hardly at all. It is in their interest to do as much flying as possible to maintain their flight skills, which progressively decline the more they instruct. They will be happy to have you just sit and watch. Be brave; ask to start the engine, try to call the tower, and attempt to taxi the airplane. Get as much from the experience as you can. The instructors will be very clear about what they will allow and will without hesitation take the airplane from you in the interest of safety. That's why they make the big money.

Everything in aviation is backwards, and the airplane is no exception. Steering is done with your feet, and the power is in your hands. You should have learned to fly before driving. Don't forget to chuckle when you head for the grass, while you wiggle the ailerons hoping to turn the aircraft. If you think your instructor is secretly laughing at you, he is. Everyone makes the same mistakes. It all becomes quite predictable after a while. Which brings on another rule: You never learn anything without making mistakes. Think about it, what can you learn from doing something you know how to do? Now think how well you remember doing something you completely goofed and will never do again.

Keep in mind that anytime you go for a demo ride you are being sold; you are not engaged in flight training. You cannot learn anything when nothing is required of you. If you are lucky, you may get to work the controls on takeoff. What an ego boost. You may be willing to sign up for lessons right now. Keep in mind also that people are capable of following specific commands and flying an airplane, without the slightest idea what they are doing. This is interpreted as learning how to fly, but is really just part of the sales process.

Flying an airplane means exercising judgment and making decisions, all of which is done by the instructor. He also keeps you on track by regularly steadying the airplane. When you reach an altitude where you can't hit anything, you can wiggle the controls all you want. You still aren't flying though. This is where you should get a trial balloon question as to your interest in taking lessons. Gauging your interest and personally tailoring the experience is part of the game. You are purposely led to believe that learning to fly is easy. Your ignorance allows you to talk yourself into lessons. You are not getting the feel for flight training as you wanted, only the feeling of flight; big difference. All too soon you will be in the airport traffic pattern for landing.

You will not do the landing. Not that you couldn't be talked through all but the final approach, but this is where your instructor gets to sell himself to you. You believe the Hollywood mystery shrouding landings; that they require such skill and mastery that mere mortals can't do it. After witnessing a great landing, you will idolize the god in the right seat. The purpose of this is to inspire such confidence that only this instructor can be trusted to teach you to fly. The bond is formed. Have you carefully picked your instructor? No, you have blundered into revering the only person you may have ever seen land an airplane. The good gut feeling you have for that person is completely based on fluff.

You will marvel at the featherlike touchdown while the instructor nonchalantly leans over and whispers, "Did you like it?" Unknown to you, your instructor is thanking his lucky stars he didn't bounce. It ruins the whole effect. I don't think I ever bounced a demo landing. The possibility of losing the student is economically out of the question. So I come in faster, roll the plane on, and use up a little more runway than necessary. Business is business.

Of course, the landing will seem graceful to you. Unless you have previous experience in small aircraft, your last landing was in some jumbo airliner, weighing thousands of tons, slamming down at 140 knots and slowing to the roar of thrust reversers. You can't compare that to an airplane that weighs in somewhere near a Volkswagen, touching down at maybe 30 knots, gliding on to the runway, and slowing with light braking. Yet that is exactly what you will do. As you taxi in, it dawns on you that the adventure is over. You want to go right back up again. You will be a little tired, a little thirsty, and excited to be alive. Back in the lobby of the flight school, you will be hit with the reality that you just flew an airplane, or so you think. Go ahead, call your friends. On your way out the door, don't forget to pay your bill and sign up for your first lesson.

Some of you will find that flying isn't for you. Bounding around in a craft that is all too small, unattached to anything tangible, is no fun. It is a good thing you took this brief sample before plunking down a large, unrefundable sum. Some people are not comfortable in small airplanes. Most of you know this, however, before you ever leave the ground. Flying looks very different observed on a big screen monitor from the comfort of your plush couch. The sensations are very different when you cram into a sardine can with big wings that telegraph every jolt. There is nothing to be ashamed of; at least you tried. Go find another activity. You may have some explaining to friends who have suffered through your potential exploits for the last several weeks.

Some of you will back right into a demo ride without the slightest intention of going flying. If the instructor is exceptionally slick and polished, this is how it happens. You drop by your local flight school to look the place over, chat a bit, read the literature (sales propaganda), and get a feel for their training environment. While trying to be casual as you explore the facility, so as not to attract attention, an instructor saunters up to make your acquaintance. You don't know this, but any potential customer is a magnet for attention. A new face in a flight school stands out like an American tourist on his first overseas stop. Depending on how much macaroni that instructor has been forced to eat the last few weeks, you may just find yourself up in an airplane today. Then you go and spill your guts out about your dreams of flight to an instructor who is sizing up how best to sell you. It is not enough to fly well, or even teach well; to be valuable to a school, an instructor has to sell well. A badly handled sales pitch will have you feeling like you are buying used cars. You will leave that school for another, where they will do exactly the same thing, only with better fluff.

"So, have you ever taken a look at an airplane up close?" croons the instructor ever so innocently. You probably have some time to kill so this seems reasonable.

"I'll just grab the key so we can get in."

Well, it can't hurt to just take a look. You are curious. After all, you can't just hop into an airplane and go. There is planning involved. Your instructor (sales professional) returns with the key to an airplane that is free for the next hour or so. It's such a pretty thing out there on the ramp. It just seems to say: "Fly me."

Next you walk around the airplane. You play with the control surfaces. It looks so small when you get close up. Light trainers are small compared to the behemoth airliners you are probably used to. They will feel bigger when you get the panoramic view from a few thousand feet up.

"Why don't we just sit in it a while, and I'll show you how everything works."

You've walked around it a few times, kicked the tires, and taken all the looks you wanted. Why not get in? Someday you may actually fly one. Not today of course, but someday.

"Here, try on this headset. It will feel more realistic. Why not have the full effect and fasten this seat belt."

So this is what it will be like in a small airplane. It's not so bad. Maybe this is just a little too realistic.

"Now that we are here and all strapped in, would you like to take a quick flight and see how this thing works? Seems a shame to waste such a beautiful day sitting on the ground."

Clang go the jaws of the trap. Stunned, you think: I can't go flying, I just came by to ask a few questions. What am I doing strapped into an airplane with a headset on? If your curiosity is stronger than your fear, you are going flying. While still in shock, you can only react to the instructor's commands. To keep you distracted, they will have you do as much as possible.

"Clear! Okay, just turn the key. I've set the controls."

Why did they yell clear? What key? The engine roars to life; as much as a trainer can roar. Oh, the propeller. You understand now. Hey, this is fun. Doubt and excitement mingle as you try to figure out how you ended up with a perfect stranger taxiing for takeoff, when all you really wanted . . .

By the time you are ready to go, you will have rationalized your predicament by convincing yourself that you really wanted to fly today anyway. You refuse to admit that you were manipulated so expertly by the instructor.

"Okay, we are cleared for takeoff. Just push the throttle all the way in."

Push the what? Jesus, we're moving.

"No, not so fast. Okay, that's good."

You didn't say how fast. How was I to know. This whole flight was your idea.

"Good, now just steer with your feet."

Steer where? Where are we going?

"Looking good."

Compared to what?

"Now just ease back on the stick and lift the nose."

Better give a good pull; I don't want to look like an amateur.

"I got it."

Oh, that's where you wanted the nose. I can even see the horizon now. This is fun. I wonder where we are going?

"You're doing just fine. Now we have to turn left, so just ease down on the wheel to the left and add some rudder with your left foot."

This feels odd, and the world is at a most peculiar angle.

"No, here let me show you."

Oh that feels better.

"Are you having a good time?"

Considering you had no intention of flying, most of you will enjoy this bit of serendipity. Some of you will not have so much fun. Instructors run the risk of people not enjoying the experience at all. They feel tricked and want to be on the ground right now, even after willingly consenting to the flight. Because of economic pressure from above, and the success rate of this technique, instructors will continue to take the risk. Except for a few individuals who warned me of their concerns and doubts well ahead of time, I never took someone up who didn't have a wonderful story to tell. What the doubters had to prove by going up remains a mystery. Most of you convince yourself ahead of time whether you will enjoy flight. The demo is unnecessary because it can only confirm what you already know.

When people come to me for an introductory lesson, they get a proper lesson. If they come for a demo, I tell them straight out that this is a joyride, it has nothing to do with flight training, so let's go have some fun with an airplane. My standard demo would begin with some 60-degree banked turns for thrills and gravity (G) forces. Just like an amusement park ride, only better. Then I would hang out all the flaps and fly as slowly as possible. My victim would then hear some technical jargon about maximum capability of the aircraft and so forth. The sky is my circus tent, and I am the ringmaster. For the strong of stomach, I would show them the most gentle power-off stall possible. Most people are breathing hard and ready to sign their lives away to me at this point. But just in case, I always held the clincher in reserve.

To my constant amazement, the best way to hook a student is to do wing-overs. People absolutely love wing-overs. This is a maneuver where during a steep climbing and descending turn, one wing appears to fly over the other. Hence the name. With the ball centered (you'll learn about this later), and the pitch and bank limited to 30 and 60 degrees respectively for legality and passenger comfort, the wing-over is truly beautiful. First-timers have never had a view of the world like this. They are surprised and proud to find themselves comfortable at such a strange angle. What an accomplishment. The wing-over always brings a smirk of glee from the left seat, as the sheepish voice says, "I want to learn how to do that." Chalk up another student.

There are schools run by unscrupulous owners with sinister greed who abuse student and instructor alike. One tactic they use is to hire ten instructors where the need is for only five. During the weeding out process, the instructors who bring in the least business are dumped. Instructors are paid only for the time they generate revenue by teaching. It costs the school nothing to let them hang around all day trying to drum up students. Such a hostile learning environment should be avoided. You will recognize such a place instantly because you will begin to feel like fishbait. If you stay around, you can take part in "the shark dance."

Think back to those opening scenes from *Jaws*. Down in the depths of swaying kelp, a string bass plays in your head. Daaaaaa Dum, Daaaaaa Dum. Now put yourself in the lobby of the flight school. As the waft of fresh prey arouses the instructors interest, magazines slowly lower to reveal beady, piercing eyes. Daaaa Dum, Daaaa Dum. You carefully step toward the false security of the photos against the wall, trying not to make eye contact. Your skin begins to crawl. Daa Dum, Daa Dum. The instructors slowly rise and begin to circle, leering as they move.

"Can I help you?" asks a fresh, eager face.

"No, may I help you?" perks a second voice, trying to sound more authoritative.

"I'm available. What can I do for you?" quips a third.

Daa Dum, Daa Dum. No person in his right mind would stay around for a demo flight. That is why it is so amazing how many people do. Now for Round 2, the marketing campaign.

"So, what kind of flying do you want to do? I can help as I have all the ratings."

"What you need is the voice of experience. In my three years of teaching, I can tell that you will be a great pilot."

"How about a tour? Having taught in this area the longest, I can show you all the highlights."

Da Dum, Da Dum. The string bass reaches fever pitch as the frenzy turns inward for a verbal assault. If you haven't left already, this would be the time to skip out as the instructors will be preoccupied with infighting. You will hear them jostle for position as they fight for who is the next in line to take a student. You are meat for the hunters, that is all. It makes no difference which instructor is best suited to teach you. If this is the condition of the staff, what shape are the airplanes in? What kind of training will you get for your money? It is very common for people to openly lie about fellow instructors to get more business. The school owners don't care. They are too busy lying about other schools. Besides, they don't care how the money comes in, only that it does.

Your first flight should be a fun experience and a wonderful memory. Find a place where you are comfortable. Take a full lesson if you want to seriously evaluate flight training. If you want to just play — take a demo ride.

4 • THOSE FIRST FEW HOURS

Okay, bucko, you've decided to go for that certificate and become a private pilot. In your possession, a new stack of books covering the Federal Regulations, handbooks on how to fly, aircraft manuals, weather books, and aeronautical charts you have no idea how to use. Having all this information makes you feel like a pilot. The trick now is to actually read all the material. Because you know nothing, nothing will be expected of you, at first.

Although the FAA takes great pains to counsel otherwise, flight training is all accomplished completely by rote. Follow the motions of the instructor and try to deal with the significance later — maybe. Your hands and feet will be beaten into reflex action by endless repetition, while your brain vainly tries to apply some meaning to those actions. You will consciously suppress your brain because you have already learned, even before your first flight, the single greatest impediment to your training — those dangerous words "because my instructor told me to." With that simple belief, you give up any chance of thinking for yourself, before you even get started.

The current system of flight instruction places total responsibility for all training on the instructor. The unforeseen, and so far uncorrected effect of this, is to remove any responsibility for training from the student. Student pilots like most people are lazy and gladly hand over not just the responsibility, but all the blame for anything a student does not know, or forgets, any mistake they make, any violation of the Regulations, any time safety is jeopardized by their actions, any low scores on written exams, any unanswered questions from an examiner, and any failed flight checks. Total responsibility according to the student, examiner, FAA and the instructor themselves, falls totally on the instructor, regardless of what the student or instructor, did or did not do. It doesn't matter.

With the burden of responsibility removed from the student, any procedure becomes a rote reflex for the hands and feet, and any piece of knowledge becomes an independent fact to be stored away. There is no effort from either party to train the brain of the student to think. There is no incentive for the instructor or student to give individual bits of wisdom or technique a purpose, fit them into the structure of training, or make sure the student comprehends how any particular piece integrates into the big picture. Learning to fly becomes a series of separate and distinct, rote, and meaningless exercises. The more structured the flight program, the more flying will be taught by rote, the less the student will be permitted to think.

With the responsibility placed totally on the instructor, they also gain total control. Many instructors like this because it makes it easier to handle students, ratchet up their ego and self-importance, and cause their students to learn only what they decide their students will learn. Students therefore take no initiative beyond what they are told, read no material beyond what is assigned, if that, ask no questions outside the current topic, or develop any of the qualities of individuality, decisiveness, resourcefulness, or action, all of which are critical to the successful development of any pilot.

This is the fundamental flaw in our flight training system. All other breakdowns can be traced directly to this problem. Time after time, students would fly with me and I would ask them why they did certain things the way they did? After recovering from the shock that there may be alternatives to the methods taught by their instructor/god, they always respond "because my instructor told me to." That is never an answer for anything, and only proves they have taken no responsibility for their training. Do yourself a favor and ask why every time you get a procedure or piece of knowledge. Make sure you understand the big picture. Give a purpose to every action. Take charge of your training. Do not settle for spoonfeeding, relying on your instructor to provide you with everything. If you establish a positive mental attitude, set ground rules for yourself, take initiative, and constantly ask questions, you will learn with excitement and vigor and get the best from your instructor.

A word about scheduling lessons. Most flight schools, especially the big ones, schedule you for two-hour blocks of time. On a typical busy day, the student before you will get back late with the airplane. You will be checking the airplane over when you should be flying. Your instructor will be late as well. You will probably be off the ground half to three quarters of an hour into your two hours.

You will still have to fly to the practice area. In order not to back the schedule up too much, you will have to head home at least half an hour before the end of your two hours. That should leave you about fifteen minutes of flight time for your lesson. This is no accident. The school makes money depending on how long they can get you to keep the propeller turning, not by how much they teach you. The incentive is to teach you as little as possible while having you fly as much as possible. The way to do this is to spend lots of time with the engine running doing things other than learning. That is why your lessons are so short.

This is where instructors with more freedom to schedule can save you much time and money. I block four hours per lesson. I schedule the airplane for three. The first hour is a cushion for the previous student, and ample time for a good preflight. The extra nonflight hour is for preflight and postflight discussions. There is no hurry, and the students get far more training per lesson.

The first thing you will learn is how to preflight the airplane. This is the check you do before firing up and blasting off. Besides being required by the Regulations, it's nice to know all the parts are still attached. Unless you own

the airplane, you never know what the last person did to it;
and in an airplane you can't just pull over with a problem.
None of this will occur to you when learning how to preflight because your attention span is short enough to have
your mind wander to the upcoming flight long before this
lesson is over.

You also delude yourself into thinking all airplanes
are perfectly maintained because anything you are about to
fly in must be safe, and the instructor with whom you trust
your life will bend the physical laws of nature on your
behalf. Well, your instructor is just a regular person trying
to scrape out a living and most likely trying to move out of
instructing, and your airplane is probably over 20 years old
with thousands of hours of flight time on it. Pay close
attention to learning your preflight.

You learn your preflight by walking around the aircraft painfully slowly as each part is carefully explained. As
your mind overloads, you get anxious to move on to more
fun stuff. After two or three lessons where you both are
there for the preflight, your instructor will assume that you
now know the procedure. From then on, you will be handed
the keys and expected to insure the aircraft is safe for flight
all by yourself. This is your first real responsibility. You
will be trusted from here on out to perform a good preflight.
Don't you wish you had paid closer attention when the
instructor was around?

Here is our first breakdown in training. You will val-
iantly try to remember the procedure for preflight from what
you observed from your instructor, even though the airplane
manufacturer has provided you with an entire preflight checklist in the airplane manual. Because you rely on your instructor for absolutely everything, you have no knowledge it is
there. If you were adventurous enough to discover it, you will
not use it for its intended purpose because either your instructor has not told you to use it or you have not seen him use it.
You will instead try to memorize the steps in the manual and
preflight by guesswork.

You will not think to use the preflight checklist for

preflight until someone cares enough to suggest it to you. I was halfway to my instructor rating before I ever used a preflight checklist. You also won't use it because you don't see other pilots using it. Most people flying the same aircraft eventually memorize a reasonable preflight. But how do they know they aren't missing the same thing every time? Sometimes the manufacturer leaves out an item your instructor deems important. This is fine; you can always add to the list, you just can't remove anything.

The reason that you learn the preflight so early is not to build your sense of responsibility. Flight training is structured for just the opposite. It is because this may be the only break your instructor gets between students. All the postflight briefings and paperwork are done while the next student is out getting the airplane ready. The school makes more money by being efficient, but your training and safety suffer as you don't know yet when things look right, let alone when something is wrong with the airplane.

Since making a good impression on your instructor is more important to you than the safe operation of the aircraft, you will go through the motions of a preflight. Your instructor, depending on time pressures, will ask a few critical questions to insure that nothing will happen during the flight. The classic question is usually, "Is there any water in the fuel tanks?" Most students will answer with authority even though they have no idea what fuel looks like when contaminated with water.

You proudly stand by your aircraft when the instructor swaggers out. All instructors swagger; it is required.

"So, did you check the oil?"

Oh no. You stop for a second. You think so. You hope so. "Yes, I did" comes your less than sure reply.

"Uh-huh. So, how much oil do we have?"

Still on this oil thing. You already said you checked it. Now you feel you did something wrong, so you try to come up with a good guess. "We have four quarts."

You aren't really sure about any of this. Human nature is such that you would rather risk both your lives in an airplane, on an oil supply for which you are not sure, rather than answer embarrassing questions from your instructor. Instructors always ask follow up questions, ostensibly to find out what you really know. In actuality, it is because student pilots are chronic and habitual liars. You know you have no responsibility for your actions, so no matter what you say, the instructor is ultimately responsible. You will be the first to blame them anyway for any consequence of statements you made in order to avoid being embarrassed. They must therefore dig for the truth, and ask you questions.

"So how much oil do we need to fly?"

What you take for sarcasm is actually a valuable lesson. This kind of questioning is most motivating. If your instructor has not told you to memorize the minimum oil quantity, or shown you where it is in the manual, you won't know. If you have any initiative at all, you will go home and read the manual. With both of you now growing impatient, you will be told the answer to this question.

"The minimum we can operate with is four quarts; so, can we go?"

What is this, a trick question? If four quarts is the minimum and you have four quarts, then, yes, you can go. "We're fine. We can go fly. I can't wait."

"Well, if four quarts is the minimum, and we have four quarts now, what happens if the engine burns some oil in flight?"

Ever feel set up? You are paying for this privilege. Good instructors often teach using this technique because it makes you think. What you have just learned is that you must always have reserves when you fly. You are developing judgment. You must because your instructor won't always be there. Don't dwell in self-pity at having been tricked; take in the lesson. Be leery, though, of the instructor who asks if you checked the oil, you reply yes, and he lets it go at that.

Communication is such a subtle and precise art that major catastrophes have been caused by simple misunderstandings. An instructor can ask a student if he checked the oil, when the instructor really wants to know if there is sufficient oil for flight. The student can respond that he checked the oil, without knowing how much has to be in the engine to go flying. Be good to your instructor. Don't make him dig. If there is ever anything for which you are unsure — ask. In this example, ask your instructor to say exactly what you are checking for.

Students are also so embarrassed when asked about something they completely forgot to check that they sometimes lie and hope for the best rather than face criticism. I have shut down engines on the ramp beside the runway, called the tower to explain the delay, unfastened myself, climbed out of the airplane, and checked something as innocent as the security of the cargo door when the student wasn't completely sure that it was locked. I almost always find it locked. The point though is to be sure. (With experience I learned to just check it myself — but I still ask about it.)

You are a student trying to master an activity with endless details. Do not ever lie to an instructor. You will build respect and trust with your instructor by immediately admitting mistakes. Checking things on the ground is easy. It is at worst, inconvenient. But it is nothing compared to the danger of rolling flight bags out an open cargo door over a city.

A friend of mine once asked a student if he had sufficient fuel for the lesson. The student was embarrassed at having not checked, so he lied and said there was plenty. The fuel gauges indicated relatively full tanks, so the lie was plausible. The fuel gauges of small airplanes, however, are notoriously inaccurate. Any pilot wishing to remain in powered flight will trust only his watch and a dipstick. Students don't know this because it is hard to tell beginning students that the fuel gauges don't work. They might leave and the school would lose business. My friend, though, had no reason not to believe the student. When they lost their engine after many practice takeoffs and landings (touch and go's) because the fuel was gone, their emergency landing resulted in the loss of a couple of aircraft and a hangar.

They lived. If you are a conscientious student, you will have your instructor periodically review your preflight inspection.

You will learn how to start the engine. Well, not so much start it as hopefully coax it into life. The airplanes out there are old, and the technology is very old. When light aircraft were being certified decades ago, the FAA determined that the engines would not be reliable without a dual ignition system consisting of two magnetos firing two sets of spark plugs. While auto engines have advanced to computer-controlled electronic ignition systems, airplanes still use magnetos. You will learn to fly the way aviators have long learned to fly because nothing much has changed under the cowl. You will find that each instructor has his own favorite way of getting the engine started. This procedure may be different than the one in the checklist. It may work better, it may not. However, the first time you fly with another instructor, you will get a new and different method altogether . . .

"Why did you do that?" quizzes your substitute teacher.

Don't you feel strange now? Student pilots hate criticism of any kind, especially when they think they are acting correctly. Because your instructor violates the checklist in one particular way, you think everyone does. You cannot base assumptions of pilot behavior on your instructor; but you will. This is logical because your instructor is the only pilot to whom you have ever paid close attention. Now you have a new instructor who violates the checklist a whole new way. When you get back with your regular instructor, you pause before starting the engine because you now have three ways to start the engine, the checklist method, and the interpretations of two individual instructors.

Another rule of aviation is that every pilot is an expert at his own technique. Since you have no technique of your own, and since you must please each instructor, you have a dilemma. To make your lessons easier, try to follow the recommendations of each instructor when he is on board. When you get your certificate, you will probably add your own variations. The best advice is to always follow the checklist; that is the way the engineers like it, and after all, they built the engine.

Learning to taxi should be a great source of amusement. Not only do you have to steer with your feet, you have to work two independent brake pedals. You steer by pushing the bottom part and brake by pushing the top. This makes for some interesting combinations. Each brake on the foot pedal works its own main wheel so you may use any combination of brakes to either stop straight or make a left or right turn sharper. You may not push both rudder pedals at the same time however. The system that swings the rudder and steers the nosewheel isn't set up that way. The power is in your hands. That control in front of you is called a throttle.

Juggling all this requires much more practice than the average student believes. Which leads to the impression that if they can't even taxi, how can they fly. Flying can be easier because there is a lot more room up there, so don't sweat your taxiing. Juggling all those new controls is going to take practice. You have to get used to taxiing on the centerline. We don't have lanes in aviation — the wings would get in the way. Okay, now try to do all this at once while your instructor babbles because he has nothing better to do. This is where you will be asked if the cargo door is locked. The secret of taxiing is to forget that you have ever been in an automobile. Never relate cars to airplanes — ever.

There is nothing that causes student pilots more anxiety than talking on the radio. You will lay awake nights in a cold sweat, not because you almost took out a high rise office building, but because you said something really stupid on the radio. You judge your mistakes not by any relationship to safety, but by how many people you think know about it. In this age of electronic phones, ham radios, and conference calls, you would think that a simple conversation over an aircraft radio would be no dig deal.

Guess again. Student pilots are preoccupied with hiding the fact that they are student pilots. To do that, you try to always sound professional. You may have no idea what

you are saying or what the response is from the tower, as
long as no one thinks you are a student on the radio. You
assume that real pilots know everything and never have to
ask questions. We can always tell the students because
they don't question what they don't understand and always
try to speak as quickly as possible. The goal here is to
communicate, to expedite the flow of traffic, and maintain
safe operations. The essence of any radio call is who you
are, where you are, and what you want to do.

Some students learn at busy fields served by an Airport Radar Service Area (Class C). With this you get radar
service on departure and arrival. To accomplish this, you
are put into a computer euphemistically called "the system." You will be assigned a four-digit number code that
will identify you on radar. All of this takes the transmission
of a great deal of information to the ground controller.

This is daunting to the fledgling aviator. Some rated
pilots are so intimidated by the radio work, they avoid such
airspace entirely. There are students who print up beautiful speeches on laminated 3 X 5 cards. Others speak as
rapidly as possible, hoping not to forget anything. The poor
ground controller misses most of the transmission and has
to start again. When students do get it right, they are so
pleased with themselves for sounding professional. When
the controller comes back with taxi instructions, the student misses it and the call has to be repeated.

Now everyone knows you are a student pilot. Lighten
up. We all had to start somewhere. Take your time. If you
forget something, the controller will simply call you back.
Speak slowly, clearly, and always ask questions. If you
listen, you will find the professionals questioning anything
they don't understand. The controllers all know who the
students are anyway. You are flying a training aircraft that
regularly departs from a flight school. Who else would be
on board?

Learning to use the radio can cause huge misunderstandings with your instructor. Many students, when
handed the microphone, go into suspended animation.

"It's just a conversation with the tower. No one else is listening. Just tell them we are eight miles south for landing." This sounds so easy coming from the instructor. You don't believe it of course, but somehow try to speak. As you rehearse your speech, you will notice your instructor twisting in his seat. You are now six miles out and about to violate the Airport Class D Airspace. You don't know that and think your instructor is unduly impatient with you. Finally you state, "Tower, this is Cessna 54321." You forget the rest and give the instructor a blank stare.

"Tell them you are now six miles south for landing."

When you finally do call, the person in the tower heard your initial transmission and pause, and is calling for you to "say again" at the same instant. This results in a squeal because only one party can talk at once. Your instructor has no choice now but to take the controls and bank sharply from the airport. You are confused and feel persecuted and punished like a baby just because you goofed on the radio.

However, in the time it has taken you to go through this process you have very nearly busted Class D; the airspace four nautical miles from and up to 2500 feet above an airport with an operating control tower. Your instructor has worked damn hard to get where he is and refuses to have a violation of the Regulations on his record that would prevent him from going to the airlines just because you don't know how to use the radio. He can't explain all this as you head for the airport. The only recourse available while you key the mike, or push to-talk switch, is to keep the airplane out of the airspace. You may resent your instructor based on erroneous assumptions. Your instructor may not know what you are thinking.

This is a breakdown. If you ever have a question about any event in your training, you must talk about it with your instructor. Radio work is one aspect of flying that must be done correctly every time. In a busy airport environment, you do not have the luxury of endless practice. Airplanes move fast; some airplanes move very fast. Practice on the ground. Role play with your instructor. When you get the hang of it, you will realize that 90 percent of the time you already know what the controllers will say back. That is why pilots sound professional on the radio.

Should you be learning at a field with no tower, you can practice all you want on the radio and not get in too much trouble. Your communications aren't put on tape either. This is the most informal radio environment; sometimes it gets a little too informal. You are ready for takeoff and your instructor motions you to call your departure on the radio and go. If this is one of your very early lessons, you may have your hands so full with flight stuff that you pass on the radio call. Instructors love talking on the radio. At an uncontrolled field, you talk directly to the other airplanes and to whomever might be volunteering on the ground station, called "Unicom."

"Smith traffic, Cessna 54321 departing runway 32, straight out."

"Is that you Bob?" comes a voice over the headset.

"Yeah, I got a new student. We're weaving down the runway as I speak."

Asshole.

"Oh yeah, don't let him kill you."

Jerk.

"No, this one will be just fine with practice."

Well, maybe you'll give your instructor another chance. Your instructor will brief you on proper radio phraseology and then do something like this. There are times to be strict, and there are times when you can get away with little chats. You don't know when those times are, so you should always be formal in your radio work. Besides, that's how we know who the students are at the uncontrolled fields.

As you improve your radio skills, you will find yourself communicating more information while using fewer words. Some students, though, when they become comfortable with the radio, go overboard with jargon. I had to break a student who used to sign off with "Cessna 321, roger wilco, over and out, good day." Bit much, don't you think?

The specific words you use on the radio can be critical.

Recently I was out with a student on a cloudy day, practicing touch and go's. I saw some lightning and called the tower. "Cessna 321 has observed an air-to-ground lightning strike, ten miles south of the field. We shall be departing the airport traffic pattern." That message communicated to every airplane in the area that conditions were changing rapidly. Our home field was 20 miles away. Visibility below the clouds was great, but lightning is lightning. On the way back there was another strike to the north. My student, bursting to make the report, called our home tower with the words "Jones tower, we've had a lightning strike!"

Oh no.

"Do you require assistance and the equipment standing by?" came the immediate reply from the tower.

After clearing things up with the tower we discussed the importance of saying exactly what you mean.

The first phase of your training is called presolo, for obvious reasons. This is where instructors hope to cram enough rote reflex actions into your hands and feet to let you safely fly by yourself one day. The FAA has clearly spelled out the presolo requirements in Part 61 of the Regulations. Part 61 covers all the eligibility and training requirements for each certificate. You won't read this, of course, because you are relying on your instructor to tell you what you need. Rather than wait for spoonfeeding, why not learn the requirements for each phase of your training so your can prepare ahead for your lessons and monitor your own progress. You will also discover immediately if you are deviating from the stuff you must do.

Clarity of meaning is not a specialty of the Regulations. Take this requirement for presolo training, for example. You must be able to perform "straight and level flight, shallow, medium, and steep banked turns in both directions." Read literally, one might wonder why only the steep banked turns have to be done in both directions. You also have to be able to perform "climbs and climbing turns," and "descents with and without turns, using high and low drag configurations."

What they are trying to get at are the four fundamentals of flight: climbs; descents; straight and level; and turns. All they had to say was: Be able to use these skills in combination. You will probably practice each of these skills separately in the practice area. If you think about it though, these maneuvers are necessary in every flight. Once the takeoff is over, you are in a climb. A climbing turn gets you out of the traffic pattern. Straight and level gets you to and from the practice area. You will descend back to the airport. Descending turns get you into the pattern. The use of flaps on final approach gets you the low and high drag configurations. You may be double covering the requirements in your training.

One big problem in your early hours is that you learn lessons separately and never apply them to the situation for which they are designed. I can take any presolo student and say, "Show me a descent." After some doubt and fumbling, the nose will tentatively lower, the power will come back, and the aircraft will descend somewhat like the instructor showed them. This is rote training.

Now, if I say, "Let's go back to the airport and land," the application of the descent is completely lost, and we will most likely fly over the airport at cruise altitude. The student knows how to descend. He just demonstrated that to me. He knows he must descend in order to land. He does not know when I want him to descend and is waiting for prompting because he does not want to make a mistake, look like a student pilot, and receive criticism. So he does nothing.

We do not teach students to think because the instructor is totally responsible for training, so the student does not have to think. Lessons are a series of instructor commands and reflex responses. These commands are repeated every lesson because the student takes no responsibility for retention and must be retaught every time. Students never apply what they have learned because pleasing the instructor is more important than learning how to make decisions. Students will learn how to do a descent and only do it when specifically told to do a descent.

This is a common breakdown in training. Instructors

are confused because they teach you descents, they observe you doing descents, and then they can't understand why you refuse to do them when you should without prompting. This is because they believe the FAA, which teaches that training is not by rote, which we all know is completely wrong. Instructors are taught that students can transfer knowledge and apply techniques, when the truth is that student pilots only do what they are directed to do, when they are directed to do it. You have to take responsibility here for your own training. Find out why you are doing each maneuver, learn its purpose, and where to apply it in flight. That is the only way you will learn how to think and make good decisions. Because students are allowed to abdicate any responsibility for training, this critical step is missed, and training is greatly delayed.

Many flight schools, especially the big academies, will standardize the lessons but not standardize the training. For example, the instructor tells you to climb. There is no clear picture in your mind for you to latch on to and go directly to a climb. Each time the instructor tells you to climb and you sit there with a blank stare, the instructor takes the airplane and tells you to put the nose "here." Without a clue as to where exactly "here" is, you hold the nose where it is this time, with no ability to duplicate it next time. This is why most students have to be retaught each maneuver, every time they do it. This is training by rote.

The FAA wants you to learn to fly by the integrated instruction method. That lofty but nebulous term simply means flying by both an instrument reference and a visual reference. Everything works backwards in flight training, so you will concentrate on those instruments because you assume real pilots always rely on their instruments. Since you haven't had the instrument training of real pilots, you have no idea how to take what you see and put it in a usable picture. You also don't develop the habit of constantly looking out the window for other airplanes. Each of those instruments you value so highly has its own particular delays and errors, of which you have no idea. This will result in you being told you are "chasing the needles," another nebulous term which has no meaning to you, even though you know you are doing something wrong.

Outside the window is the best reference ever created to observe what the airplane is doing — the horizon. This immovable, unerring, constant indicator is always there for the visual student. It requires no interpretation. Your perception of what real pilots do, along with your fascinations with the dials, will prevent you from learning how to fly by visual reference. There you are being told to climb, when you don't remember where "here" is, while you stare at instruments you can't read. That is why you have trouble in your early hours.

There is of course, a cure. Standardize the training. You will cut your training time in half if you can get your instructor to write down specific pitch and power settings for every configuration in normal flight. Memorize them. For each particular pitch and power setting, you will get the same airspeed and rate of climb or descent. You can climb, descend, and fly straight and level, using a particular power setting and a visual reference where the nose appears relative to the horizon, and do it the same way every time without ever thinking about it. When you don't have to think about how to do it, you can learn to make decisions when to do it. I have yet to fly a trainer that doesn't take off and climb at full power, cruise about 2300 revolutions per minute (RPM), descend and fly downwind about 2100 RPM, and approach for landing at 1500 RPM, touching down at idle. So much for all the power settings you need to know. Your instructor or aircraft manual may require some variation. However, the point is to use the same setting for each particular configuration every time you do it.

For the pitch settings, we have a very sophisticated method. Fingers; that's right, we need your fingers. Most of my experience is in Cessnas, so they will be used throughout the book. You can adapt my examples to your own situation with a little creativity. Okay, take the Cessna 172. When you takeoff, raise the nose until it appears to touch the horizon. The plane will stabilize at the best rate of climb speed (Vy) of 73-76 knots depending on the model. Your height in the seat, however, will cause slight variations in the picture you see out the window.

Now, if you lower the nose enough to slide two fingers horizontally between the horizon and the nose, you will stabilize at a cruise climb speed of 85 knots. Lower the nose further to slide four fingers under the horizon and you will be straight and level. Set cruise power and trim, and you are done.

When ready to come down, set descent power and put six fingers under the horizon, and trim. You will descend at your cruise airspeed at about 500 ft/min. If you reduce the power and maintain the pitch, you will keep the same rate of descent but cover less ground in the process. If you plan it right, you won't have to touch anything until downwind where you raise the nose to three fingers, leave the power alone, and trim. For your landing, set approach power, and put the nose back down to six fingers, and trim. It works every time. Except for the actual takeoff and landing, you can now fly a 172 very accurately using only the horizon, your fingers, and a tachometer to measure the RPM.

You may think that this is rote training. True; learning the pitch and power settings is a rote exercise. However, the human brain can only contain so many thoughts and carry out so many actions at once. If you have to dwell on remembering how to configure the airplane every time a change is needed, you do not have the space in your brain to worry about the radio or watching out for other airplanes. Once your configurations become instinct, you free up space in your brain to learn the things that most students learn much later in their training. The big one is "situational awareness," a glorified way of saying what the hell is going on in the airspace around you. It also takes students a while to realize and accept that level flight does not mean putting the nose on the horizon. If your instructor does not have specific settings for you, then get him to provide pitch and power settings. This simple system removes much of the confusion and delay from your early hours.

Learning to fly is completely different from school learning. In our education system you learn a subject, take a test, and forget everything you learned. Flight training is all cumulative. You must not forget anything. The four fundamentals will be performed every single flight. The problem is that you will think they are only for demonstration in the practice area, to be promptly forgotten as you move on to something else. This is why you cannot apply what you are learning.

Climbs; descents; and straight and level are one group of fundamentals because the wings stay level. Turns, are something else. Your experience in cars demonstrates that turns, especially ones which are too fast for the road, create great side forces. In flying, that is really gauche. I have to mention autos here even though I said never to think of them, because you will do it anyway. To avoid any side forces, you are told to remain "coordinated" in flight. This term, usually applied to proficiency in sports, is pretty much meaningless to the new student.

Many aviation terms are nondescript and misleading, so throughout this book I shall rename them. Whenever you hear or read the word *coordinate*, replace it with the word *balance*. For any given angle of bank, you need a certain amount of rudder to balance the side forces, and a certain amount of increased pitch to maintain your altitude. To accomplish this, you will be told to keep the ball centered, stop chasing the needles, and remain coordinated. You may spend many hours of flight time trying to decipher what that means. To state this in English: Put the airplane in a bank relative to the horizon, add enough rudder to keep you centered and balanced in your seat, and when you find the picture where the nose on the horizon maintains the altitude, hold everything still. You have to learn how to fly the airplane despite how you are told to fly the airplane.

The *ball*, as mentioned earlier, is nothing more than a steel or agate sphere, sitting in a curved glass chamber filled with kerosene, located on your turn coordinator. When the forces of turns are in balance, the ball will be centered between those two lines. Keeping the ball centered is what your instructor wants when he tells you to coordinate. Too much or too little rudder in a turn will cause the side forces to pull the ball out of the center. You have to push it back. Coordination in flight has nothing to do with proper hand/ eye movement.

There is another set of forces which work against remaining coordinated. For reasons better left to your flight manuals, you will see that the slower you fly and the more power you use, the more the nose will wander to the left and the ball gets pulled to the right. Call it *torque* or "left turning tendencies," these forces will cause you constant problems.

Say you are climbing out to the practice area. Your instructor continually tells you to stay coordinated, so you feel you should do something. The airplane doesn't feel all that strange to you, because students, unaware of these forces, are always unbalanced (uncoordinated) in flight. You sense a conflict in logic, but push the right rudder anyway to kick the ball back and humor the instructor. Your instructor, satisfied that you understand the concept, which you do not, goes back to looking out the window. You think you have learned something, even though being a student you don't care about its meaning, purpose, or application. You feel you have corrected the problem so you release the pressure on the rudder. The force hasn't gone away just because you pushed the rudder temporarily.

In a few minutes, the instructor will look back at the ball, see that you are unbalanced again, and tell you in a voice like an American tourist trying to make a foreigner understand English, to put in right rudder. You did it once, why do it again? You never make the connection with the force, and the instructor never provides it because he has been told you can transfer knowledge, which you cannot. This game will go on until the cycle is broken.

As everything is taught backwards, you are told to remain coordinated by centering the ball, this balances the forces and maintains your heading. Try it the other way. If you hold your heading with the rudder while climbing out and hold a ground reference with the nose, you will always be balanced. The ball will never wander from the center and your instructor will wonder how you mastered that trick hours before he did in his training.

The propeller spins to the right, which causes the airplane to turn to the left. As long as the propeller is spinning, you have to correct for this force, so pushing the rudder and letting go accomplishes nothing. The plane is rigged, usually, to take care of this force itself while in cruise. This is why you forget it when you aren't in cruise. Train the brain. Understand the force and the correction will take care of itself.

Steep turns will be especially challenging. Although your first solo will consist of low-altitude medium turns, during your presolo time the FAA wants you to master steep turns way up high. A steep turn is, by definition, where the upper wing moves sufficiently faster than the lower wing because of the bank angle, to generate more lift and keep the airplane constantly increasing its bank. Nice effect, no? There is much emphasis on your proficiency at steeply bank turns.

Shallow turns are of no use in training to the FAA and most flight schools because the aircraft's stability keeps you out of trouble by trying to bring you back to level flight. They are difficult to perform well and build a fine sense of touch while not making you feel strange or uncomfortable, so you will of course not practice shallow turns.

Medium turns are where the stability of the aircraft is neutralized by the overbanking tendency of greater bank angles. The airplane can fly by itself in a medium turn until the fuel runs out. You would learn how well the plane flies without your help and how to work with the airplane rather than against it. These being valuable lessons, they must be ignored.

No, what you have to master is the steep turn; a maneuver where the forces of nature conspire against you to throw the airplane out of your hands as rapidly as possible. You never have to do a steep turn in normal flight, however; only for an emergency or evasive maneuver. Since your early solos will take place in the traffic pattern, and steep turns should never be done at low level, and you don't need them for cross-country flights, there is no reason to perfect steep turns until you are getting ready for the checkride. You may wonder why you have to get so good at them this early in your training. Better that you experience them early, understand the forces, learn the dangers of steep turns at the wrong time, and work on them throughout your training. But that is not the way it is.

The FAA wants you to be able to fly at various airspeeds from cruise to the minimum controllable airspeed to maintain altitude. This is good. It gives you the full range of airplane capability, differing control pressures, and handling qualities. Unfortunately, most instructors skip all those speeds in the middle and go right from cruise to the slowest possible speed. I'll ask a student on a phase check to fly at some arbitrary speed like 63 knots. In a Cessna 172 there is nothing you do at 63 knots, so it is a good speed to use. Most students cannot do it. When I tell them they can fly slowly like their instructor taught them, the rote reflex kicks in, big relieved smiles cross their faces, and we go immediately to minimum controllable airspeed. This demonstrates only that their brains are dead.

Slow flight and minimum controllable airspeed are terms often confused and used interchangeably. Slow flight is any speed less than cruise. Minimum controllable airspeed is any speed where an immediate stall will result from a pitch increase or power reduction. Other than a flight test (because it is required), can you think of any time where it is practical to fly an airplane as slowly as possible on the edge of a stall? However, there are lots of times to fly slowly: approaching for landing, downwind, overflying an unfamiliar airport, aerial photography, dragging the area for an off-field landing, search and rescue, etc.; so most instructors pass up this technique, too.

Instructors like minimum controllable airspeed because it is an extreme condition and it demonstrates their superiority. Students like it because they think it is great training, for what they haven't a clue. There you are way up high, the engine is surging, the flaps are hanging out, the stall warning horn is blaring, and you sit there with a stupid grin on your face thinking, look what I can do and doesn't this feel strange. For safety you should never do this without an instructor. You do this without any practical reason in your head other than that is what your instructor told you to do. Fascinating. Like steep turns, this is a technique you master before you solo, don't do during your solo flights, completely forget during your cross-country flights, and have to learn all over again for the flight test.

The reason for experiencing minimum controllable airspeed is not to see how well you can keep an airplane somewhat stable in an unstable condition, but is to show you why you never want to be in this condition especially at low altitudes. All this will be completely lost during your lessons as you endlessly repeat reflex actions. Many instructors teach without purpose because that is the way they were taught. It was good enough for them, they got their ratings, so why change? Passing bad characteristics continually from instructor to student is like letting abused children become parents.

After you are comfortable with and accept without question steep turns that are too steep and slow flight that is too slow, you will get stall training. No one is ever quite ready for stalls except those iron-nerved rocks of character that laugh at danger or people who are lying to themselves and the instructor. My instructor said he loved stalls, but to this day I have never enjoyed stalls. It is just something that has to be done. To the new student, stalls appear as instants of uncontrolled flight where his stomach leaves his body.

Everything is taught backwards so before you solo you will have to master stall entries when you should really be learning stall avoidance; recognition; and recovery. You will spend much time and money learning how to get into something you must avoid on your own. You desperately try to learn a technique but succeed at first only in alternatively filling the windshield with blue sky and mother earth.

What are you learning from this? You are learning that stalls are difficult to enter the way you want (whatever that is), they feel really strange, and flight training is more difficult than you were led to believe. You should be learning that stalls need never be entered if you know how to recognize them, they are more easily prevented than recovered, and that stalls entered at low levels can kill you. Our current system gives the impression that you will be a great pilot if you can expertly stall the plane, that if stalls cause you concern you are a baby, your instructor will always be there to save you, you can practice these on your own and nothing will happen, and that stalls always occur at safe altitudes when the pilot induces them. None of this is true.

Since the stall/spin accident continues to claim pilots, no student should be allowed to practice stalls without an instructor, unless they have had spin training. To be legal, you need either a parachute or to be going for an instructors rating, by which time you will know what to do with stalls. The best you can hope for as a student are spin entries and recoveries. You must get this training or spins will always haunt you. The FAA dumped spin training because they wanted to promote general aviation without scaring students by requiring spins.

Students today scare themselves to death by inadvertently entering spins from stalls they were taught very well how to get into, while practicing stalls by themselves, without the slightest idea what is happening. Wouldn't you be better off with spin training from a competent instructor? Most of you approach stalls with apprehension, putting up a brave front because you think real pilots aren't afraid of anything. Confess your feelings. Stalls and spins are scary. This is why we learn to recognize when they are coming upon us. Not being able to recover, however, can be fatal. The choice is yours.

Much of your trouble grasping the concepts in your early hours can be traced to that fact that you don't know why an airplane flies. Your instructor will try to tell you, but they may not know. The examiner who finally tests you may not know why an airplane flies. I'm not convinced that anyone really knows exactly why an airplane flies.

I do know that every pilot is convinced that he knows the correct words to say to prove that he knows why an airplane flies. Learning aerodynamics becomes simply a matter of learning the words that everyone else expects you to use to describe why an airplane flies. When you use those words, other pilots think you know why an airplane flies, even though you know that you do not know why an airplane flies. But you will convince yourself that you do know, because people who you think do know why an airplane flies have told you that you are using the right words. This is how we create generations of pilots who have no idea why an airplane flies, even though they are convinced that they do. Students don't question things that they are told that they know, so the cycle continues.

This is where you should put down those fuzzy handbooks and read Wolfgang Langewieche's *Stick And Rudder*. This book is a classic, which means no one reads it anymore. If you want to know why an airplane flies, go read that book . . . Welcome back. For those who didn't read Wolfgang, here is a summary of why an airplane flies. It may not be exactly accurate, but it is comprehendible.

An airplane flies because the wing is an inclined plane, which when it flies, meets the air at a slightly higher angle to the direction it is going. This is called the angle of attack. It is the most important concept in aviation. Air strikes the bottom of the wing which forces the air downward. Because of the curve (or camber) of the upper surface, air is accelerated over the top and then downward as it leaves the wing. The wing pushes air down from below and pulls it down from above, which pushes the airplane up. If you want to see this effect, put your hand flat out your car window like a wing, then tilt it up slightly. The faster moving air on top of the wing exerts less pressure than the slower air striking the bottom. The difference in atmospheric pressure helps to hold the airplane up. Should the air meet the wing at an angle too sharp for the air to flow over the wing, it stalls. Lowering the nose to reduce that angle will restart the airflow. So much for aerodynamics.

This is my example. Air is a fluid. Water is a fluid, too; it's just closer together. Think about what happens when you water-ski.

When you start, the ski digs into the water at a sharp angle. It mushes along gaining speed and moves progressively to a flatter angle on the surface. You plane across the water. An accelerating wing is similar to a wa-ter-ski except that it is far more efficient because of the airflow over the top surface. However, a wing still launches from a steep angle and flattens as it gains speed. The wing planes through the air. That's why its called an airplane, silly. Thank you, Wolfgang.

Back to the water-ski. When you release from the boat, the drag of the water slows you down. To stay up, you must continually increase the angle at which the ski meets the surface. Eventually you will reach an angle which can no longer hold you up, and the water-ski stalls. An airplane stalls pretty much the same way. A water-ski behind a slow moving boat creates drag like an airplane in slow flight. It is not an exact description, but the comparison does help you to understand fluid dynamics.

You will not find this example in any flight school or text book that I know of. What you are expected to believe, is that a wing is just like a garden hose. Uh-huh. Your instructor is trying to teach you about Bernoulli's theory, the official explanation for why an airplane flies. The theory is that water flowing through a garden hose will go faster and exert less pressure at the point where you pinch the hose, because a venturi is created. The same thing happens when a wing goes through the air, and that is why an airplane flies. . . . What an amazing leap of faith. You can actually say that to an FAA examiner and pass a checkride — really! So if a wing is like a garden hose, is a propeller like a fire engine? No person in his right mind would ever accept such an explanation, except in flight training.

Remember the flaws of student pilots. You accept no responsibility for your training and do only what you are told by your instructor. The logical extension is that your instructor has earned a certificate to teach, so everything he says must be true. You do not question, therefore you rationalize. You mold your thinking to fit the teaching rather than question the teaching so the explanation makes sense. You also know that if you say what you are told to say, you get your certificate, regardless whether you know what you are talking about or not. You will sit with your instructor and believe that a wing is just like a garden hose. You will accept this without question because your instructor told you so. You will say the right thing during your flight check and you will become just another pilot who has no idea why an airplane flies.

Question: There are those who say the wing holds the airplane up by creating negative pressure. Since negative pressure exists only in a vacuum, how does the wing remove the entire atmosphere above it? What is pushing down above the wing to create a garden hose type venturi? Why aren't you asking questions like this to your instructor? When you truly understand why an airplane flies, all the maneuvers and procedures will make sense. I had to go to the NASA Ames Research Center and talk to two of their top scientists before I finally understood why an airplane flies. The knowledge is out there; however, under our current system the standard explanation is so poor that it is effectively useless.

Once your instructor is reasonably confident that you have learned the rote procedures for high work, you get to come down to earth for some fun. It's time for ground reference maneuvers. They should be called ground track maneuvers, for it would be more accurate. Besides giving you a great view of the neighborhood, they are designed to teach you how to correct for the effect of wind while tracking paths on the ground and dividing your attention so you don't fly into the ground you are tracking, or hit another airplane doing the same thing. You will fly between 600-1000 feet above the ground to clearly see the effect of the wind. Flying this low is not only exciting, with rare exception, it's legal.

This training maneuver still exists despite the fact that low-level flying is a major cause of both accidents and noise complaints. Are we teaching you that flying low is all right as long as you are practicing ground track maneuvers? Don't do these on your own. With minor distractions 600 feet can very quickly become 400 feet and lower. You fly where you look. There are obvious safety issues, regulations, and your instructor's certificates to consider. Your instructor is totally responsible for you even when he is not with you. You fly on his certificate until you receive your own, so be good to him.

Why do these maneuvers? The logic is that you can't stay lined up with a runway if you can't correct for the wind. This is true. But ground track maneuvers won't teach you how to do that. You can't transfer knowledge anyway so there is no purpose for these procedures, except that they are required for the flight test. What ground track maneuvers will teach you is how to do ground track maneuvers, period.

You may be confused by the difference between airspeed and groundspeed. Air moves over the ground. Your airplane flies through the air that moves over the ground. The airspeed indicator measures the speed at which you fly through the air that moves over the ground. The airplane has no idea what it flies over, only what it flies in. That is airspeed.

Groundspeed is how fast you fly between two fixed points on the ground. Wind, or the air moving over the ground, will help or hurt you depending on its velocity and angle to your path over the ground. The time it takes you to traverse two points on the ground with the wind acting on you is groundspeed.

Any wind that crosses your path other than from directly ahead or behind will change your path over the ground. Turning the aircraft or *crabbing* into the wind enough to counter the force of the wind pushing you off your track is called *wind drift correction*. You supposedly learn these maneuvers to hold a ground track, like when flying a traffic pattern. However, when you get to actually fly a traffic pattern, you will be so busy that you will forget everything you ever learned from ground track maneuvers.

The one maneuver you can almost relate to something useful is the rectangular course. A traffic pattern is a rectangle. You pick a rectangle out there on the ground that approximates a traffic pattern. You fly at a traffic pattern altitude. You fly at traffic pattern airspeed. The difference is that you must keep your reference rectangle inside your flight path, on the left where you can see it. With your reference inside your flightpath, you are never lined up on anything. All this can teach you is how to fly outside a traffic pattern and parallel to the runways.

This is why you cannot transfer this knowledge to a traffic pattern. After perfecting this maneuver, you are expected to be able to line up on a runway in front of the airplane, not inside and to the left. Why not just fly a real traffic pattern and maintain altitude. You can line up, follow, and track away from a runway, just like a real traffic pattern. That would do you some good. When you can correct for the wind and fly a decent pattern, add in the descent for landing and climb for takeoff. The next logical step would be touch and go's. Instead, we will warn you of the dangers of low-level flight, and then have you fly outside a simulated traffic pattern, while dodging hills, power lines, and other aircraft.

Time for the Twilight zone. S-turns across a road. How flying nice, round, curves over a road is significant to your training is a mystery. You won't cross a runway perpendicularly unless you plan to land on its width. For this maneuver you will need a long straight road, canal, power line, railroad track or something; preferably a road for that is what the maneuver is called. You will also need a perpendicular wind, otherwise the explanation in the handbook won't work.

It never happens. In practice with an unknown wind and nothing to line up on, you will never know how much of your nice, curved flightpath is because of wind, drift correction, or luck. The only time you are lined up on anything is when you cross the road. At that instant you have to be perfectly square to the road and can't use drift correction. What can you learn from this except how to make nice turns? Have fun with these. You may be low enough to see people on the ground pointing at the weaving airplane, apparently lost. I would love to get a jetload of passengers and do some of these, just to stay proficient.

What they are trying to teach you is this: With a tailwind your groundspeed increases; with a headwind it slows down. When you fly across a wind between points on the ground, you turn into the wind to counter its force. The faster your groundspeed, the steeper you must turn to maintain your track. That's it.

I should take a moment here to clear up some terms that most pilots use in conversation but really can't define. I know, I've asked them. They don't have to because every other pilot thinks he knows what the words mean. They are *maneuverability, controllability,* and *stability.* The technical definitions are too confusing, so try these. Maneuverability is what the airplane can do. Controllability is what you can make the airplane do. Stability is the airplane's resistance to what you try to make it do and its ability to fly just fine without your help.

So much for the early hours, on to the next phase . . .

5 • THE FIRST SOLO

Somewhere in the early hours of your training, the realization hits you that you will actually have to fly this thing by yourself one day. Up until now, you have been happily following your instructor's commands to the letter. He has been making the decisions; you have been moving the controls. Monkeys can be taught to fly like this. You begin to realize that it is the act of making decisions, not moving the controls, that actually flies the airplane. You must train your brain to make decisions, you must develop judgment, and you must change your attitude toward flight. This is the goal of flight training.

The problem is that the system is not set up to accomplish that goal. You must do it on your own. You will see this change when you start trying to solve the problems that arise. You try to resolve the traffic conflicts. You begin to analyze why your technique doesn't work. You try to think ahead what any procedure will require. You will see that learning comes in breakthroughs, followed by long plateaus of nothing. It is not a smooth and consistent progression. The thought of soloing is frightening because you know you have to make all the decisions yourself, and they have to be correct. Up until now, you have been a passenger. You become a student pilot when you accept that you can fly by yourself and start preparing for it. This is the first step in accepting your ultimate responsibility, the authority of *pilot in command*. That is what flying is really all about.

Although you have crossed the threshold and are starting to think like a pilot, the FAA still has to be notified before it is official. Nothing is real in aviation unless it is on paper. You are not a student pilot until you get a student pilot certificate. The FAA will now know who you are, where you live, and your current state of health, because all you have to do to get a student pilot certificate is pass a third class (the easy one) medical exam. This is to insure that anyone who operates an aircraft is healthy enough to not constitute a threat to safety. When you fly with an instructor, you also fly on his medical certificate as well as his pilot certificate.

You want the privilege of solo flight, so you now need your own medical and student certificate. Your medical is good for 24 calendar months, and it expires at the end of the month two years after your exam. Finding an FAA medical examiner is easy. Look in the Yellow Pages. Usually your flight school will have a particular doctor to recommend. Sometimes this is fine. Sometimes though, the person doing the recommending is getting free medicals for steering over the business. "Go to Dr. So and So, he's easy," is a common phrase around flight schools. Do you really want the easy exam? If there is a question of fitness for flight, don't you want to know? It can be heartbreaking to have a health problem stand in the way of a dream, but you owe it to yourself, your future passengers, and everyone you fly over. The good news is that there are pilots flying with one eye, who have had cancer, heart attacks, and strokes. Perfect health is not required.

However, the further you carry your dream, the stricter will be the requirements. If you plan a career in aviation, you might consider taking a first class exam (the hard one), to see if you can pass the one you will need for the airlines before investing your life savings in flight training. The thoroughness of the exam, however, depends more on the individual doctor than the class of exam. Some doctors poke and prod every cavity. If you survive the exam, you are fit to fly. Others will hand over a medical certificate if you can see well enough to avoid the furniture and have the strength to write a check. To avoid any conflict of interest between your school and their doctors, you can always pick your own.

The medical certificate serves two purposes. On the front side is your bill of good health. On the backside is space for a series of instructor endorsements. The top spaces are for four models of aircraft you can be endorsed to fly solo, although most students have enough trouble just flying one. The bottom is for your cross-country training endorsement. Have you even read the back side of your medical? Anyway, endorsements are your official permission to operate in solo flight. You cannot even breathe on an aircraft by yourself without the correct endorsement.

You must understand that the student pilot certificate is a real pilot certificate. It has privileges and limitations to which you must strictly adhere. Most students haven't a clue what their legal rights and responsibilities are because all responsibility ultimately rests with the instructor. Your blind faith and irresponsibility are what gets so many instructors in trouble. They know that, which is why they have to control you so closely. This, however, prevents you from learning to think for yourself. Everything is backwards.

Many a student has come before me for a cross-country sign-off or to discuss the current weather for a possible practice flight. I am now responsible for that student, having given him my recommendations, so the first things I check are his endorsements. Many students have expired or inadequate endorsements, making their solo operations illegal. Had the FAA shown up for a ramp check, both the student and the instructor would have their butts hauled down to the FAA office.

Most students have no clue that they must insure their own legality for flight because they feel they operate with impunity under their instructor's wing. Many instructors neglect to keep up on the expiration dates of the endorsements they give and get themselves in trouble because, even though they told the student to watch the dates, students don't care because they don't have to care. I have seen instructors fired for lapsed endorsements. You must take the responsibility to make sure all your endorsements are current and correct.

There are two components for any privilege or certificate: an aeronautical knowledge requirement and one for flight proficiency and experience. The instructor who teaches you in a particular make and model certifies that by endorsing the back of your medical certificate as previously mentioned. They also have to endorse your logbook stating that they gave you the training (which you haven't read) in Part 61 of the Regulations. It certifies that you are competent in those procedures and may operate in solo flight for 90 days. This is the winning offender for most neglected endorsement from student and instructor alike.

Should you wish to land at another airport, you will need an endorsement from the instructor who goes with you to those other airports. Should you want to fly through a Terminal Control Area (Class B), you will need an endorsement for that, too. Most students have one instructor who takes care of all those endorsements and knows exactly what you are capable of doing.

Instructors other than your primary instructor are understandably reluctant to give you recommendations for any privilege that they have not specifically endorsed. Some schools exacerbate this by switching you around to different instructors every few lessons. Sometimes your instructor is on the golf course when you want to fly. Sometimes the instructor who gave you all your endorsements takes off for the airlines, and you have to start all over with another instructor. You can only get an endorsement from the instructor who specifically gives you training in that procedure. Sometimes you have endorsements from several different instructors, so no one instructor can give you recommendations on any given day for all your privileges because he didn't give you your training. Endorsements can be a mess, and that is just the flight endorsements.

Back in the good old days that never were, the rules were simple. You learned to fly from a small field, you soloed in a few hours, and you practiced in the vicinity of your airfield, where you couldn't get in too much trouble. Flying is now far more complicated. The rules are stricter and more confusing. The airspace is more complex, getting worse, and has been renamed. The airplanes are heavier and faster.

Students, unfortunately, are still the same. To operate safely in the current environment, the FAA has wisely required a presolo test to confirm that you have the necessary aeronautical knowledge. The presolo test is where you discover what you have retained from your training so far. When presented with new material, most students simply nod their heads and say, "I understand." For the moment, you probably did. Your instructor, having covered the topic, figures you understand and he need not mention it again.

Because you accept the material without knowing how to use it, the information will not be with you for long. That's not good enough in aviation. All knowledge is cumulative, and you have to have instant recall and know how to apply all your knowledge to various situations. For reasons you already know, the presolo test will be a cold slap of reality because you do not have a working knowledge of the material.

The presolo test must cover the relevant regulations in Parts 61 and 91 which apply to student pilots and the information from the aircraft manual to operate the aircraft. This is not a traditional test where you can pass or fail. This is open book, and the goal is to teach you where to find the information you need and then teach that material to you. It may appear difficult at first, but this is the easiest test in aviation. The challenges increase in difficulty as you advance, but so does your capacity to handle them. That's the idea anyway.

The content and form of this test are purely at the discretion of your instructor, who has the opportunity to tailor a creative test that really makes you think. One instructor I know based his test on problem scenarios his previous students had run across. It is a great test. Many schools unfortunately standardize some multiple-guess, rote exam where all the students have copies of the answers ahead of time.

The beauty of this test from the instructors' point of view is that it relieves them of the accusation of personally holding up your first solo. This bureaucratic obstacle says that you cannot solo until you know this stuff and how to use it. In our society, the anonymous decree on paper carries more weight than the professional judgment of a real live human being. It is this test which convinces you to learn and know the material because you want to solo, not the word of your instructor who has being trying to convince you all along to learn this stuff. That takes care of the aeronautical knowledge requirement.

Before any certificate can be issued, you have to meet eligibility requirements, even for the student certificate. You can take lessons anytime you want; it is only when you desire solo privileges that you need to be certified. You must have reached your 16th birthday. You must be able to read, speak, and understand English, unless you are a foreign student with lots of money, learning to fly at a Part 141 academy.

This isn't an official exception to the Regulations; it's just the way it is. Anyone who has flown within 100 miles of an academy will hear student pilots with dangerously little English at their command. There is nothing like the sound of thick Asian, European, and Arabic accents on a crowded frequency around a busy terminal to increase the tension. Learning to fly is difficult enough without having pilots who speak poor English trying to comply with air traffic directives. Foreign students make great pilots. They are well-educated and capable people. That does not, however, relieve them of their duty to communicate in proper English in return for the privilege of training in the United States. Anyone who has ever had to take evasive action around a student who did not understand what was happening will agree.

Anyway, the third requirement is the medical. Since

most students have no idea that the student certificate is in fact a pilot certificate, I would recommend that the medical become a separate certificate and a card be issued to students; one side of which carries space for all the required endorsements, the other side lists all the Regulations pertaining to students, clearly spelling out their privileges, responsibilities, and limitations. If you agree with any of the suggestions in this book, please write the FAA.

Flying has bureaucratic walls designed to take all the fun out of flight training. Scaling those walls is just part of the ritual. Some of you will find the things you must comply with are just too much hassle and use it as an excuse to get out of flight training before ever having experienced the joy of solo flight. Hang in there. As with many endeavors, the moments of ecstasy require hours of frustration beforehand. Some of you will drop out anyway. Well, there is always bowling.

It takes balls (figuratively speaking) to take an airplane up by yourself. Somewhere before you solo, the initial shock of just being up in an airplane will wear off, and you will start to be driven by a desire for precision and perfection. Your reward will not be the accolades of peers; rather you seek the fulfillment of meeting and beating a challenge. Flying well is a very personal thing. Call it "the right stuff" for lack of a better term. Something happens as you prepare to solo that adds a new dimension to your personality and confidence to your spirit. That is what carries you past the bureaucracy.

Up until now most of your training has concerned things that you should not do: stall the airplane, turn too steeply, fly too slowly, and maneuver too close to the ground. You have also been taught how to get out of the things that you should not do. Before you can solo, you have to demonstrate that you are good at the things that you should not do, just in case you do them, which you should not. You will now be taught more practical things, like how to takeoff and land properly. During this time, you will completely forget how to recover from the things that you should not do, which is why you should not do them.

Normal flight consists of climbs, descents, turns, and straight and level. The other essentials for normal flight are takeoffs and landings. Every flight begins with a take-off and should end with a landing. This will be the most demanding, intensive, and frustrating part of your training. Touch and go's are busy, and things happen fast for the presolo student. When you can takeoff and land consistently and relatively well, you will have accomplished something. Things change when you operate near the ground. The air is soft, but the ground is very hard, making this the most critical phase of flight. Being close to the ground, people, especially your fellow students, can now watch you fly. You cannot slide all over the sky when you operate in the tight confines of the traffic pattern.

Your preconceived notions of takeoffs and landings will initially delay your getting the hang of them. You think takeoffs are no big deal. Sure, you start on a narrow runway, but you get to blast off into the wide open blue sky. Landings are definitely a big deal because you have to come from the wide open blue sky and land on that narrow runway. There is an obsolete saying, "Any landing you can walk away from is a good one." That may have applied back when flimsy biplanes operated from fields dotted with potholes.

Today the standard is much higher, especially if you are training for a career. Nobody wants you to bruise their airplane. There will be many landings during this time you wish you could just walk away from. You need to put a realistic perspective on landings. It takes about 300 hours of flight experience to be able to evaluate all the variables, runway conditions, and weather to consistently make good landings. What you are trying to achieve before you solo is to learn to land reasonably well and safely under ideal conditions.

Everyone reaches a plateau in training where no improvement is happening. You will think you are the only pilot who can't learn to land an airplane. This is just a normal phase in your training. You must keep going. What will happen, when you let it happen, is a breakthrough. Lightbulbs will go off when you cross that magic point, and suddenly takeoffs and landings make sense. Aha! You won't always land well, but you will understand what you did wrong. The reason for the 300-hours experience is to develop the judgment to know what to do ahead of time to make a good landing.

There is a reason you think you will never get the knack. You spend almost all your time watching someone who does not yet know how to fly — you. The only other person you see is your instructor, who does know how to fly. With only these examples, you figure that all pilots are like your instructor, and you are the only hopeless student. If flight schools could make videos of the students, and the instructors when they were students, learning how to land; if some Friday night complete with popcorn, those videos could be shown; everyone would have a good laugh and students could remove the burden from their shoulders and simply concentrate on learning how to land properly. Just remember, if flying were easy, everyone would do it.

There is a tendency among students to refuse inflight demonstrations by the instructor. You will have long since forgotten the landing demos from your first lessons when you go hard at the touch and go's. Any model you had in mind to follow is gone from your memory. You refuse demos for two reasons: One, you are paying for the airplane, so damn it, you are going to fly it; and two, you really want to please the instructor so you know you need just one more chance and will get it right the next time.

Many instructors are put off by this attitude and have two ways to deal with it: One, you are paying the bill, so you can call the shots; and two, if you don't want their help, they will just sit and happily take your money. If you encourage your instructor to model the first touch and go each lesson, you will have a fresh example in your mind to follow. Once you know what to look for, you can more quickly and easily make the same picture. When you get the rote stuff out of the way, you can concentrate on developing skill and judgment.

The best thing you can bring to these lessons is a good sense of humor. The instructor is there to insure safety, and trainers are constructed to let you see some amusing effects without doing any real damage. Sit out on the approach end of a runway if you think you are the only one who can't land an airplane and count the tire squeals. Remember that airplanes are designed to cruise, not to land. The greatest contributing factor to the perfect landing — is luck.

Your attitude toward takeoffs is too frivolous. Point the plane down the runway and go, no big deal. Your complacency will be tolerated for a while. Before the instructor can let you go alone, however, he must have confidence in your ability and judgment. Much of this can be determined from your attitude. Do you voluntarily check the weather ahead of time and discuss how that will affect your takeoff? Or do you just show up at your lessons with the look of a child on his first day of kindergarten?

If the winds are strong, do you want to cancel and go home? Or do you say to your instructor that the winds are too strong for solo flight, but if possible it would make a good challenge and learning experience for a dual lesson. Do you ever check the windsock right before you go? Or like most students, do you consider the decision set in stone way back in the lobby when the instructor first said you will go.

How actively do you participate in the process? How close are you to accepting the responsibility of pilot in command? Do you look to see if any airliners are on final even though the tower has cleared you for takeoff? Just three weeks after the incident in Los Angeles where an airliner landed on top of a commuter, I was sitting on a runway ready to go when a King Air was cleared to land on my runway. While reaching for the throttle to get off the runway and the radio to announce my intentions, the King Air called "traffic on the runway" and went around for another approach. My student, a rated pilot, had no idea that it happened. He had only been listening for our call sign, so anything else was superfluous.

We do not expect a high degree of situational awareness from soloing students; that's why we send you out under ideal conditions. You must be aware though that anything can happen, so it is critical to have a certain degree of competence beyond the basic mechanics. Flying is thinking — any idiot can move the controls.

The weakest link in the chain of learning is the application of the correct technique to the proper situation — without prompting. The FAA requires you to be proficient in normal and crosswind takeoffs. Most students by the time they solo can handle a light crosswind. They know the technique to use in a crosswind. They never use it, however, unless specifically told to do so. I have watched countless students waffle and limp into the air using the normal takeoff technique, even when they knew there was a crosswind. My students got to solo after they said something like, "We have a crosswind out of the west. This is what I'm going to do ..." Bless you.

You know that the takeoff is much more than the mechanical description of the roll, rotation, and transition to normal climb described in the handbooks. You cannot fly without thinking. Since the system is designed only to teach you the mechanics, your training will be greatly extended. If you find yourself continually repeating the same mistakes in flight, get out of the airplane and spend time on the ground with your instructor and analyze your thinking.

Here is an example of what I mean. A student came to me for a phase check. That is where a student flies with an instructor other than the primary one just to make sure the first instructor didn't miss anything. This is required by the Part 141 schools and is a good practice anyway. Well, there we were sitting on the ramp with the engine running. The student diligently copied down the Automatic Terminal Information Service (ATIS), which is information on current conditions at the airport necessary for takeoff and landing. I noticed they had drawn this neat tic-tac-toe board where all the information was put into its own cute little box.

"Why do you do that?" I asked.

"Because my instructor told me to," came the half-answer/half-question.

That told me everything I needed to know. "Well, what does it mean?" I asked.

"This is my ATIS information."

Hmmm. We taxied out. The student stared directly ahead. I rephrased my question and tried again. "So, how does this affect your takeoff?"

There was no response to this question, only the blank stare of a student who could not think beyond rote, reflex actions. I had better get his confidence back. "What are the winds?" I asked.

"The winds are 240 degrees at 10 knots."

That was true, but I had to find out how much he actually knew about flying. "If the winds are 240 degrees and we are using Runway 30, what does that mean?"

"It means we have a crosswind."

Hallelujah. There is hope after all. I really thought that after such discussions I would see a crosswind takeoff, but like with so many other students, we lurched into the air sideways using the normal technique. There was no correction for the wind at all. I inquired what the crosswind technique was. He knew exactly what it was. There was no answer when I asked why he had not used it. The reason was simple. Like so many students, this one had always been told exactly what to do, every single time, without ever having to think for himself. His next takeoff had pretty good crosswind technique.

Back to the original question, I asked why write down the ATIS information if that information is never used?

"Because my instructor told me to," was the answer, accompanied by that same perplexed face you always see from anything other than a yes/no question.

I give up. This is so typical it is scary. Why bother writing down the information if you have no idea why you are writing it down? Why do you just accept tasks without any purpose for them in your mind? If your instructor showed up one day and told you in a perfectly serious voice to go practice your evasive maneuvers by flying backwards around the pattern, would you do it?

If your takeoffs feel sloppy, they probably are. Don't be satisfied with thinking you slipped something by the instructor; find out how to fly better. I frequently let sloppy techniques go briefly without comment, just to see what students will tolerate from themselves. Most takeoff errors happen during the transition of weight from the wheels to the wings and during the initial climb out of the ground effect. Ground effect is the change in airflow off the wing because of the close proximity to the ground, which results in more lift at a lower speed. The common mistakes of lack of crosswind correction, torque correction, rapid or excessive pitching of the nose, all boil down to not having an adequate plan in mind before you go. Get your instructor to help you develop good takeoff planning.

Landings you will treat completely differently. Your takeoffs are bad because you don't give them a thought. Your landings are bad because you think they are the most critical thing on the face of the earth. A good landing is the hallmark of a good pilot. That only the velvet touchdown of the proverbial greaser is acceptable technique. Why not, no one wishes you "happy takeoffs."

So why is it so hard to learn how to land an airplane? One reason is that every landing of every day is filled with so many variables that no standard rules will ever apply all the time. You cannot learn to do something consistently well when there is nothing consistent about it. The change in temperature, the wind, the personalities in the tower, the flow of traffic, and the unexpected, all conspire against you. Learning to land well means learning to handle all the variables, and like everything else it takes time and practice. There is no magic secret to landings.

Your lack of skill works against you as well because your fluctuations in altitude, airspeed, and flightpath mean that you are never starting your approach from the same place. Each landing becomes a new experience. Learning to land means learning to compensate for all your fluctuations.

There is a way to help yourself while you are learning. Before you go for a session of touch and go's, get with your instructor on the ground to discuss the particulars of the current conditions and how you are going to handle them. Same old story, make a plan, and you will be ahead of the game. Typically though, you have to botch a few touch and go's before the instructor decides to tell you why the conditions are fouling you up. It's too late then. Take heart though, you will never solo in less than ideal conditions. Your instructor will tell you this is to boost your confidence so you will enjoy the experience. In reality, they don't want you to get them in trouble by screwing up and hitting something.

There is another problem. The reason you have so much trouble touching down with the airplane is because you have no idea where the air ends and the ground begins. The closer you get to the ground, the less you are aware where it begins. You only know where the ground was when you started the transition from the approach to the landing. Anything beyond that is at best a guess, as you try to round out the airplane based only on where you think the ground might be. You cannot be ready for contact with the ground when you have no idea when that will occur.

Since the traditional round out or flare is dependent on precisely timing your arrival at the landing attitude, so that it is just above the ground, when you run out of lift, at the slowest possible speed, you can't possibly do it if you cannot see the ground. That is why you flare too soon and end up hanging in the air or too late and fly into the ground. If you ever time it right, it is sheer luck. You think you have learned something but are frustrated when you cannot repeat the performance. That is why it takes so long to learn how to land an airplane.

Your instructor compounds the humiliation by stating the obvious about your landing ten feet high or ground plowing. They might come back with something useless to cure it, like telling you that you were "behind the airplane," which means you didn't think far enough ahead. That doesn't deal with the problem of you not knowing where the ground begins. You can talk all lesson about the dangers of flaring too high or trying to snap off the nose wheel, and you can try a landing again. You can botch it a different way because you are overcompensating for where you thought the ground was last time to where you think it might be this time.

After a frustrating lesson, you can sit with your instructor and try and remember what you did during the lesson. You can't because you won't remember not seeing the ground. Even if you did remember, you won't think that is the problem. If your instructor only deals with the standard teaching methods, he won't know why you don't know where the ground is either. How can the instructor solve a problem that neither of you understand? None of the standard stuff will work because it does not identify and deal with the real problem.

This forces you to adapt and compensate for the inadequacy of your training rather than challenge the system to help you. You will eventually teach yourself some way to approximate your distance from the ground; sometimes your system will work, and sometimes it won't. When it works more than it does not, you will solo. But you will carry a flaw with you that almost every other student carries, because the reason no student knows where the ground begins is because the method we use to teach landings to students is wrong and does not work.

All the coaching in the world won't stop the mistake that every student makes. You fly the airplane down near where you think the ground is, raise the nose to where it totally obscures any forward vision, idle the engine, release all control pressures which neutralize your flight controls, stare straight ahead at the blue sky over the nose, wait, pray, and hope for the best. All students give up flying the airplane the minute the ground disappears from view. From then on the landing is just a guess.

The FAA handbooks, advisory circulars, noted aviation authors, and magazine experts all carefully document where to look. The problem is that they have so much experience, over so many years, that they cannot think like a student anymore. What they state is perfect for the accomplished pilot. It is useless for the beginner. Where they tell the student to look is the wrong place to look.

A student pilot does not have the discipline and confidence to take his eyes off the nose nor has he acquired the skill to use his peripheral vision, so he stares straight ahead directly over the nose and completely gives up flying when the ground disappears. I ask students with landing problems where they are looking. They cannot tell you. They cannot believe that there is any connection between where they look and how they fly, preferring instead to blame their lack of flying skill or their instructors. Instructors cannot tell where their students are looking because it is extremely difficult to observe that when flying so close to the ground. The instructor is too preoccupied with the safety of the airplane to take his eyes off the ground and see what the student is doing.

There is help for you. While you are making your final approach, look at your desired touchdown spot. Hold your six-finger pitch to control the airspeed. Adjust your power to control the flightpath. As many books will tell you, if you hold your touchdown zone in an imaginary target just above the nose, you can adjust your flightpath by observing whether your target spot appears to move toward you or away from you. You will overshoot if the spot moves toward you and undershoot if it moves away from you. That is a good technique; it works and it is common knowledge.

Also common knowledge is when to start your transition to landing. There is an optical illusion where instead of appearing to descend toward the ground, the ground appears to be coming up to meet you. This is where the round out begins. This is also where all the standard methods fall apart and lead you astray. Here is my difference. As you begin your round out, look at the far end of the runway. Make it your new target just above the nose. As the speed bleeds off and the airplane descends during the round out, start to put the nose just under the far end of the runway. Do whatever it takes to hold it there. The airplane will automatically slow to the landing speed and assume the correct landing attitude as you approach the ground. Because you always have a ground reference directly in front of you on the nose of the aircraft, which is the only place you want to look anyway, the ground will never be out of your sight.

The beauty of this technique is that you can know throughout your entire landing exactly where the ground is and still not have to exercise any judgment so long as you keep the nose just below the end of the runway. The sight picture assures the landing. Think about it. The end of the runway is a fixed reference. Your eyes are a fixed reference as far as you are concerned. That leaves the nose of the aircraft as the movable variable. As the airplane descends, the end of the runway appears to move higher in relation to the nose. In order to maintain the sight picture keeping the nose on the end of the runway, you have to progressively raise the nose. The end of the runway serves exactly the same function as the horizon for a visual reference, keeping your training consistent. If you raise the nose too fast, it will obliterate the end of the runway. You will have to lower the nose to bring it into view. If you have a tendency to flare to high, you will lose it because the procedure forces you to lower the nose until you are closer to the ground. If you tend to fly into the ground, the procedure makes you raise the nose to the end of the runway before you get too near the ground.

Try the explanation another way. From our chat about fingers, you know that six fingers below the horizon is a descent, four fingers is straight and level, two fingers is for a cruise climb, and the nose on the horizon is the takeoff attitude. It is also the landing attitude because that is where the threshold of lift allows you to either takeoff or land. On landing, replace the horizon reference you use in flight with the end of the runway and you have the same transition. When you touch down, they will merge to exactly the same reference, which is why the picture works.

Try it yet another way. Imagine a straight line from your eyes to the end of the runway. The nose is way below that line in descent, somewhat below the line when transitioning through level flight, and fixed on the line before you reach the landing attitude. That is your round out. There is only one pitch attitude for every altitude above the runway that will keep the nose lined up with the end of the runway. You will always achieve the landing attitude at the instant you touch the ground, because when the nose is lined up with the end of the runway at ground level, it is on the horizon.

You will always make a reasonably good landing using this technique. You will never find it in any other book that I know of. You will probably have to modify the sight picture slightly, depending on your height in the seat and the type of aircraft you fly. This technique is especially valuable at night when the ground references have disappeared. By keeping the nose on the red lights at the end of the runway, you accomplish the same thing. This technique was developed to help students land at night. One day I just tried it in the daytime. It worked. The only catch though is that having never flown a taildragger, I have no idea if it works for them. Find someone who does know, if you are curious.

Most instructors and the handbooks stress the wrong things when addressing landings. The landing does not begin until you arrest your descent to the ground. Up until then, you are still on the approach. Instructors and examiners make a big fuss about maintaining a *stabilized approach*. Since no two pilots share an exact definition, this is yet another meaningless expression to students. What they are trying to have you do is maintain a consistent airspeed and approach path.

Now, ask your instructor how do you maintain a stabilized approach in unstable conditions? Any landing approach, especially a long straight in, will encounter changes in temperature, terrain, wind, ground heating, engine power or manifold pressure (I'll explain that one later.), and turbulence. Nothing is stable. If you just remember to line up with your touchdown spot on approach and the end of the runway for landing, you can't miss. Since landing an airplane is so overwhelming to you, hearing a fuzzy platitude like "maintain a stabilized approach" gives you nothing to latch on to. After some experience in this technique, you will begin to see the ground using your peripheral vision, while still looking down at the far end of the runway.

Once you can accurately judge where the ground is, *now* is the time when you can begin to learn how to land with the nose blocking your forward vision. Everything is backwards. Why does the standard method teach you to land with your nose in the air *before* you know where the ground is? Ostensibly the goal is to land as slowly as possible and stall at the instant of touchdown. Why? Any student pilot will tell you this takes too much judgment and skill. Stall anywhere above the ground and you drop in. Wait to late to stall and you nose it in.

This technique is supposedly taught to save tires, brakes, and general wear. Watch any touch-and-go session and see how much wear and tear is inflicted on an airplane with a student attempting the full stall technique, and you will see that there has to be a better way. Why not challenge the premise that an airplane ever has to be stalled during the landing.

When you use my method, come in just a touch faster than the recommended approach speed; five knots should do it. With your nose on the end of the runway, idle the engine. You will roll on to the runway under control and with a full field of vision. You must not come in too fast as you will float. However, by maintaining the sight picture, you will just achieve the target speed and attitude further down the runway, that's all. The extra speed is needed to maintain flight control effectiveness throughout the landing. Controls lose effect with declining airspeed. The full stall technique requires the greatest control input at the very time students are losing sight of the ground, giving up, neutralizing the controls and hoping for the best. That is why it doesn't work.

By gently rolling the airplane on to the runway, the student has full use of the controls, the ground is in sight, and he is still flying the airplane. When you can see the ground, you see what is happening with the landing, you can analyze your faults, and learn something for the next attempt. No student attempting a full stall landing really knows what is happening. The worst part is you don't know why you can't learn how to land and feel you have no choice but to blame yourself, when really it is the system that will not allow you to learn to land properly.

When you can see down the runway, you can also see if you are lined up with the runway. Students invariably are not. When the nose blocks the runway and you stare blindly at the sky, the airplane can change direction without perception by you, because you have no reference with which to sense the change in direction.

This is especially evident in a crosswind. Yet another fault of the full stall landing. Students can make a passable crosswind approach, only to lose sight of the ground, give up, neutralize, pray, and land sideways with a screech, with no idea that the airplane had changed its heading, but not its direction of flight. What happened? When you give up control of the airplane, say five to ten feet off the ground, the wind hitting the tail immediately starts to weathervane the airplane into the wind. The momentum of the airplane, however, keeps it moving pretty much the same direction. The screech comes from the wheels as they turn to match the direction the rest of the airplane is going.

Since you could not see the ground after the nose blocked your forward vision, you could not know that the nose had swung because you had no reference with which to perceive the movement. You are told for crosswinds to come in with a lower attitude and land a bit faster. This isn't specific enough. You will also be told to add half the velocity of the gust to your approach speed. Well, you can have strong crosswinds without gusts, and being such literal people, you will not change your technique. The biggest problem is that the way you first learned how to land was to block out the world with the nose, so that is where you will always return. By using the end of the runway as your target, you will always have the reference to perceive drift, and you won't have to change a thing in your technique.

If the crosswind won't misalign the airplane, you will surely do it yourself. It comes about from something totally unrelated, something we in the biz call a negative transfer of learning. It happens because you are taught when you taxi to always keep the nosewheel on the centerline. Since you can't see the nosewheel, you have to guess where it is from a parallax view over the nose to the spinner. By attempting to put the nosewheel on the centerline from a view of the spinner from the left side of the airplane, you acquire a habit of centering the airplane by looking across the nose at an angle. This will permanently screw up your landings. It will forever distort your perception of where the airplane nose is pointing. On the ground this isn't much of a problem.

In the air however, looking across the nose at the spinner to center the aircraft subconsciously makes you line up the aircraft spinner with the runway centerline. The only way to put the spinner between you on the left side of the airplane and the runway centerline is to swing the whole airplane around to the left. You will always be crooked when you land. Because of the habit of looking across the nose at the spinner you acquired from your instructor insisting that you taxi with the nosewheel on the centerline, you will swear that on landing you are lined up correctly even as the wheels screech every time you land. Don't ever let an instructor make you taxi with the nosewheel on the centerline!

Instead, put *you* on the centerline. If your nose, not the airplane's nose, is situated say ten inches from the exact longitudinal center of the airplane, taxiing with your nose on the centerline puts the airplane a whole ten inches off center. Big deal. Most taxiways can handle it. Most runways can as well since the average trainer wheelbase is about eight to ten feet and the runways that students will use are at least 25 feet wide. When you are putting your touchdown zone and end of the runway just above the airplane's nose, put the centerline directly in front of your nose. When your reference on the nose is the part of the cowl directly in front of you, you will be looking exactly where the airplane is going. You will land *your* body on the centerline, not where you think the spinner and the center of the aircraft will hit the centerline.

There is no regulation that states you have to screw up all of your landings by trying to guess where the center of the aircraft will touch down by looking across the nose at an angle. I have drawn gunsight targets with a grease pencil on the windshield of the airplane directly in front of the students to break them of looking across at the spinner and force them to look directly ahead to put themselves on the centerline. Forget the other half of the airplane. If you look straight ahead, you will land with the wheels pointing straight ahead. Forget putting the airplane on the centerline; always taxi and land with you on the centerline. This will allow you to fly from either the right or left seat without any problem, because wherever you sit, you are always the center of the airplane.

We have now built a landing where the ground is always in sight, you always have a fixed reference at the end of the runway, the controls remain effective, you don't give up, you know exactly where to look, you are always lined up correctly because you are the center of the aircraft, and you don't have to think about any of this for it is all automatic. Not bad.

What about those crosswind landings that are driving you nuts? You have probably figured out that the standard way you are taught to land aggravates the effects of a crosswind. Since everything is backwards, the way you are taught to handle crosswinds only makes things worse. The two official ways to correct for a crosswind are the *crab* and *wing down* methods. The crab method is great for the approach; simply turn the airplane into the wind and track towards the runway. You can even draw on your ground track experience if you happen to have parallel runways and have been assigned the runway on the right. From your rectangular course, you have learned how to fly out-side a track to your left, so just track outside the path of the left runway and you should be fine.

I'm just kidding. Unlike ground track maneuvers, when you descend for the approach, the winds continually change. The windsock gives you wind information, but it indicates only the surface conditions, which won't help you because that is where you raise the nose to block your vision. Any-way, there you are a 10-20 hour student; you are expected to crab into the wind, fly down the runway at an angle, and kick the airplane straight at the last possible second before touchdown to avoid drift and side forces on the gear. Don't even try it.

You don't know where the ground is or where the nose is pointing, so you haven't a prayer of judging this one correctly. After the full stall is beaten into you, it is damn hard to adjust to the lower attitude and attempt to kick around an airplane just above the ground. This wouldn't be an issue had you started your touch-and-go training with the sight picture, rolling landing. When you have profi-ciency at the rolling technique and have developed a sense of where the ground is through your peripheral vision, you will be able to try the crab method and see if it works for you.

I never use it because it is based purely on guesswork, and that is no way to fly an airplane. The only time the crab is useful is on approach. That is because it does not cause anywhere near the drag of the other method. If you are low, it is perfectly acceptable to crab; just don't try to land that way while you are still learning.

The other way is called the *wing down* method. All it consists of is a sideslip into the wind at the same rate as the force of the wind tries to push you off your desired track. Since it is just a sideslip, it should be called the *sideslip method* so students would understand what was going on.

Calling it wing down causes one of the main errors in the technique because students think of the ailerons and lowering the wing as leading the crosswind correction, when it is the rudder and directional control of the airplane that is most important. Everything is backwards.

The folly is compounded because this method is taught so badly. Even though it is the most effective way for you to control the aircraft, you will be driven by frustration to the crab method, which does not work. You will be asked by instructors, examiners, and written tests where to put the airplane controls in, for example, a left crosswind. The answer you are supposed to give to say the right words so the officials think you know what you are doing is this: If the wind is from the left, then the left wing has to be down, which means the left aileron has to be up, which is caused by lowering the left hand, which puts the plane in a bank to the left, which means the opposite rudder has to be pushed with the right foot to correct for the bank and remain longitudinally straight.

The answer is technically correct. It is the question that is all wrong. You can't possibly use this mental process while flying an aircraft on final approach; it is just too complicated. This thinking leads you to arbitrarily put in whatever amount of aileron and rudder strikes your fancy, point the plane in some direction other than where the crosswind had it, but not line it up with the runway, and call it the wing down method. Half the time you will get your aileron and rudder reversed and correct the wrong way.

There is another problem. Since the beginning of your training, you were verbally abused anytime the "ball" is out of center and the airplane is "uncoordinated." You have also been told never to use opposite aileron and rudder. Now you are told to do them both. This goes against all your training. It also takes away all meaning from the word "never."

Well, if you can be uncoordinated on landing, what is the big deal on takeoff? This inconsistency will bug you. There is no pleasing your instructor. The fault, of course, is in the training; your instructor is just passing on the abuse. Try this explanation. An airplane is a weather vane. When you try to fly one way over the ground during your approach, and the wind is moving another way, it strikes the tail just like a weather vane and attempts to point the airplane into the wind. Let it. The more the nose deviates from your flightpath, the greater is the component of crosswind. You can see what you are dealing with and make decisions. One option is to find another runway or airport. You have immediate information as to your situation before you get close to the ground. Since the nose points into the wind, you know exactly where the wind is coming from, without having to do the usual mental exercises from the ATIS while you are still flying.

Put in enough rudder to put *your* nose by looking directly ahead, not the nose of the airplane as seen over the spinner, on the centerline. The most important thing in a crosswind is to have the airplane pointed straight down the runway when it touches, so it is the rudder that leads this method. Now, put in enough aileron into the wind to hold your position on the ground track. The ailerons control the degree of sideslip necessary to offset the drifting force of the crosswind.

Simply put: The rudder keeps you straight; the ailerons keep you positioned. You will never get your controls reversed. If the nose is to the left, the right rudder brings it back on center. You will know instantly if you put in the wrong aileron for you will turn. The beauty is that you also know how much control input is required because you are using the strength of the crosswind component as your guide. When the wind changes as you descend, vary your controls; rudder first, as necessary to keep you both straight and positioned properly. When you sideslip into the wind with the same force that the wind tries to make you deviate, technically you are uncoordinated, but really you are *balanced*. Change your words, change your thinking. So much for the approach.

The full stall landing ruins you yet again. In order to

land on one wheel out of the sideslip method, you have to be able to put yourself, not the spinner, on the centerline. To do this, you have to be able to see the ground. Whether you crab or sideslip, the whole effect will be sabotaged when you lose sight of the ground because the airplane will drift as soon as you neutralize the controls. You give up at the worst possible time.

The airplane is at its slowest right before touchdown. The slowest airspeed requires the greatest control deflection to maintain direction and position in a crosswind, the very point where you are staring at the wild blue, waiting and hoping for the best. Use the target directly in front of you on the cowl and center it on the end of the runway. The sight picture will put you automatically at the proper attitude, airspeed, direction, and line up the proper wheel for touchdown. I have seen reasonable crosswind approaches blown in the round out because the standard methods give you no appropriate sight pictures. To sum up: Using the standard methods, when you have learned enough on your own to compensate for your faulty training, you will be ready to solo.

Where does all the pressure to solo come from? Certainly not from the instructor or the flight school. I have never heard an instructor say to a student, "Gee, it's taking you a long time to solo." You are much more likely to say that to your instructor, as if it is his fault. We know the standard methods extend your training. You also know how to fix that.

The problem here is with you. Led by public perception and Hollywood, you believe that real pilots solo in some arbitrary figure, like 15 hours. If you pass whatever mark you set for yourself, you consider it a personal failure and blame the instructor because you won't take any responsibility for your training. Your fellow students, who should know better, and your friends, who you got into this to impress, will keep asking, "Have you soloed yet?" No one outside aviation has a clue as to what soloing is all about. They just think they know that you will be a real pilot when you do it. Hardly, it is just a phase in your training. You know that, but the outside world does not.

To avoid a stigma that exists only in your mind, you will pressure your instructor to let you solo when you reach your arbitrary limit, even if you know deep down you are not ready. Some students actually bully their instructors into letting them solo, but if anything goes wrong, guess who gets the blame.

Another problem: Back when life and airspace were simple, pilots were limping around the pattern by themselves with 7-10 hours of flight time. Although a badge of honor for the aviation elders and goal for young pilots, the old-timers will usually acknowledge the folly and danger of what they did not know when they soloed. Students today still want to solo in the same amount of time appropriate to a country field from the '40s, while learning at a congested terminal in the '90s.

You can postpone your solo until you have more experience and still get your certificate in the same amount of hours. Concentrate on the certificate and the solo will take care of itself. Having said that, I must confess that your first solo will be one of your most memorable and enjoyable experiences ever, provided you have the skill and experience to be safe.

Your first solo and the privilege of solo practice flights are very different. You are gearing up only for the first solo. However, the student pilot certificate and instructor endorsements that allow your first solo also permit you to operate within a 25-mile radius of your airport. The first solo, and the next couple of solo flights after that, will be supervised by your instructor who will wait for you helplessly on the ground below. (I sometimes duck into the tower should the student want to chat with a friendly voice.)

Your first couple of solo flights will consist of three touch and go's. You will be briefed by your instructor, signed off, and let go. You know the solo is only three times around the patch, so why did you have to master all that other stuff? The other procedures are designed so that you will avoid trouble when you operate within your 25 miles on solo practice flights.

It is unfortunate that the way those procedures are taught has the opposite effect of encouraging you to try them. During your dual lessons, you wouldn't get the impression that a lot can happen in 25 miles, but it can. You can get lost. All it takes is an unfamiliar view of the landscape from an angle that you hadn't seen on your lessons. You can get mixed up by Air Traffic Control (ATC) without the benefit of your interpreter aboard. The weather could change during your hour or so of solo practice. There is actually an endless list of variables that can combine and contribute to your misery. Also, your instructor immediately corrects your mistakes. On your own, one mistake can quickly unravel into a series of bad decisions. This is why you must be ready to operate 25 miles from home before you can go three times around the pattern by yourself.

Many students have done very well on presolo tests only to forget all their teachings as soon as they are let loose to practice on their own. Most students do not perceive how quickly 25 miles can be traveled, even in a trainer. Your lessons may have frequently taken you to airports and practice areas that extended beyond 25 miles. Since you went there with your instructor, you feel safe and assume you can go there on your own. Better draw a 25-mile circle on your chart just to be on the safe side. Knowing the rules is for tests; applying the rules is for real life. You won't be able to solo until your instructor is sure you won't get *him* in trouble by your actions.

There is some flexibility in the rules. For instance, you can get a repeat solo cross-country endorsement to fly to and from an airport beyond the 25-mile limit. In my experience though, few students have any idea of their rights, responsibilities, and limitations. Any compliance with the Regulations becomes a matter of previous experience with the instructor and luck. I asked one student on a phase check where he could fly with his solo endorsement.

"Wherever my instructor says I can."

Typical.

How is it, you ask, that procedures and methods get passed down through the decades, without any consideration or improvement? It happens because no one questions what they are doing. Flight instruction is such an individual thing. No one watches your instructor teach you. No one will watch you teach your students. Bad habits are free to pass down the chain. Only the results of teaching are observed, and then only on phase checks and checkrides, hardly an accurate representation. The specific way you learn procedures will become for you the only correct method, even though in reality there are an infinite number of correct methods.

During your checkrides you may or may not fly with someone who learned procedures the way you did. Techniques are not judged to be correct by examiner's because they work; they are judged to be correct because they are familiar. This entire book is designed to question how and why we do things, to spark the debate that reviews our whole system of flight training. You are not expected to agree with everything I say; you are only expected to ask your own questions. That is how we will spark the long overdue debate.

Case in point. The *FAA Flight Training Handbook,* my bible, states that when approaching to land, the descent for landing begins on the base leg. It does not begin abeam the touchdown spot. You are expected to maintain your pattern altitude until commencing the base leg. Most of you will be taught by your instructors, because that is the way they were taught, to reduce power abeam the numbers. This is so prevalent in flight training, that what is really a myth, has become a fact in most pilot's minds. Such myths abound in aviation because no one takes the responsibility to check what they are taught against the published Regulations and procedures.

For anyone who thinks he "knows" that the descent should begin abeam your touchdown spot, I refer you to the *Flight Training Handbook,* Chapter 9, pg. 95, paragraph 2, which reads: " The base leg is that portion of the airport traffic pattern along which the airplane proceeds from the downwind leg to the final approach leg and begins the descent to a landing." Any questions?

My chief flight instructor came to the U.S. with the intention of following our manuals to the letter. Consequently, we all learned to initiate our descent 45 degrees from the touchdown spot as we began our base leg. I only learned the other way when my students told me I was wrong. Such arrogance. Such is the power of myth. Those of us who teach by the *Handbook* method are in the minority. The majority thinks we are wrong simply because we are different and in the minority. Truth in aviation becomes a matter of voting power. The FAA examiners who learned the other way think we are wrong, but allow our procedure because it is safe, although to them it is unconventional. The only reason we all learned the *Handbook* method is because the chief instructor did not go through our system. Both methods work. Like all instructors, however, I believe my way works best.

Now, how did this myth originate? I have a theory. Back when light, fabric, two-seat taildraggers dominated the general aviation skies, airplanes were slower and glided better than today's trainers. You could fly a tighter pattern and initiate your descent earlier with no problem. If you tried to hold your pattern altitude until your base leg, you would be up there all day and get a lovely view of the airport as the runway passed way beneath you.

Airplanes have changed. Even our light trainers are faster, heavier, are all metal, and descend more quickly than earlier aircraft. The lengthened pattern which results from holding your altitude longer gives you a little more time to plan, gives you a steeper descent which improves your visibility, and makes extending your downwind a snap because you don't have to climb back up. So for all of the above reasons you develop better situational awareness. Since instructors pass on what they learned despite the rest of the world moving on, teaching students to descend abeam their touchdown spot remained in the standard repertoire. How many myths start out this way? Which method of landing do you use? Which one do you prefer?

How will your first solo go? I don't know. There is no

way anyone can know. Since you are the only one in the
airplane, you are the only one who will ever know. That is
part of the charm. You will solo when you have met all the
requirements, done all the paperwork, passed the tests,
acquired all the endorsements, met the standards, and
have the ability to safely go around the pattern on that
particular day. Your instructor will then build you up towards safe flights out to 25 miles. Sometimes students get
so anxious and excited to solo that their ability to fly leaves
them. You can't solo when you are more concerned with
soloing than flying.

I can't tell you about your first solo, but I can tell you
about mine. A week after my 16th birthday, a cool, crisp
wind blew snow crystals across the front yard of a very
excited young pilot. A bright sun arose to light the world
with rainbows of snow dust hanging in the air. The dream
of soloing on the magic 16th washed away the previous
week in a swarm of torrential hailstorms. The night before
had been frozen and still. What had been water the day
before had now encased the wheels out on the aircraft
ramp. This New England morning, an anxious teenager,
after leaving the cereal in the refrigerator and the milk in
the cupboard, hopped into the family sedan so Dad could
drive to the airport. It would be a year before I would
acquire a driver's license.

The sky reflected an iridescent blue in all directions. I
had my own reflecting to do. That first demo ride back in
Melbourne, Australia, at the tender age of 12 got me started.
After flying a Cessna 172 atop phone books, I was hooked,
yet it would be a year before my real training would begin
in the United States. An unsettling move, massive culture
shock, and much parental begging would intervene. There
was an often-repeated five-mile bike ride from my home in
Lexington, Massachusetts, to L. G. Hanscom Field in
Bedford. An old Air Force base sharing space with general
aviation and several electronics defense contractors,
Hanscom was most intimidating.

It took much cajoling, but one school was satisfied that

an overeager 13-year-old had the knowledge and maturity to begin flight training. All I needed was my parents' permission. That night the folks received many papers to sign. What a family dinner to remember. It all came together over several years from a series of odd jobs, savings, and scraping the money for a lesson every month or so.

The last few weeks had seen very concentrated and very expensive flying as the solo was imminent. Upon arrival on the big day, the flight school appeared so terribly normal. Not the setting at all for such a momentous occasion. I was the only one feeling momentous; to everyone else it was just another cold Saturday.

Unlike many students, I knew when I was to solo — it had been planned for weeks. Many instructors play this game with students where they hide the solo right up until they hop out of the plane, so that nervous jitters won't overcome the student. This works sometimes, but not for me. Planning for an upcoming first solo is a great way to rapidly accelerate the maturity level of a 16-year-old.

The bulk of my preflight consisted of chipping the aircraft out of the ice and draining ice crystals from the tanks. My aircraft today was my favorite: a 1965 Cessna 150, with a pull starter. 8353J started out life a majestic purple and had since been reduced to faded lavender by the sun. We affectionately called it "The Jet." Some years later I returned to find that a merciful god in heaven had painted the jet a striking red, white, and blue.

There was no doubt about the weather; however, the jury was still out on the confidence. Our destination was Boire Field, Nashua, New Hampshire. After the obligatory warm-up touch and go's, I longed to dump my instructor on the snow and get on with this. I didn't want him to leave either because then I would have to fly by myself. I am just a kid, why are they making me do this? Only because I have longed for this moment since I was five years old, that's why. I was an old man now by comparison.

My instructor said a few words about possible emergencies, mentioned something about the airplane being lighter without him, signed my logbook, donned a warm hat, and hopped out on to the frozen ground near the runway. After briefly reviewing emergencies in my head and praying none of them would happen, I just sat for a minute. The first time you look at an empty right seat is one of the loneliest moments you will ever experience. Your mind goes blank.

What do you do? What would you do if the instructor was there? You would get the checklist, start the engine, make a radio call if necessary, and head for the runway. That is how a solo begins. Serenity comes from entering the world of the familiar. There is nothing here that you have not done before; you just have no one looking over your shoulder. You may not have your instructor on board, but his voice will be in your head. While busy with my tasks, I swear my instructor was talking to me. Only his physical body was out there on the snow.

There is another pause just before you takeoff. It becomes very real from here as you realize that once you takeoff, you are going to have to land all by yourself. I paused to look at my instructor for a visual clue that I could call the whole thing off and come back another day. He just looked off somewhere in the distance and paced impatiently, never giving me a glance. Let's do it.

Once the power came in, I was too busy to worry about anything. Something magical happened. I started to feel like a pilot, like I could do anything, meet any challenge, and fly to perfection. I swirled in contradiction; invincible yet fragile, wanting it to last forever while hoping to get it over with, wishing the world could see me yet glad to be alone. The only respite to evaluate comes on the brief stretch during downwind. I looked at the empty seat beside me . . . Time to get busy again.

The first landing was textbook perfect. I wish it were on film. My newfound confidence would then be tested. A Piper Cub had slid in front of me. It flew slow enough to close the gap between us. On final, a combination of slow flight and S-turns helped, but didn't resolve the conflict. I initiated a go-around at the instant the Cub popped back up after touching down. I had been taught to continue any go-around once commenced, but this was the exception. The Cub did not answer my calls and our flight paths were converging. I idled the engine and proceeded with a landing again. It wasn't pretty, but it was safe. It wasn't near the approach end of the runway either. It was, however, the best I could manage under the circumstances.

On downwind for the third approach, I noticed that tiny snow crystals were floating around inside the cabin. This airplane leaked. My third landing was fine and slow enough to have me off by the first taxiway. With newfound arrogance, I taxied painfully slowly to my waiting and nearly frostbitten instructor. My instructor spent many minutes warming up before offering any congratulations. We flew home uneventfully save for the fat grin slapped across my face.

Upon arrival there were no marching bands, no fanfares, and no fireworks, although I certainly felt worthy. It was still terribly normal at the flight school. I wanted to scream out to those Bozo's what I had done. Then I stopped to think. They hadn't done it today, I had. It was mine to revel in, not theirs. Each of the pilots in that room had his day once and probably felt as I do now. There was nothing I had to say to them. I spoke not a word.

Soloing is a personal achievement that every fellow pilot understands. Anyone who is not a pilot will never understand, so there is not much you can explain. My solo was a quiet, personal victory; which is how it should be. When the pressure was off and the effect of the adrenaline had subsided, I got the most massive headache of my life.

My father also drove me home that day. I had done something he had dreamed his whole life, yet never made a reality. Although as proud as he could possibly be, our relationship changed that day, especially as far as flying was concerned. You separate yourself from the general public when you solo, never to return. You seek the comfort of the sheltered world of aviation the further you advance in your ratings, for only like pilots will understand you. You can talk to the friends who encouraged you to get into this world, but you will never speak the same language. They won't understand the glory of your sweat, the landings that jarred your fillings, the stupid things you could not later explain, the bond with your airplane and instructor, the frustrations of weather and maintenance, the humiliation of phase checks, and despite it all the need to persevere and fly. If you survive all this with your body and soul intact . . . if you have put in the study, practice, and work to get through your solo . . . if you have done all this, then you deserve it.

Now, go and write your own story.

6 • FARs AND AIRSPACE

Flying an airplane is relatively easy. What will cause you far more grief is trying to learn all the rules and procedures that allow you to fly legally and safely. No one likes to wade through pages of regulations in preparation for flight and written tests, yet that is the prime motivation for learning the material. There is a good reason for almost every one of the Federal Aviation Regulations (FARs) although it may not be apparent to you as you study. You need a working knowledge of the FARs (what I have been calling the Regulations) so you can apply the correct regulations to your situations. That is how you stay legal.

The way in which this already dry stuff is generally taught is awful. There is nothing more boring, dull, tepid, dreary, miserable, tasteless, flat, and sleep-inducing than studying FARs and Airspace. My job, therefore, is to make them live.

The first thing we need is the proper perspective. You arrogantly assume that you are the center of attention in these rules and they have been put here to harass and impede your piloting. Wrong. If you look at the FARs in total, you will see that they are written to protect people and property on the ground first, the pilots and passengers in other airplane's second, your passengers next, and finally you. With this disillusionment fresh in your mind, let's look at the highlights and major sources of confusion in Part 91 of the FARs, the flight rules.

Responsibility and authority of the pilot in command: This is the rule that allows you to be a pilot. It is your raison d'etre (look it up), so to speak. "The pilot in command is directly responsible for, and is the final authority as to, the operation of that aircraft." Mortals cannot get any more power than that.

We have created a society designed to remove all responsibility from the individual and his actions. You have spent your life learning how not to take responsibility. It is no wonder that pilot training requires such an attitude adjustment. Think of the world today. Insurance covers your accidents. Lawyers take the responsibility from you and charge someone or something else. Presidents don't even take responsibility for their administrations. Ship captains and pilots are the last truly responsible people left. Your entire pilot training is supposed to prepare you to accept this responsibility.

Think about the implications. When you are pilot in command of an aircraft in flight there is no power on earth that can second-guess your decisions. Notice I said — in flight. Since the FAA can't just drop through the window of an aircraft in flight, especially after normal business hours, they are content to wait until you land. Notice also where it says under your emergency power that you "may deviate from any rule of this part to the extent required to meet that emergency." You really didn't expect the federal government to give any individual ultimate power, did you? If you thought what I said earlier sounded too good, you were right. Those words, extent required to meet the emergency, are very dangerous to you. Who judges what is necessary and what is excessive? As long as the aircraft is in the air, practicality dictates it is the pilot.

However, after the flight you will have to answer for anything you did and any decision you made. The chilling effect of knowing that you will be held totally accountable is what actually controls all your actions in the aircraft. It is you who will do the second-guessing. Any crisis that arises will cause you to make sure you can justify your actions to some official on the ground before you act to save your aircraft. You will not declare an emergency unless you think someone else will agree that it is an emergency. Such hesitation can be fatal. This is why that whenever a pilot has a problem, his first emotional reaction is not fear; it is embarrassment and the dread of an inevitable inquisition.

The psychology of emergencies is very interesting because of this rule. For example, under certain conditions of temperature and humidity, a Cessna 150 carburetor will instantly ice up causing the engine to quit, should you have a smart-ass student trying to impress you with his stall recovery technique by pushing the carb heat and throttle in simultaneously. It's happened to me a few times. Always push the throttle first like the procedure says, please. Anyway, the engine quit at 4500 feet right above the 8000 feet runway of an old army base. I like practicing maneuvers within gliding distance of something useful. With the carb heat on, and a descent steep enough to keep the prop windmilling, my first reaction was: Oh God, how am I going to explain this. I'm a two-year flight instructor — why me? My student just sat puzzled, thinking: This isn't what is supposed to happen. You folks don't even acknowledge anything that varies from the handbook description.

I could, possibly should, have immediately declared an emergency. I had no engine power. The radios were immediately tuned to both local and emergency frequencies. Then I thought of the inevitable questions. So I waited. If the engine did not come to life by 2000 feet, I would let the world in on my secret. Welcome to real life. Time passes slowly when spiraling down on a dead engine with only what's left of carb heat and a prayer for company. The engine sprang to life at exactly 2200 feet, mike key in hand, and off we went. I never said a word to anyone.

This rule is misleading. You are certainly directly responsible for anything that happens from the moment you approach that airplane with the intention of flight. However, you are far from the final authority. Since this is a direct contradiction, and since this rule is the foundation of all other rules in this part, the flight rules must be based on direct contradictions. That is why they protect the people and property on the ground first. Everything is backwards.

Civil aircraft airworthiness: The pilot in command, that means you, must determine if the airplane is both safe for flight and airworthy. I'm not quite sure of the difference because the FAA won't define the terms. Anyway, the preflight isn't just a really good idea; it is a legal requirement. Of course, whenever the FARs mention pilot in command, you know you are going to be responsible for everything on the airplane whether you check it or not. How much can you check in the normal 20-minute preflight? What if you blow a tire on landing, career off the runway, and walk across the ramp right into an FAA inspector who is always on hand for such occasions?

"Did you check that tire on your preflight?"

"Why, yes, of course. I even rolled the wheel through an entire revolution to reveal all parts of the tire under the wheelpant, and it looked just fine to me."

How else would you answer a question like that? You should know that you are responsible for all the things you rely on others to take care of for you. Take the maintenance logs of rental aircraft. How do you know that airplane is legal just because the person behind the desk gave you the keys? We assume an awesome responsibility when we accept an airplane. You can't check everything that you are responsible for because there just isn't time. Most of us get by with luck because nothing happens. Just remember, though, that it is luck, and if something happens that you should have checked, you are responsible.

Careless or reckless operation: This is the rule that always gets added on to any violation because anyone who violates a rule must have been either careless or reckless, or both. It also bumps up your fine to whatever the going rate is for such things. Check out the language: "No person may operate so as to endanger the life or property of another." I told you that you could go kill yourself and they wouldn't care.

Consider the finer shades of meaning. If you operate within the limits of the airplane and the FARs, you should never be guilty of being careless or reckless. Well, maybe. What if one day you are flying at 600 feet above ground level (AGL) in a sparsely populated area, perfectly legally, when some farmer with binoculars calls in your numbers to the FAA. Can you be both legal and careless? It's all a matter of interpretation. I knew a pilot who flew a short approach inside the tower to a runway. Most folks stay outside, but some go inside. On this day an FAA inspector happened to be in the tower. The inspector violated the pilot for reckless operation and told him to report to the FAA office in three weeks after the inspector returned from a vacation. Priorities, I guess.

The pilot, in the meantime, did the exact same approach the following week. The folks in the tower, who are also FAA personnel and had seen the same maneuver many times, requested that the pilot repeat the approach so it could be filmed for the lead-in to an FAA promotional video they were working on that day. Why not! My pilot friend reported to his violation appointment with a copy of the FAA video under his arm. Same maneuver, different interpretations.

This rule is purely subjective so even complying to the best of your ability, there is no guarantee that someone will not consider one of your actions as either careless or reckless. As for my friend, it would have been a little embarrassing for the FAA inspector to fine the pilot who starred in one of their own films.

Dropping objects: You may drop objects out of airplanes with reasonable precautions; as long as you are not careless of course, or do damage to persons or property below. Guess who is left out again? I can't think of any need to drop things out of airplanes. My theory is that this rule came in when Curtiss Jenny biplanes carried mail in the 1920s, and it was just too much trouble to land every time. Oh well, once a rule — always a rule.

Alcohol or drugs: As if there is a difference. Everyone knows that you cannot legally fly within eight hours of consuming alcohol. However, there are two more clauses to this rule. You cannot fly while under the influence or when your blood/alcohol level exceeds .04 percent. Now, is .04 the threshold of under the influence or could a hangover be considered as influenced by alcohol? No one defines this. Why are all three levels mentioned? The rule for drugs is more straightforward. Anything that affects your "faculties in any way contrary to safety" will keep you on the ground. Why couldn't they just say that for alcohol and make life easy?

Preflight action: Just in case you weren't convinced that you are totally responsible for every nut, bolt, and rivet before you fly, the FAA provides yet another infinitive statement. You are responsible to "become familiar with all available information concerning that flight." Well, what do they mean by "all"? All is a big word. Not only are you directly responsible and the final authority, but you have to be all knowledgeable as well.

To be absolutely responsible and absolutely knowledgeable is absolutely impossible, yet that is the standard we work under. The pilot can always be judged at fault when the standard is perfection because pilots are only human. Even though this rule has good guidelines for things you should check before you fly, it is the omnipotent generalizations tucked in the fine print that allow the FAA complete latitude to interpret the rules any way they want. That puts you and your pilot certificate, especially if your livelihood depends on it, in an incredibly precarious position.

Here is an example. Familiarizing yourself with all information includes getting a briefing. Do you know any pilot who gets a briefing before every single flight? Especially if he has multiple flights from the same airport on a sunny day? But what is the first thing they ask you at the Flight Service Station (FSS) when you call? They want your aircraft number. Is it to prove you are a pilot and not a civilian getting weather for a Sunday picnic? No. It is to use against you should anything happen on your flight and you didn't get a briefing. The first statement on any accident report informs you if the pilot got a briefing and about the weather when the accident occurred.

Operating near other aircraft: You can't operate "so close to another aircraft so as to create a collision hazard" according to this rule. Then in the very next sentence you are allowed to fly in formation as long as everyone agrees. Isn't a formation of airplanes by definition a collision hazard? Isn't a collision hazard careless and reckless? How close a formation are you allowed? There is no consistency to these FARs because of the constant contradictions. Like so much of our society, what is legal is determined by what nobody hears about, what you get away with, and what does not go wrong.

Right of way rules: For the most part, it makes sense to have the aircraft on the right take the right of way. However, the head-on rule needs some work. Picture a scene where an aircraft is on a collision course with you. It is slightly to your left. Students who know the rule about altering their course to the right in the face of a nearly head-on aircraft do what they think is the right thing and put us directly in the flightpath of the oncoming aircraft. Suicide isn't what the regulation has in mind.

The overriding rule is see and avoid because the regulations can't cover every situation. If a course to the left will avoid an aircraft that is nearly head-on, then get out of the way as quickly as you can. This rule only works if both aircraft see and alter course to avoid. If you see the other aircraft altering course to the right, it makes sense for you to do the same. You should never turn into the flightpath of an oncoming aircraft, putting it under your nose and out of sight, when you have no idea if it has seen you. Yet this is exactly what students trained by rote always want to do.

The classic muddling of right of way occurs when de-

termining the priority for landing. The right of way goes to
the aircraft on final approach and the aircraft which is
lower. So there you are on base in a light single when a big
twin calls in on a five-mile final approach. You are lower,
they are on final. Hmmm. The experts have argued this for
years. No solutions yet. They can cure this problem by
simply stating that either the lower aircraft, or an aircraft
on a final approach of three miles or less, will have right of
way. When two aircraft have equal priority, the larger
aircraft will have the right of way.

One quick note. You are supposed to pass slower aircraft to the right. What do you do in a right traffic pattern?
Another thing to think about is what happens when ATC
directs you one way for traffic avoidance, and it goes against
the right of way rules? Talk to your instructor about all
this stuff.

Minimum safe altitudes: You will be legal if you
keep above 1000 feet AGL over urban areas, 500 feet AGL
over less congested areas, and 500 feet from people and
structures out in the boonies — except for takeoff and
landing. The exceptions are always the key to these regulations. From the regulation covering control zones (Class D),
you know that you can take off or land with 1000 foot
ceilings and three miles of visibility. You also know from VFR
rules that you have to stay 500 feet below the clouds in
controlled airspace. Now, if you know the rules, you can use
them to your advantage. In many cases, they are written to
give you the most latitude in exercising your authority.

Case in point. You are 20 miles from home, the clouds
are at 1400 feet and the area between you and the airport is
urban and controlled airspace. How can you fly home VFR?
If you fly at 1000 feet, you are clear of the urban area but
too close to the clouds. If you fly at 900 feet, you will be too
close to the urban area — except for takeoff or landing.
Fortunately, there is no official definition for where a takeoff ends or a landing begins. When I faced this situation, I
determined that my landing would begin 20 miles out. I

also heard from the controllers on "approach" to fly at an appropriate VFR altitude. That means 500 feet below the clouds. I can take a hint. We flew home at 900 feet AGL.

Also in this rule is a requirement that if a power unit fails, you have to be able to land "without undue hazard to persons or property on the surface." They still don't care about you. Fortunately, there was a highway leading from our location to the airport. You have to be in compliance with all the rules, all the time.

Compliance with ATC clearances and instructions: Along with saying something stupid on the radio, doing something stupid from what you thought you heard on the radio is the next greatest fear. Student pilots want to comply with ATC directives without question and to the letter. So much for that final authority bit. Yes, you must do what they say, but here is the key: "unless an amended clearance is obtained." The rule takes great pains to point out the omnipotence of ATC, except in an emergency, where you will have to defend yourself later. This is how pilots are coerced into giving over control of the aircraft to ATC.

From the exception in the rule, I have learned to just keep asking for new clearances until I get what I want or I need. The folks in ATC are only human. I know, I've met some of them. Sometimes they make mistakes. The system only works when both sides work together. Just remember that it takes a lot more to become a controller than a private pilot. Those folks in the tower will most likely bail you out many more times than you catch them in a goof — although one controller did try to land a King Air on me.

The rule is: Follow the instructions if you can; if not, get new instructions. I was departing an airport served with an ARSA (Class C). The controller gave me a vector (heading to fly) and an altitude restriction. The frequency became jammed after that, he forgot me, and I was heading directly for a line of hills. I could not call back. I could not get new instructions. I was VFR and there were fast jets to sequence. After violating the altitude restriction and changing my heading about 40 degrees, I got the controller back on the radio.

"You forgot me. I'm now at 2500 feet heading north."

"Oh — sorry. Resume own navigation." (Translation: I have other things to do, good day.)

No problem.

So much for the obvious conflict. But what do you do when you are vectored toward airspace that you need permission to enter, like someone else's ATA (Class D), or a TCA (Class B), or a Restricted Area? Remember that even though you are under ATC control, you are still responsible for your own navigation. If a controller gives you a vector to maintain that will take you into a TCA, you must get a either a TCA clearance, a new vector, or just turn the airplane to remain clear and explain your action when you can. I have had to do them all. Remember the controller works for the FAA and is never officially at fault. Time after time, I have heard pilots told to call the supervisor upon landing when they flew into a TCA, at the direction of a controller, without a clearance. Pilot in command means that no matter who screws up you are responsible.

Light signals: You of course have to just memorize your light signals. But how many of you have actually seen a light signal, such that if you lost your radio, you would know what to look for? I was out one day and asked for a light signal for my student. The tower folks hadn't used it in so long they had to read the instructions.

Operation at airports with control towers: You know you can't takeoff or land without being cleared by the tower. But what does being cleared really mean? When you study your regulations, you should visit the tower folks. They have their own set of regulations. Knowing what the controllers have to do will put your rights and responsibilities in perspective. Their primary responsibility is to make sure aircraft do not hit each other. Everything else is secondary.

When you are cleared to takeoff, it means that if you

want to go the runway will be clear of traffic. That is all it means. If for any reason you do not want to go, tell the tower and they will cancel the clearance. When you are cleared to land, it means the same thing. It also means you are clear to go around, you never have to land. Cleared for the option gives you real flexibility. This is where you can exercise your pilot in command authority because a clearance to go is never an order to go. Should you be cleared for an immediate take-off, the runways will be clear only if you go right now. Work with your tower people for the good of the system. We fly the airplanes, they keep us separated. We keep us separated, too.

Visual flight rules: So long as you are a private pilot, you will operate within these rules. The FARs have this confusing chart that is supposed to help you understand your weather minimums. Try this chart. The secret is to think of controlled and night as one group, uncontrolled above and below 1200 feet AGL as another group, and anything over 10,000 feet mean sea level (MSL) as the third group.

Altitude:	Controlled & Night:	Uncontrolled:
<1200'AGL	3 st. mi. visibility, 1000' above, 500' below 2000' beside	1 st. mi. visibility, clear of clouds.
1200'AGL — 10,000 MSL	" "	1 st. mi. visibility 1000' above 500' below 2000' beside
>10,000 MSL	**All Airspace** 5 st. mi. visibility 1000' above 1000' below 1 st. mi. beside	

Isn't that easy? The rules we have to live by actually make a lot of sense even if they are absurdly complicated. The trick is to redo the explanations so that you can understand the information.

A word on night because there are three definitions for night. At sunset the lights go on. At civil twilight you can log night time. One hour after sunset you have to have three takeoffs and landings within 90 days to carry passengers.

It is virtually impossible for the FAA to enforce the cloud clearance rule. Consequently, pilots develop peculiar ideas about what 2000 feet from a cloud looks like. A length of 2000 feet is over a third of a mile. Student pilots, who should know better because this stuff is fresh in their minds, sometimes want to snuggle right in next to the clouds. Clouds look so pretty, all fluffy and white — how can they be anything but safe? It's not the clouds you have to worry about. Even at 2000 feet, a Learjet popping out on approach can close that distance in seconds.

The rules of VFR are designed to protect you from what you cannot see, namely IFR traffic in the clouds. It is designed even more to protect the IFR traffic from you. They do this by giving you a minimum distance to stay clear of the clouds. I have burst out into the sunshine many times on an IFR flight plan only to have some VFR idiot sitting right in my path.

The folks on the ground can't see clouds on radar until they are severe enough that you will probably be on the ground. Therefore, they can't enforce cloud clearance rules like they can for airspace, which is marked right on their screen. Besides, it isn't their responsibility. It is the in-between weather where IFR and VFR mix that causes the problems, so you must maintain your cloud distances to keep everyone safe. Work with your instructor until you can accurately estimate both your distance from clouds and your visibility. Almost all pilots can recite this rule, but over time fly progressively closer to the clouds because nothing happens to them, right up until the day . . .

Inoperative instruments and equipment: The procedure that you must follow in order to fly with something not working is so confusing and intricate that you might as well fix it. Can you guess the intention of this rule? You may have a minimum equipment list for your aircraft, but probably not. From the title you would think that it would list the minimum equipment you need to fly. Wrong. It lists the things that do not have to be working before you fly. Which is why it should be called the inoperative equipment list. The best thing you can do to wade through the other requirements is to make a simple checklist for yourself for all your future flights and checkrides. This rule isn't even necessary because the aircraft manual and a set of other regulations prescribe lists of required equipment. Shouldn't the pilot in command determine what is necessary beyond that? So much for the FARs.

Nothing is guaranteed to irritate you more than the study of *airspace*. All those colors on the chart blur together. Airspace is not designed from a master plan. It has evolved over decades. Unfortunately, the obsolete stuff is still around. The good news is that the reclassification will take some of the problems out. The bad news is that pilots will have to learn to match the old airspace name with the new class letter. For the airspace that remains unchanged, the problems will remain.

Despite the early previews, you won't really know what is going on until your specific sectional and terminal charts are issued. No matter where you are as a pilot, you can regroup all the airspace into new and logical categories such that no matter what happens, you will have the mental flexibility and agility to adapt to any changes, as well as what remains the same.

The problems in understanding airspace originate in the names we use. Controlled airspace is confused with air traffic control when it really refers to weather, which then mixes with positive control which refers to weather and traffic, add in uncontrolled airspace, mix in all the different weather minimums, both day and night, the special use airspace, and the other airspace, and you have a mess. Let's organize the mess.

The first group of airspace we will call *weather controlled*. For the VFR pilot, the only function of this airspace is to provide higher weather minimums so you can safely see and avoid IFR traffic. The only requirement to enter is the controlled airspace weather minimums. Include in this group: Transition Areas, Control Areas, and the Continental Control Area (Class E). As you know, the world changes at 10,000 feet, making the Continental Control Area obsolete. But it was on the books so I had to include it.

Call the second group *traffic controlled*. In this group, we include airspace where there are or were no weather requirements at all, but you either must communicate with FAA controllers or FAA personnel are available to serve you. Include in this group: Airport Traffic Areas (Class D), Airport Advisory Areas, Military Operations Areas, and Military Training Routes. Airport Traffic Areas will always be controlled by control tower people. The other three are monitored by Flight Service Stations and can give the VFR pilot advisories when they are within 100 miles of the FSS.

Make your next group *radar controlled*. Radar now covers much of the country to help the VFR pilot in a variety of ways, like flight following, for example. This group is where the congestion of air traffic dictates separation through radar control, in airspace with controlled airspace weather minimums and very specific operating procedures. Include in this group: TCAs, ARSAs, and the Positive Control Area PCA (Classes B, C, and A). The PCA is the odd one out. I'm not sure if it all is radar controlled. But that is the ceiling for VFR operations. You need an IFR clearance to enter, so here it is.

Take what is left of the Special Use category and call it *military airspace*. In this group if you want to avoid the hazards or get permission to enter, in most cases you will deal with a military controller or civilian folks who are coordinating with the military. Included in this group: Prohibited, Restricted, Alert, and Warning Areas. There are no weather minimums associated with this group nor general procedures for all operations. They all have differing levels of danger and restriction to the VFR pilot.

The Control Zone was the only one left. It didn't go anywhere else so I called it *the misfit*. This is the one that used to confuse you. It was that dashed blue line on the chart that you mistook for an Airport Traffic Area, except that ATAs had no weather minimums so they were not charted. Control Zones weren't always five miles across either. They were whatever they were on the chart. They had controlled airspace weather minimums and extended up to the Continental Control Area. Over Alaska and Hawaii there was no limit, even though all the weather minimums change at 10,000 feet.

There was no requirement for communication to enter a Control Zone even though the Control Zone was not in effect unless you could talk to some controller. If the weather was below minimums, you needed a special VFR clearance to enter, which was not available at night without an instrument rating, when you wouldn't need it anyway. You knew the weather in a Control Zone because when it was in effect, there had to be FAA-approved weather observers on the field. The people in the tower usually served this function. However, some airports without a tower had Control Zones. You didn't even need a radio to land at those airports being uncontrolled, so even though you needed a certain weather minimum, it was your guess as to the current conditions. Flight Service Stations, if they existed on such fields, gave you the weather, but not a special VFR clearance if it was below minimums because they were only able to give advisories. Control Zones are now Class D.

Our airspace is far too complicated, and there is no legitimate reason for it. Here is what I would do to simplify. VFR anywhere below 10,000 feet would require three statute miles visibility; 1000 feet above, 500 feet below and 2000 feet beside the clouds. VFR above 10,000 feet would be five statute miles visibility; 1000 feet above, 1000 feet below, one statute mile beside the clouds. Pilots with a commercial certificate or an instrument rating could operate at and below 1200 feet AGL, unless restricted by other airspace, during the daytime, with one statute mile visibility, and clear of clouds.

By standardizing the weather, you could remove all the controlled/uncontrolled airspace designations. This would eliminate all airspace from the chart except TCAs, ARSAs (Classes B and C), and the military group. Military airspace could then be reclassified into two groups, Restricted and Alert Areas. Anything to do with training would be an Alert Area, including training routes, and anything that presents a hazard to nonparticipating aircraft would be Restricted. Color military airspace bright green.

Make it simpler still. Since the distinction between ARSAs and TCAs is rapidly closing, why not combine and redesignate them Positive Control Areas (Class A). Color them bright blue. The rules would be the TCA rules. All VFR aircraft would need a clearance. At our busiest terminals, make a chart notation that VFR clearances are unlikely or unavailable between certain hours. You could then publish Airport Traffic Areas and not clutter the chart. Color them bright red. With the weather standardized, printing the Victor airways would be optional. Imagine a chart with only Airport Traffic Areas, Positive Control Areas, Restricted Areas and Alert Areas. If you have a better system of your own, tell the FAA.

The whole idea of uncontrolled airspace fascinates me. It is leftover airspace where the FAA didn't know what else to do with it. Imagine, allowing private pilots to operate in the lowest visibility and cloud clearance, close to the ground, around sometimes very busy airports, without a tower, while talking to each other on a frequency designated for any number of other airports. The current system of controlled airspace usually gives out at either 1200 feet or 700 feet AGL, and the FAA doesn't want any responsibility for traffic below that. Unless something goes wrong, of course. Putting control towers out there is not the answer because pilots can regulate their own flow quite well.

However, allowing pilots of all designations to operate at one mile visibility and clear of clouds over busy airports is nuts. Instrument pilots are taken care of because they have their own special requirements for each airport with an approach procedure. By permitting commercial pilots to operate during the day at one mile visibility, clear of clouds below 1200 feet AGL, pilots could still earn a living and the commercial certificate would actually mean something. That would leave private and recreational pilots to have three statute miles visibility, 1000 feet, 500 feet, and 2000 feet from the clouds anytime below 10,000 feet.

The problem with the current low visibility is the closing times between aircraft. Two aircraft converging at combined closing speeds of 200 knots, or 230 miles per hour (mph), will cover one statute mile in 16 seconds. That is not much time to see and avoid. For high performance aircraft closing at 260 knots, or 300 mph, the time goes to 12 seconds.

Uncontrolled airspace, as well as all airspace, has not come about through any comprehensive strategy. It just kind of happened. Courtesy of the Aircraft Owners and Pilots Association (AOPA), comes this little history. Airspace evolved because of various perceived needs in different parts of the country. New airspace and air traffic control procedures were implemented after accidents, usually mid-air collisions. Airport Traffic Areas governed by control towers started it all in the late 1920s and spread slowly across the country. In those days, en route traffic separation was nothing more than noting departure and arrival times. Accidents occurred when aircraft failed to meet their expected times.

During the 1940s and 1950s, our national route structure and en route traffic control developed. In this time we got the control zones, control areas, and transition areas, as well as the increasing use of radar. With the introduction of jets in the late 1950s, everything changed. Out of the old airline climb corridors came more control zones, the Continental Control Area, and the Positive Control Area.

The late 1960s saw the introduction of the Terminal Control Area. In modern times where fast jets mix with light general aviation aircraft, positive control is achieved with the Airport Radar Service Area, the mandatory version of the old, voluntary Terminal Radar Service Area. Not enough of you volunteered.

This is not an exact history. I am still looking for a really good book on the history and development of our airspace system. It might explain how we got into this mess.

The new airspace will be in effect in September, 1993. The purpose is to simplify the understanding of airspace. However, since what you learn first lasts forever, the FAA can attach whatever letters they want to various airspace, and the vast majority of you will always translate it back to the old name in your mind so that you will understand what you are flying through. The result is that you now have to learn two names for all the new airspace which doubles your mental workload.

You would think that new students who start after September, 1993, would have it easier. However, since a name like "Class B" doesn't mean anything by itself, the instructor will have to teach the student that this is an area of controlled airspace around a terminal, which we used to call a Terminal Control Area, thus necessitating that new students learn both names to keep all the letters straight in their minds, which doubles their workloads as well. Only the FAA would think of this as a simplification.

The justification for the new airspace is to standardize nomenclature and regulations for international travel. Since the rule changes from our current system are so minuscule, weren't we already at an international standard? How many of you fly internationally anyway? If you did, wouldn't you learn the airspace of the country where you were traveling? Also, flying cross-country as far as airspace is concerned is greatly simplified by flying on an IFR flight plan, where positive control will usually clear you right through all the different letters. How many pilots actually take off VFR to foreign countries or come to the U.S. to fly?

The easiest way to transition to the new letters is to look at the system from the top down, where the first letter has the most restrictive airspace and it loosens up and descends from there.

* "Class A" is at FL 180 where you have to be operating IFR.
* "Class B" will top out about 10-12,000 feet MSL and you can operate VFR with an ATC clearance.
* "Class C" should end about 4,000 feet AGL and all you need is to establish communications.

These three also require that your transponder be turned on.

* "Class D" will be around four nautical miles wide and 2500 feet AGL high. Give the tower a call here.

So much for the technical stuff.

* "Class E" is an amorphous blob of controlled air space below A, above G, and beside and above B, C, and D.
* "Class G" is uncontrolled airspace; what we used to call — uncontrolled airspace.

The Continental Control Area is gone. This makes sense because present regulations render it useless. However, you will notice that all "Class G" airspace ends at 14,500 feet MSL. Since only those who learned the old system will know why, new students who want to understand will still have to learn the old name of obsolete airspace, superseded now by present regulation, because it still governs the top of uncontrolled airspace even though it doesn't exist anymore.

The most noteworthy change is that "Class B" cloud requirements have been lowered to just clear of clouds; a novel idea from the FAA where private aircraft will now be able to operate VFR with the minimum cloud clearance in the vicinity of overcrowded terminals with the greatest concentration of 250-knot airliners.

The great unknown of the lettered airspace is how

radio communications will go since B, C, D, E, and G all rhyme. The whole reason for the development of a phonetic alphabet was to avoid such confusion on the radio. This will be especially crucial where Airport Traffic Areas are over-lapped by ARSAs which border TCAs. Picture a poor student on a training flight with a fuzzy radio, a babbling instructor, and a command from ATC to remain clear of B, traverse C, and call the tower for D. However, if we go to a phonetic alphabet to name the airspace, you could have a radio call like this: "This is Delta heavy, with information Delta, approaching Class Delta."

You should be comforted by this new change in airspace as it is purely for your benefit . . .

7 • CROSS-COUNTRIES

———————————————————————————→

This is the most fun you will have in your training. This is the answer to every romantic dream of flight you ever had. Finally, after all the toil, you will get to seek adventure over routes and territory unexplored by you. Out into the great unknown goes our intrepid aviator. A wandering soul still not quite yet a pilot. So why is this such a high point in your training? Is it because you visit new and exciting places? Is it because you leave behind those tired, old, all-too-familiar practice areas? Is it because you finally get to enjoy the view instead of the usual endless repetition of tedious presolo maneuvers?

No, it's because you get to leave your instructor back at the airport. For literally hours at a time, you get to enjoy your new environment in relative silence. You can build your confidence when things go right. You will discover your resourcefulness when problems arise. You will learn to control your panic when you get the sickening feeling that you are not quite sure of your position and then reach back to your training to solve the mystery.

Before you can do any of this on your own, you have to do a few dual cross-country flights with your old instructor. You have to prove you can get from here to there, and then back to here again. When you can do that, you will get an endorsement on the back of your medical certificate that says you have received cross-country training and are proficient in the requirements of Part 61 pertaining to student cross-countries. Make sure you get this endorsement. I have sent many students home for a beer instead of sending them on an adventure because both they and the instructor forgot the endorsement. What a waste. Beer always tastes better after a long flight.

Once you have the primary endorsement on your medical from the instructor who gave you the cross-country training, any instructor can give you the second endorsement that certifies that your planning is good and the conditions that day are safe for you to make your flight. It goes in your logbook. The logic is simple. Say for example that you fly somewhere and the weather changes such that you cannot safely return. Your instructor won't have to come fly out the next day and endorse you for the trip home; any instructor on the field can do it. They are not taking responsibility for your cross-country training. Your primary instructor has done that.

This endorsement only covers what an instructor can verify himself, namely your navigation log, and the weather information. This is a great convenience when your primary instructor has another lesson and you want to take off cross-country. They may (heaven forbid) have a day off, a hot date, or a day on the beach . . . who knows? The point is that you need a separate endorsement for each cross-country flight. Some instructors are reluctant to sign off a student they have never flown with despite the supposed safeguards. Sometimes I have refused to sign off students when I did not trust their flying ability or their instructor's ability to teach. Sometimes, though, unscrupulous instructors will take advantage of students who don't know their rights.

"It's too bad your instructor isn't here. I'd be glad to sign you off. But I have never flown with you. Maybe we could do a quick review flight first."

This is garbage. It also happened to one of my students after I left a school for greener pastures. If you have a valid solo endorsement, a valid cross-country endorsement, your planning is good, and the weather is safe, you are legally entitled to fly. Each instructor controls his own endorsement so he can always refuse. Many students unaware of these details get the runaround. Many instructors don't understand the endorsements either. With a little effort and responsibility, you can save yourself a lot of grief.

Another thing to watch is phase checks. Student pilots have a lot of phase checks in some schools. I send my students up for a check before they solo and before the checkride. Some schools add in a check before you go on a solo cross-country. That's three flights with different instructors who all have different things they look for, different methods of instructing, and different things to teach. Varied experience is good, but you can have too much of a good thing. Phase checks can cost several hundred dollars of your flight money just to see how things are going.

In the Part 141 schools, you can fail the phase check and have to go back for lots more training. With good instruction, you should never fail a phase check. You can also use them to check the school, your training, and your instructor, as they are checking you. I had a chief instructor who used to keep students from all the instructors for a few lessons after a phase check by telling students that their regular instructor was not very good and with just a few hours of personal instruction, he could fix everything. What a scam.

So what is a cross-country? Talk to your nonflying friends and they will think you are flying from San Francisco to New York. Sounds like fun. As a student, you are restricted to 25 miles for solo practice. Anything more than that is a cross-country flight. However, to count for your certificate you have to fly over 50 nautical miles from your point of origin. This is a straight line distance point to point. Zigzag courses with familiar short stops don't count.

As with any new privilege, you have aeronautical knowledge and flight proficiency requirements to fulfill. First the cross-country knowledge. The backbone of all VFR navigation is the sectional chart. Few students spend anywhere near the time required to become intimately familiar with all the information available on these charts. Consequently, you waste lots of time in your lessons, miss easy solutions to navigational problems, overlook obvious hazards, try to run through all kinds of airspace, and sit perplexed with the information staring you right in the face. You will spend your entire flying career working with these charts. New symbols, new hazards, now completely new airspace; there are changes every issue of the charts.

The last couple of years have seen the inclusion of three-letter identifiers for every airport and the designation of a Common Traffic Advisory Frequency (CTAF) so all aircraft would know which frequency to tune in at an airport for the traffic information. Look at the chart. The colors give you a mental picture of the terrain. All the symbols, ticks, stars, and numbers around an airport mean something significant. Be close friends with your charts. Please always call them charts. Maps are for groundlings.

Students often miss the big picture when it comes to understanding navigation because you are focused in your training on nitpicky, tunnel-visioned techniques. This leads to some rather peculiar manifestations of logic, common errors, and interesting misconceptions. Do some research on your own, or take a course on basic navigation. It doesn't have to have anything to do with aviation. In fact, a course in ship navigation may be even more useful since that is where we stole all our navigation wisdom. Reach beyond your training whenever you can.

Long ago sailors and explorers divided the world into a grid using lines of longitude running from the North to the South Pole and lines of latitude running parallel around the Earth, perpendicular to the lines of longitude. Lines of longitude converge at the poles and expand to their widest at the Equator. The base line is the Prime Meridian. It runs through Greenwich, England, where we get Greenwich Mean Time (GMT) or Zulu time from. On the other side, it is the International Date Line. Lines of latitude are consistently equidistant. The base line is the Equator.

The lines of longitude and latitude are measured in degrees and are numbered for identification. Those are the big numbers at the side and bottom of your chart you haven't yet identified. Each degree is divided up into 60 parts called minutes. A minute of latitude, because it is always parallel and equidistant, is also a nautical mile. Those are the little lines between the degrees. Each minute is divided up into 60 parts called seconds. A second is 1/60th of a nautical mile, or 1/60th of approximately 6000 feet. Therefore, a second is about 100 feet. With degrees, minutes and seconds of longitude and latitude, you can find any point on earth within 100 feet. On your chart the lines of longitude are printed straight, even though they curve toward the poles. This makes your chart a projection and not entirely accurate. However, the distances involved do not make for much of a distortion. Besides, you need straight lines to plot your courses.

The fundamental problem with VFR navigation is the constant conversion from true to magnetic directions. Your plotter is used to find your true course, usually from the lines of longitude, which are really curved. Your true course is your angular difference from true north. All courses are really angles measured in degrees from true or magnetic north. The problem is that we have no instrument in the airplane that can be used to fly any true direction. The only directional instrument we have is the magnetic compass, and this is where the problems come from.

Magnetic north and true north are in two different places. Unless you are on the agonic line, (look it up) if you draw a line from your location to true north, and then from your location to magnetic north, the resulting two lines will make an angle. That angular difference is your variation. Fortunately, this job is done for you and printed on your sectional with magenta dashed lines where the variation angle is included. True directions serve no function in the airplane unless you have to plot a new course. The magnetic course is the only number you can use with the only directional instrument in the airplane, making it your only known course.

The magnetic course also determines your hemispherical altitude. This makes it by far the most useful number in

VFR navigation. Knowing this, you might be rather amused to find that many of the official navigational logs you are given by your flight school and put out by reputable professionals in aviation never include the magnetic course on the navigational log sheet. What is going through their minds? You have to learn to fly with this stuff. I had to design my own cross-country form because nothing else worked well enough for my students.

To find the variation from true north on the chart to magnetic north so you can navigate, you will have to add or subtract the angular difference. You have heard the expression "east is least, west is best." Now, if you know the big picture of navigation, you know that refers to whether your variation is to the east or west from the agonic line, which is where magnetic and true north are in a straight line with no variation at your location (see, I told you anyway.) If the number on the isogonic line where variation is constant is followed by an "E," you have easterly variation and will subtract the variation to get your magnetic course. No big deal. However, a shocking number of students have made no attempt to understand the big picture. They are the ones who come up with such contortions of reality.

"So like I'm here on the West Coast, I should like add my variation 'cause west is best, right?"

We have to talk about reading a chart.

"Well, I'm flying east today, so I should subtract my variation."

I suppose we can always change the earth's magnetic field when you decide to fly west.

We haven't talked about the wind. Finding wind correction angles is heavily stressed during most student's training. It is covered in written exams and flight tests. The fact that winds aloft forecasts are notoriously inaccurate seems to have escaped everyone's notice. Winds aloft are measured in true directions (here we go again) because the winds that cover untold thousands of square miles cannot be broken down by individual isogonic lines, airports and routes you might take by the computers at the National Weather Service. In theory, you use the winds aloft information to try to plan a heading that will correct for the force of the wind so that you can maintain your magnetic course. Whether you find that angular correction by converting the winds aloft to a magnetic direction and compare it to magnetic course, or you take true winds and true course to find the angle and then apply that to your magnetic course, it does not matter for the angle remains the same. As long as you always derive your magnetic heading from your magnetic course, you will have numbers you can use in the airplane.

True course is measured from known longitude and latitude. Magnetic course is an angle from true course, the difference being your variation. These are measurable numbers. True winds are at an unknown angle to your true course because that number is only as precise as the accuracy of the forecast. Therefore, magnetic heading is anybody's guess from magnetic course because the wind forecasts are so bad. Got it? Keep in mind while you learn all this that you cannot mix something true and something magnetic in the same calculation and hope to find anything.

The above theory limitations and all is not what most of you are taught. The FAA and Jeppesen/Sanderson standard VFR planning log takes you from true course, through true heading, to magnetic heading. This is truly ludicrous. True heading is a concept that must be banished from your mind if you are ever to find your way. Think about how it is derived. True heading is an angle from true course, neither of which you can fly by any instrument in the airplane. It is derived with winds aloft that are notoriously inaccurate. From this you are supposed to calculate your magnetic heading and fly cross-country. This is madness! How can you expect to navigate from a number that can't be measured in the airplane and is based on erroneous information?

Question: If the magnetic heading you calculated from true heading takes you off the crisp pencil line on the chart

you so neatly drew on the ground, how would you recalculate where to put the nose? You can't because there is
nothing to base it on. This is why most students end up
arbitrarily picking a heading that somehow gets them over
their checkpoints or just tracks the VOR and forgets the
navigational log. Now, if you knew your magnetic course
when you got off track, you would have an immediate
reference that would at worst put you parallel to your
desired track.

Magnetic course is a real number. It is the magnetic
direction from point to point based on the earth's magnetic
field, measured on a chart and read directly from your
magnetic compass. From your magnetic course, you can
deal with the real winds aloft and from subsequent checkpoints find a heading that holds the course. Should you be
lost, when you locate yourself, you will see where the winds
blew you and either correct for the wind or replot a course
to the next checkpoint or your destination. True heading is
not a real number. Without exact wind information, true
heading does not exist. If the true heading does not exist,
then neither does the magnetic heading you calculated from
true heading. You cannot find your way with this system. You
cannot navigate from numbers that do not exist.

The oldest form of VFR navigation is called "pilotage."
This is a glorified name for flying between prominent landmarks to your destination. Think of this as your aviation
history lesson. There was a time when daring aviators
carried people and stuff from old Smith Field (when it
really was a field), past Bob's farm, down the railroad
tracks, over Lake Bluebell, turned left at the grain elevators, looked for a town five miles ahead, and on the far side
of town was your destination. That is pilotage. You have to
learn pilotage because nothing is ever thrown out in aviation, we just keeping adding on new stuff. Also if all your
instruments fail, how will you get home?

The problem with pilotage is its lack of selectivity.
Students tend to head for the first landmark they see that
looks somewhat like their checkpoint, regardless of little

details like distance, course, and groundspeed. On the chart landmarks look very distinct, and a logical course is easy to plot. As soon as you leave the comfortable confines of your sheltered practice areas to navigate with pilotage, all the towns look the same, all the highways look the same, and the railroad tracks, power lines and private airports all completely disappear. The easiest way to get lost is to start dashing off to where you think the destination lies based on landmarks you think are your checkpoints. Never go for any checkpoint that takes you more than a few degrees off your magnetic course. Pilotage without a back-up will quickly lead you astray. It is most useful to confirm what you already know. If you can verify something measured or calculated with a prominent landmark like the big, bold letters of an airport name, then that is verification.

Dead reckoning is the coolest name for any procedure in aviation. Picture yourself at a cocktail party.

"Yeah, I'm a student pilot. We navigate using dead reckoning. If you screw up your reckoning — you're dead!"

The truth be known it is a misspelled contraction of deduced reckoning. Deduced reckoning is navigation by measured course (don't use true heading!) and calculation of time, distance and groundspeed.

Students, though, love to go all out and create these Leonardo da Vinci masterpiece logs for the instructor. Why do you do this? I have seen the most elaborate, detailed, sophisticated, artful, exquisitely planned navigation logs that are of absolutely no use whatsoever in an airplane. The checkpoints, set every ten miles or less, are neatly drawn perfect circles, enclosing sideroads, rivers, tiny hilltops, and other invisible objects. Individual magnetic headings for each checkpoint are carefully calculated with their own independent wind correction angles, methodically recorded in the proper space. Airport diagrams become perfect copies of pilots guides that have already done the job. How are you going to read the fine print when you bounce? How are you going to maintain heading and altitude, watch for airplanes, and both listen to the radio and your babbling instructor, while constantly having your head buried in the cockpit solving endless problems of time, fuel, and distance with the wonder wheel (so called because you wonder how to use it when you leave the ground)?

This assumes a level of inflight competence hardly approachable by your basic student. It assumes you can hold a heading within one degree. It assumes the winds aloft forecast is precise. It assumes your magnetic compass is accurate or that you have factored in the deviation. It assumes that the directional gyro will not precess, or if it does, you will immediately notice. If you can guarantee all this, then those navigation logs make sense.

The biggest problem with student planning is too much information. You are not going to fly all that far, nor all that fast, so you don't need to plan like it is a transatlantic crossing. VFR navigation just isn't that precise. When you can't read the log, you are going to get the trend of the wind from the checkpoints, guess at a heading, try to hold it within ten degrees, put the airplane somewhere near the destination, look for the airport, and then fly directly to it once it is in sight.

To avoid going overboard in your planning, do this. Unless there is at least a 10,000-foot mountain in your way, draw a straight line from your point of origin to your destination. Plot and calculate your magnetic course. Pick an appropriate hemispherical altitude from your magnetic course that will clear all the airspace and obstacles in your path. For the highest true airspeed in a low-performance aircraft, try to fly around 6000 feet. Find a few good checkpoints at least 25 miles apart either on or near your course. Get a weather briefing. Calculate one wind correction angle for the whole trip, unless a significant change is forecast along the way, and apply it to the magnetic course. Calculate the time, fuel, and distance. Calculate a descent point in distance and time that will allow you to descend from your cruise altitude to arrive 1000 feet over your destination. Draw a diagram of the runways, label them with their numbers, and include the direction of the patterns. Do a weight and balance. Go fly!

There are schools and instructors that send you on these short, just-over-the-minimum-distance, cross-countries. That's just beyond some of your practice areas. They have you calculate arrival times for checkpoints so close you can already see them. Your workload is so heavy that you build up resentment for what should be the most fun time in your learning. This is a waste and you know it. You want to be able to look out the window. You learn that you can point the airplane on course, follow the sectional for 51 miles, and be pretty sure of finding your destination. You are reminded of your instructor who has flown you all over the sky and never once calculated a thing. This is not a good standard. If you are not having fun or do not have a chance to look out the window because you are always calculating something, talk to your instructor.

Most of you are so busy with the excessive workload that you never learn the reasons and logic behind deduced reckoning. Your flights become drudgery; your efforts purely for your instructor and the checkride. When overused during training, deduced reckoning gets abandoned after you get the rating. A false sense of security is built when on short solo cross-countries you find you can get there and back just fine without all the work. Those elaborate cross-country plans go to the bottom of your stack of papers and you fly with the sectional and track the VORs. You know from your regular 1.5-hour lessons that a 30-minute cross-country flight is well within the reach of the airplane. The effect of all this is to build contempt for critical procedures through teaching them without demonstrating need or purpose. When everything is going fine, there is no apparent purpose for deduced reckoning. The reason for all the calculations before you go, and to monitor your progress en route, is that deduced reckoning gives you the very first warning when everything is not going fine. It allows you to fly close to the limits of range, fuel, and weight, and still remain safe. It allows you to fly great distances where you do not know where you are going. You can't learn any of this from flying trips that are too short. Many of you consequently have a rude awakening on your long solo cross-country.

If you want a great experience, go on a 200-mile or more trip with your instructor. Deduced reckoning will save you when your conditions change. If a checkpoint shows up too soon, too late, in a different location than expected, or not at all, you will immediately know something has changed. Start asking questions. If your ground-speed slows because the wind shifted, will you have the fuel for the destination? This is where you learn why you carry a reserve. What headings will get you back on track and then keep you on course? On a long flight the checkpoints will assure you that all is well because each checkpoint becomes a little cross-country. You want your checkpoints close enough to notice a trend before it becomes a problem, yet far enough apart not to be a burden. For the student, 25 miles should do it. A commercial pilot flying an 800-mile trip may use checkpoints that are 100 miles apart.

Should you observe a change in the weather, deduced reckoning allows you to change your plans in flight. You can receive and interpret weather information, pick an alternate airport or return to base, plot your new course (don't use true heading), calculate your arrival time, and compare that to time in the air so far to see if you will safely get to your alternate. Don't trust the fuel gauges. Many students get too little practice at inflight diversions and after certified are reluctant to change their plans, preferring instead to proceed into deteriorating weather. Other students fail to develop these skills because they are so saturated with unnecessary calculations in training, that whenever the instructor suggested a diversion, the student wanted to throw the charts and flight computer at the instructor. Rather than concentrate on how elaborate you can make your cross-country logs, learn how to include only the most useful information. Rather than calculating arrival times for checkpoints you are about to fly over, learn how to properly divide your attention in the cockpit and the proper priorities for your energies. That is the secret of deduced reckoning.

When learning to plan cross-countries, you stumble at the first checkpoint because you have no idea how to figure in the climb. I have known students who try to calculate a groundspeed from the climb speed of the airplane and some bizarre, revolving wind correction average from the airport to the cruise altitude. From this they try vainly to get some figure for the time and fuel used. The problem becomes complicated when the altitude is reached after the first checkpoint.

Some airports have overlying airspace or ATC directives that force a stair-step climb, making life really miserable. To get around this, many students try to make the top of the climb the first checkpoint. If you don't know when you are going to get there, and you have no ground reference for verification, what good does this do? What is the basis for all future checkpoint calculations? Rote training and an inability to analyze the big picture usually causes you to solve the wrong problem and still be confused.

Consider the climb. Everything in a climb is constantly changing: the wind, the temperature, the power available in the engine, the fuel burn, and your flightpath at the whim of ATC. In the grand scheme of things, the climb is a small part of your overall flight, unless it is too short, such that minutely detailed climb planning is of no significant importance. Go to the airplane manual and find a climb chart that gives you the time, fuel, and distance to your altitude and use those numbers. Make your first checkpoint beyond where you should get to your cruise altitude. Checkpoints close to home are absurd. You don't need a checkpoint when you know where you are. Anyway, figure the remaining time from where the manual says you will reach your cruise altitude, to the first checkpoint, at your estimated cruise groundspeed. End of problem.

Understand that figuring in the climb is only for overall planning purposes to make sure you have the fuel to make the journey and that you have your arrival time correct. When exactly you reach your altitude for cruise is irrelevant. I have included the cross-country log that my students use. It puts the columns in a logical order from a planning and flying standpoint. It gives you a trip-planner to figure in your time, fuel, and distance from both the airplane manual and your calculations. You may use it or take it and make your own improvements.

A word about the magnetic compass. It is more than that thing stuck in your window that you use to set your directional gyro. It is your only directional instrument. Even the most sophisticated airliner carries a magnetic compass. Why? Because when all else fails you can still navigate. It will work as long as you don't fly in thunderstorms, you keep fluid in the compass, and the earth retains its magnetic field. I got caught near a rapidly building thunderstorm once and observed my compass sitting comfortably at a 45-degree angle. Anyway, most students think of the directional gyro as a directional instrument. It is not. All it does is to measure the angle the airplane has turned from where the gyro was last set. It only has meaning when set to a compass. Since it turns with the airplane and is easier to read in turbulence, you learn to rely on it and think of it as having directional properties. This is how the gyro can precess without your notice.

Back in the old days, there was only the compass. Pilots were confused at first because of its inherent errors for which they either corrected or waited for the compass to settle. Once you learn the peculiarities, it is not a problem. Standard teaching today does not introduce the magnetic compass until instrument training, when pilots are under the hood and have no visual reference to understand what they are learning. Besides the fact that this training comes way too late, wouldn't it be nice if student pilots could navigate with a compass should the vacuum pump fail and take out the directional gyro? Get your instructor to teach you about navigating with your compass before your solo cross-countries like it says in the FARs. Soon you won't even need the directional gyro. As a student, I once flew a whole cross-country with the directional gyro covered up. I've never had a problem with the compass since then.

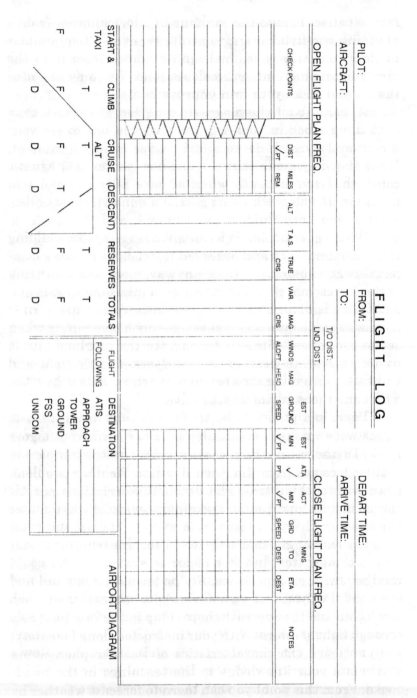

Weather bores most students. I don't know why, but anytime an instructor wants to talk about weather, student pilots give that face reminiscent of kids eating vegetables they don't like. Weather, however, can rivet your attention should you happen to find yourself in some when it is, of course, too late for classroom study. What happens is that students are bombarded in ground schools with texts written in excruciating detail and weather charts filled with complicated symbols that pilots never see in real life because they are jealously guarded by FSS personnel like monks in an abbey clutching sacred manuscripts. Students rightly see no point in learning something they cannot use.

Weather symbols are a complete language known only to meteorologists after years of study. Since we have weather people who understand the charts, why not draw on their expertise rather than try to make haphazard meteorologists out of pilots? The study of weather is tedious if all it means to you is multiple-guess questions on written tests. You must get out of the rote mode, stretch your instructor, fly in interesting weather and compare it to the forecast, and talk to flight service briefers to see how they do their job. Make the study of weather live.

There are plenty of books for pilots to learn what is important to them, particularly *Weather Flying* by Robert Buck. To understand how weather works, the best book you could get comes from the FAA. *Aviation Weather* is a book filled with cute pictures and detailed descriptions put together in a way that allows you to understand what drives our weather.

We cannot be teaching weather properly as it still claims the greatest number of private pilots. Understand weather through experience. Fly on blustery days and feel the wind. Put yourself on a dark night between the bright city lights and the overcast above. Play in the rain by flying through light showers. With your instructor along for safety, you can learn the characteristics of weather phenomena and relate your knowledge to the teachings in the handbooks. From this point you can learn to forecast weather by knowing the cloud types.

Now when you see the charts you will be able to picture the weather in your mind. When you have flown in marginal visibility, you will develop a new appreciation for VFR warnings and realize the folly of Special VFR. Should you ever fly with limited visibility at night, you will learn that night VFR can be an oxymoron. The FARs require you to have training in estimating visibility. See how your estimates of visibility compare to your instructor's and the reported weather. The best way to see if you can estimate three miles visibility is to go flying in three miles visibility. Taking responsibility for knowing the weather is part of being pilot in command. Your experience will allow you to develop judgment. You will then want to learn more about weather to make better judgments. Those boring books will suddenly become valued guides to greater understanding.

Many pilots have trouble accepting marginal weather when it confronts them, especially if it wasn't forecast in the briefing. They have this mental plan. They want to get somewhere and this weather is getting in the way, besides the briefer said it shouldn't even be there. This is what delays or eliminates the execution of an alternate plan. By not knowing the characteristics of weather, should you have spent your training flying on sunny days, you will underestimate the danger. All pilots have to have a plan B for each flight and a plan C for plan B. If you ever have to get there, take a commercial airliner.

I was flying with a student on the West Coast when the famous coastal stratus obscured our destination. When asked about this, the student responded with many options including turning for home, heading for any of three airports we could clearly see in the sunshine, and slowing to conserve fuel while we pondered the possibilities. That is great thinking. I wondered what the conditions might be at the destination. Students can be too quick to give up as well. You want to try to make your goal as long as it is safe. It turned out the destination was well in the clear; it was just that we could not see it.

Now, student pilots cannot fly without visual reference to the surface so we talked about how on a solo flight a diversion would be required. The FARs also require knowledge in radar aids and radio directives. Our destination had an ARSA (Class C) so we just got vectors to a place where we could see the airport. With a little imagination, you can cover a lot of different training requirements in a lesson.

When you fly, you are only as limited as the options you consider. By understanding weather and the system for reporting weather, you will develop the wisdom to evaluate your options. When you have seen marginal weather with your instructor, you can evaluate the merits and hazards of various courses of action. When you are released for solo cross-countries, you will have more experience to draw on should you be faced with your own unexpected weather dilemmas. It will happen.

You will learn procedures to locate yourself when you are lost. But will you use them when you need them? The psychology of emergencies causes a reluctance to accept situations because of embarrassment — and the questions. Student pilots, therefore, rationalize and never think that they are lost. They just don't happen to know exactly where they are just now. If you do not know where you are, you are lost. That is a distress situation that must be dealt with or it will become an emergency. The sooner you accept it and deal with it, the sooner you will be back on course.

If you have any doubt about your lost procedures, help is just a radio call away. Make the request. The controller is not going to ask you a bunch of stupid questions about what happened; they don't have time. They will get you on track the fastest, most efficient way possible for the safety of the whole system. However, should you fail to make the call, you, and particularly your instructor, will be questioned thoroughly on why you weren't trained to ask for help. Some students, especially the overly careful types who are acutely embarrassed at any human flaw, will not only try to solve the problem without help, they may do it using only one method of navigation. The instructor said use deduced reckoning on the way up and radio navigation on the way back. Taken literally, anything other than strict deduced reckoning on the way up would be cheating. Take this to heart: There is no such thing as cheating in an airplane — use everything you can when you are up there alone! All forms of navigation should be used all the time for verification. The best way to fly would be to track to a VOR, knowing the time you will cross the station by dead reckoning and confirming your progress with prominent landmarks.

Radio navigation is the third major form you will learn. That means learning the VOR. Sounds Latin, doesn't it? It stands for Very High Frequency Omni Range. Like most instruments, it has its own idiosyncrasies. However in its teaching, most students become hopelessly confused because you learn to use it only through examples given by your instructor, from which you on your own try to form guiding rules. The nature of the instrument leads you to form the wrong rules for all the different examples. After many lessons of forming different rules for all the situations for which you are presented, you get frustrated because no matter how hard you try to understand, your rules are always in conflict. It is also possible to go through the motions of a memorized procedure, get the correct answer with the VOR, and still not understand what you did. You temporarily fool your instructor who thinks you understand until you get a problem where you don't understand at all. Now your instructor is confused.

You should be taught the automatic direction finder (ADF) first, for it is far simpler in VFR conditions than the VOR and is not restricted by limited needle deflections. That would make learning the VOR afterwards much easier. The problem is that even though you have to know the ADF for the written test, only a handful of primary trainers have one installed. When you understand the principals by which the VOR operates, in other words the big picture, you will abandon your incorrect rules and go directly for the correct solution to navigational problems.

What is a VOR? It is a radio station that broadcasts a signal in all directions. Unlike stations on your FM dial, all it plays is Morse Code or a voice recording. We break up the signal into the 360 degrees of a circle and call each degree a radial. The signal always goes from the station. Therefore, anything you do with that signal is based on its direction from the station.

There are also 360 degrees in our compass for navigation. The "0" radial of the VOR is always set to magnetic north. What you are told to call the "omni bearing selector" is really an omni radial selector. Renaming it would cut out much of the confusion. A bearing always refers to a direction *To* a station. Things that radiate always go *From* a station. You know by now that everything works backwards, so it should come as no surprise that to go *To* a station, you have to dial in a radial on the other side of the VOR that is going *From* the station. That concept will drive you nuts.

What you have in the airplane is nothing but a receiver. It is a direction finder that identifies radials. The "course deviation indicator" (CDI), or "needle," will center when the airplane is on the radial coming from the station, and the VOR will indicate *From*. When you have identified the radial going *To* a station by centering the needle, you have actually identified the radial broadcasting from the other side of the station. You fly to the station by flying to the radial on the other side.

Because the radials of the VOR are set to magnetic north, you can think of the radial selector as a rotating, direction-finding compass. Just as your true course is measured from known lines of longitude and latitude, your radial navigation is measured in known degrees from magnetic north. You identify and set your magnetic course by radio and fly it by magnetic compass. The limitation of the VOR is that you only know whether the airplane is within ten degrees of your selected radial because ten degrees either side of center is the limit of the needle deflection. Work with your instructor until you understand.

So what can you do with this thing? You can fly directly to a station by identifying the magnetic direction that will take you there by centering the needle with a *To* indication. You can go directly from a station by identifying the magnetic direction with a centered needle and a *From* indication.

The situation is complicated by wind. Wind will blow the airplane off the track. Wind can't blow the radial, since it is an electronic signal, so it stays where it is. We are talking about "tracking" a radial. The radial is a fixed, immovable reference like a runway in the ground is fixed and a magnetic course is fixed. No matter whether you are tracking a ground reference, a magnetic course, or an electronic signal, you are looking for a wind correction angle. They are all the same thing even though they are taught separately.

When you are blown off your radial, the needle moves off center. You can tell, up to ten degrees, how many radials you have been blown off course. That's what the dots are for. The doughnut (white circle) and each dot are two degrees apart. You can also twist the radial selector and confirm where you are. Your job is to get back on the radial. If you are taught to track a radial by turning left or right when the needle has a left and right deflection, you will set yourself up for endless problems. It will take time, but you must never think of correcting to the left or right when talking about a VOR.

Always think of your course, how many degrees you have been blown off course, and what magnetic direction will put you back and hold you on course. If you are five degrees off course, the needle will be halfway to the side since it only indicates a maximum of ten degrees from the radial selected. The easiest way to get back on course is to double the number of degrees off course, in this case ten, and turn to the magnetic direction that is ten degrees over on the same side as the needle. This is the "same side safe" rule that I learned from an old chief instructor.

As you already know, the VOR cares not where the airplane is heading; it only cares about the position of the airplane in space. Therefore, talking about left or right correction is meaningless. Left or right from what? No matter what direction the airplane is currently heading, if you have a five-degree right needle deflection, turn to the heading that is ten degrees to the right of the radial selected. Depending on your heading, either a left or right turn will be the shortest path, and that is why you can't think of correcting left or right for a left or right needle. It is a very subtle but important difference.

The same goes for intercepting a radial beyond full deflection when you know not what radial you are on. To intercept a radial, any radial, *To* or *From*, regardless of your heading, simply twist the radial selector to your interception radial, and turn to a heading on the same side as the needle. Your only decision is how much of an interception angle you want. For a direct interception, turn to the direction that is 90 degrees on the same side as the needle. For a normal interception, use 45 degrees. You can prevent overshooting by simply reducing the angle to 30 degrees when the needle comes alive (moves from a full deflection.) Go to 20 degrees when the needle is halfway across, ten degrees on the doughnut, and five degrees of correction only when inside the doughnut. Work with your instructor using this method and you should never have trouble understanding how to identify, intercept, and track radials.

The "same side safe" rule is so simple and effective you might wonder why only a select few instructors know it. That is because in this age of information and high technology the FAA still hasn't set up a national flight instructor data bank for new and innovative improvements to instructing. Once you relate new techniques to something you already know, like courses, radials, headings, ground reference, and wind correction, you realize they are all about the same thing and that nothing in flying is all that complicated. Our current system, however, teaches you in a fashion that makes everything seem new and different. This is the fault of your training, not your ability to master what you might consider impossible procedures. The cycle goes on because the teaching never changes.

On the communications side of the radio, you will be exposed to individual frequencies for various situations, but you will begin to notice certain trends. The *Airman's Information Manual (AIM)*, one of the most useful books in aviation, groups most frequencies on various charts. I hope one day they make one big chart with all the communication frequencies. Here is a grouping of the common ones you will encounter:

Unicom: At an uncontrolled (which should be called either pilot-controlled or nontower) field, if anyone is home, you can get weather and airport advisories. Frequencies; 122.7, 122.8, 123.0, 123.05.

Multicom: At airports that are unattended on the ground. All communications are between airplanes. Frequency: 122.9.

Flight Service Stations: They have a wealth of services to the VFR pilot. For weather and pilot reports, call flightwatch 122.0. For all other business, any FSS can be reached on 122.2. They also have a separate discreet frequency. All traffic advisories when an airport advisory area exists will communicate on 123.6.

Ground: Ground frequencies will almost always be 121.something.

Unicom: For airport information at a tower-controlled field such as fuel, parking, or particular FBOs. Frequency: 122.95.

Multicom: For airplanes to communicate strictly air to air. Frequency; 122.75.

Before you soloed, you practiced pattern after pattern at familiar airports. You knew all the local landmarks in which to make nice patterns at those airports. This is good in the beginning because it builds confidence. However, it sets you up for a later problem. You were teaching yourself how to fly that airport rather than to fly good patterns by applying specific procedures which would work at all airports. This will be glaringly obvious the first time you take a dual cross-country to an unfamiliar airport. In your effort to get a really good look at the strange airport, your pattern will likely be very close to the runway, all the legs will be too short, you will end up rushing the procedure, you will mentally fall behind the airplane, come in high, make a steep descent, build excessive speed, and land halfway down the runway. Welcome to your first dual cross-country.

You can make your life easier by making a conscious effort to learn the pattern, not the airports. Runways give an illusion of distance depending on their width. You will fly closer to a narrower runway than you are used to and further from a wider one in your effort to make a familiar picture. The way out of this is to always put the runway centerline on the same reference point on your airplane. That way you will always fly a pattern from the same distance. Halfway up the strut works on the Cessnas. The second or third line of rivets in from the end of the wing works for the low wings. No rivets? Find your own reference. Your consistency will help to deal with this illusion. Local terrain can also distort your perceptions.

Your pattern legs may not be the pretty rectangles you are used to flying. This is where the directional gyro can be of service. If you have a heading bug, it gets even simpler. Either mark with a bug or note the runway heading on the directional gyro. Your downwind heading will be the reciprocal. The base and crosswind legs will be at the 90-degree points. For a perfect 45-degree entry, put the runway heading in the lower left 45 position for a left pattern and the lower right quadrant for a right pattern.

Now all you have to concern yourself with is the wind. Always verify with a back-up. The compass backs up your visual reference. This is really useful in a high wing when turning base at a strange airport. The directional gyro is your greatest aid if the runways change while you are mid-pattern. By knowing the new runway and finding it on your gyro, you can figure a new pattern entry in seconds.

Another thing most pilots have a tendency to do, especially at new airports, is to fly away from the runway as soon as they have passed the numbers on downwind; right or left pattern, it doesn't matter. Hold your wind correction heading whenever you extend your downwind and your pattern will be consistent.

The slope of the terrain around an airport can cause interesting errors. I used to fly out of an airport where the hill off the end of one runway would rise or fall at the same rate as a Cessna 150 climbing out or descending for landing, giving the illusion that you were flying level. Only with your altimeter and vertical speed indicator could you measure the altitude change.

Should you fly into an airport with a sloping runway, you are in for a thrill. If you are used to the sight picture of a level runway, by flying the same sight picture on an upsloping runway, you will be really low on approach. You may fly right over a down-sloper. The best you can do here is to have altitude checkpoints in the pattern. Try to turn base to final at a checkpoint altitude, like 600 feet AGL, as consistently close to the same distance from touchdown as you can. That way you will fly to the same markers for every pattern. If the picture looks funny at your usual checkpoints, you know there is an illusion. In most flight regimes, there are many references to help you verify your situation. Chances are, though, you will forget all this when you are so pleased that you found the new airport.

If you fly a pattern the way I do, you won't be descending until turning base. This will give you plenty of visibility and time to find your way. You won't be concerned with descending on downwind so you can hold your course better. If you get in the habit of always turning base at the 45-degree point every time, you will do it at the strange airports as well. You will not depend on familiar landmarks at home nor miss them when you explore new airports because you have learned how to fly "the pattern." Each airport looks different and has its own landmarks. Your consistent procedures will help you handle inconsistent situations. If you can plot a flightpath in your mind from the base turn to the runway, if you can find a landmark that keeps you on that flightpath, if you can verify that path with a heading, and maintain that path with wind correction, then you should fly a pretty good base at any airport.

Rolling on to final in a high wing at an unfamiliar airport should give both of you some giggles. You can't see the runway because of the wing in your way, and as you roll out, initially the curvature of the window will distort your direction. You will see where you really are when looking through the flat part. You will have virtually no reference for this final turn. You won't want your head on the directional gyro when you get that low, so if you do not know the landmarks, the best thing you can use is your sense of timing. Turn at the same angle of bank every time (I use 20 degrees), and you will develop a sense of when to begin the turn so that you can roll out on final when you want to. Once again the wind will cause you grief, so it must be taken into account.

The FARs have you reviewing stuff that you were taught before you soloed. Even though you are about to venture forth cross-country, you are still a neophyte pilot. There is so much that you do not know — you do not even know what you do not know. Which is why you have to have a good grounding in the basics. The problem remains, though, that the basics are not taught as well as they could be taught, so you end up reviewing stuff that still won't help you.

For example: Along with being taught to bring back your power abeam your touchdown spot on downwind, which results in rather tight patterns in the average single, many of you are taught to add flaps, change your power, and then have to trim *after* you have established yourself on the base leg. Throw in a radio call at the pilot-controlled fields, and you are doing everything on base but what you should be doing; that is flying the airplane, watching for traffic, and planning your turn to final. Then you are told to fly a stabilized approach. Sure.

In my pattern, you set the configuration for the whole approach before even turning base. Just before you arrive at the 45-degree point past the touchdown spot, set the approach power, 1500 RPM or 15 inches of manifold pressure should work for anything with reciprocating engines. Let the nose fall. Set the flaps at two notches or 20 degrees, and the nose will rise a bit. Set the descent pitch attitude of six fingers and trim off all control pressures. Now ease into the base turn. When you become proficient, you will do them all at once. You will roll out on your flightpath, on airspeed, in trim, with a full field of vision, and all you have to do before landing is make a few minor power adjustments and flare by putting your nose just under the end of the runway while looking straight ahead. Should you have a short-field approach, simply set final flaps before rolling onto final. The loss of vertical lift in the turn compensates for the nose-up tendency of the flaps, making the turn the best place to lower flaps and remain stable. By configuring before each leg, you can spend the entire leg simply flying.

All of that goes against conventional wisdom. However, it is contradictory for instructors and examiners to insist on stabilized approaches while at the same time making you constantly reconfigure the airplane in the middle of your base and final legs. You roll on to the base leg, change the power, change the flaps, and then try to trim before final. Your head is buried in the cockpit of a destabilized airplane. Now, if you stabilize the pilot with a simple, logical, and consistent procedure, the approach will take care of itself. If you can set your power, flaps, and trim before you turn base, the airplane will practically fly itself to the runway with the minimum input from you because you have already done everything. All you have to do now is adjust the flightpath to the unstable conditions.

Many of you are taught to put the flaps down in the middle of your pattern legs because your instructors have bought into a myth without questioning it. There are enough myths in flight instruction to fill a book. Let's follow this one through. You will destabilize your approaches because you are taught never to drop flaps in a turn. This leaves only the middle of the pattern legs. For those astute enough to question this practice, you will be told that dropping flaps in a turn is dangerous if you get an asymmetrical flap situation. That should sound impressive enough to satisfy most students.

With any luck, you folks are now as skeptical as I am, and that just won't do for an explanation. Always ask why. Why is it any more dangerous in a turn with asymmetrical flaps? I was then told that in a turn you will roll faster toward a more dangerous angle because of the bank. Why? Does the airplane wait until a turn before an asymmetrical flap situation occurs? The airplane will roll regardless of its attitude. Besides, you have a 50-percent chance that the airplane will roll level, thus making it safer to drop flaps in a turn. The most you should ever bank in the pattern is 30 degrees, which gives you lots of time to correct for the flaps before you go vertical. No matter what the attitude, if you felt something strange when the flaps were deployed, wouldn't you immediately bring them back up? Chances are strong that you already have some flap down from the approach before starting your landing. Unless something happens from downwind to base, the flaps should work just as they did some few minutes previously.

Another thing you can do to alleviate your fear is to watch the flaps coming down when you lower them for preflight. Many students check only to see that the indicator works during extension or stare at the flap handle, so they never know if the flaps will come down together until landing when they are watching the runway. Believing that you should never lower flaps in a turn is like believing that you should never fly fixed gear airplanes because a bird might hit the exposed tires and puncture them.

There are some airplanes where the flaps are below the wing and you can't see them from the cockpit. Why not get someone on the ramp to watch your flaps as you lower them? During my checkout in the Cessna 310, my instructor insisted that I never touch the flaps in a turn. The 310 can have asymmetrical flaps because they aren't welded to a single metal tube like some airplanes. The interesting part of this is that in flight asymmetrical flaps will let you know immediately where you will likely be flying at least pattern altitude and have a chance to deal with the situation. You also extend the flaps one setting at a time so they shouldn't be more than one notch apart. However, should you extend the flaps fully for preflight in a 310, strap into the airplane, start the engines, try to raise the flaps and have only one come up, they would be as far apart as they could get. You would never know it unless someone on the ground brought it to your attention until lift-off, which is the most critical phase of flight in a twin. This concern was never expressed by the instructor. Such is the power of myth.

I'll bet this happens on your first dual cross-country. You are sitting in cruise flight fat and happy. The checkpoints line up reasonably well and show up about when they should. As the last checkpoint arrives, you become just a bit anxious because you will soon be preparing for a landing. Your greatest concern is finding the airport. You come upon your descent point and halfheartedly start a descent that quickly levels off as you get back to searching for the airport.

Your instructor has probably been to this airport many times before and has had it in sight for the last 30 miles. That is why you don't see him looking for the airport with you. Most instructors keep the same destinations for cross-country training for reasons of laziness, predictability, familiarity with the route so mistakes show up easily, and the fact that it is abhorrent for an instructor to be lost or even to think he might be lost.

If you want a thrill, insist that you be allowed a destination that the instructor has never seen before. Your training will be more realistic because you won't get that logical progression of questions designed to lead you to a predetermined answer. When you go to a strange airport, the instructor will not have all the answers and will have to navigate along with you. Chances are he will be just as pleased as you to find the destination. That will show you how navigation is supposed to work.

Anyway, back to our story. So there you are about 10-12 miles out and the airport comes into view. You relax thinking your job is over when really it has yet to begin. You sit there pleased with yourself while drawing ever closer to the airport. You initiate some feeble descent that will certainly not get you down to pattern altitude in time. Your instructor starts twisting in his seat, doing the famous "instructor wiggle," where he physically tries to will the airplane into something other than what you want to do with it by the power of his butt muscles. The variation on this is the "instructor lean," where he tries to use "the force" to put the airplane on track when you let it drift in a crosswind.

But I digress. Anyway, you see your instructor twitching and feel that you should be doing something. He is prompting you without actually saying anything, hoping you will take corrective action. You are in an unfamiliar environment and do not realize how close you are getting because there are no familiar landmarks to guide your actions. The new airport moves closer, but you aren't coming down. There has been no prelanding check. The mixture is still set for cruise. You have not checked the ATIS or gotten any advisories. You have no plan for your approach. But you found the airport, oh boy! What a rationalization.

The reason so many students do this is because cross-country training concentrates so much on navigating and locating the airport that it kind of leaves off what you have to do to get there. Many instructors try to compensate by having you draw an airport diagram. This is where some of you duplicate the pilot's guide picture in its entirety. If you are about to punch through an ATA (Class D), there is no way you can sift through your diagram or the pilot's guides to find your way to a runway for it is far too late. You need this information long before you ever start your approach. Having the information is half the job; you have to be taught when to use it for it to be any good.

What was so short a time before a gentle cruise with nothing to do is now about to be a very busy situation. But you don't see it. Reveling in the glory of finding the airport and unaware of its close proximity, you are shocked when the instructor in a very insistent voice states that there is much to be done before landing and not much time to do it. Now the light bulbs go on at how far behind you are and the rush begins. Without the familiar checkpoints giving you all your cues, you bury your head in the cockpit trying to figure your course of action from the stack of papers on your lap while trying to do all the landing stuff with the airplane.

It may not occur to your instructor to present any options because he is perturbed, having been there before, and is preoccupied because he knows what you should have done and when you should have done it. Had he been new to the area, you can bet you would see your options. However, in this case, his impatience causes him to try to get you to do everything quickly, so that you at least will look professional coming in to the airport. Flustered, you will now do everything badly. You will completely forget to descend or set up such a slow rate that you will never make the pattern.

Student pilots without known landmarks are notoriously bad judges of distance, altitude, and descent rate. You are caught by complacency because you thought you were learning how to fly, when in reality you now realize you were only learning to fly at certain airports. You will now be visibly upset by a situation you had no idea was coming. Your instructor will be barking out orders for you to follow. He is making the decisions; you are moving the controls — sound familiar? You are trying to figure out what happened at a time when action is needed. You will be directed to the pattern by your instructor. You will be asked if you see the runway. Chances are you have forgotten how to find the correct runway and set up for an approach while you seek to understand your predicament. You will bumble into the pattern and come whizzing in for a landing.

On the ground your instructor will ask what were you thinking to get so far behind? You were not thinking about anything because you never saw the problem, so you will not have an answer. Here is why. Most new pilots believe that the airport is much farther away than it really is. You think the airplane can descend far more steeply than it does for the amount of ground covered. You also think the speed over the ground is far less than actual and are surprised to find that the closer you get to the ground, the faster the aircraft seems to move, the higher you realize you are, and the worse things seem to get.

One problem is that this will likely be your first flight above the 3000-foot level used in most practice flights. You have never had to descend from a real cruise situation in a familiar area, let alone one that is new and different. The cure is to get comfortable with higher altitudes during earlier lessons.

The other problem is that you are not exposed to descent profiles, normally reserved for high-performance training which comes long after you have ingrained bad habits from private training. You need exposure to descent profiles from day one. You won't see any need to plan descents in your early training, however. Here is why. An aircraft traveling 90 knots and descending 500 feet per minute will require nine nautical miles to lose the 3000 feet of altitude where you are used to flying. That works out to six miles to descend into a 1000-foot pattern. You get the idea that you can descend for the airport just outside the ATA and everything will work out. It will, as long as you fly at 3000 feet. Should you approach your first unfamiliar airport from 6000 feet, the 5000 feet of altitude loss you will need to get into the pattern will take 15 nautical miles. That is assuming you hold a 500-foot-per-minute descent.

Most students cannot do this because it is the one place where you don't have your head buried in the cockpit. You are looking at your destination, nothing else. This creates an interesting effect that contributes to your being too high. Many students put objectives in a certain place on the windshield. When you are many miles out, your pitch will be only slightly below level, so no real descent occurs. As you approach the destination, you have to pitch down progressively more steeply to keep it in the same place in the windshield the closer you get. This can be really funny if the instructor lets it go for a while.

Just like on landings, if you're using a target like the touchdown spot, you maintain the picture by varying the approach path with power, the pitch keeps your descent attitude and airspeed. Should the destination not be in sight when the descent point is reached, students are reluctant to descend. Depending on the visibility and how far out you see the airport, you could have a very steep approach.

You can see once again how rote teachings establish faulty thinking and bad habits. By always flying to the same practice areas, practicing at the same altitudes, turning at the same landmarks, and descending at the same places, you have been kept from learning any of the thinking that you need for unfamiliar areas. The time to learn this is during the lessons that come after you solo while you are building the skills for your first dual cross-country.

The FAA should require in your cross-country training a whole section on exploring options. When faced with a rapidly approaching unfamiliar airport, what can you do? You could circle a prominent ground reference while keeping the airport in sight and work out an approach plan. During that time, you could refer to your airport diagram. You have been taught to overfly pilot-controlled airports, but probably not tower fields. So, you won't think to overfly the ATA (Class D) and plan a sensible course of action from there. Sometimes there is an ARSA (Class C) which prevents it, in which case you can get radar vectors to the runway, just like the professionals do.

If you find yourself too high, why not turn around and come back in at a lower altitude? You could call the tower while over the ATA, report over the field, and spiral in from there to the runway. At nontower fields you can talk to the Unicom person, if available, and find out the best route for your approach. You could also talk to pilots who are familiar with the area. If you want a lot of time to plan, you could check with your radar flight-following controller, they might know the runway in use from pilot reports.

The Flight Service Station people might be able to help if they know the airport. If no FSS is nearby, you could look for Remote Communications Outlets (RCOs) for a distant FSS or try to reach one through the voice feature of VORs or even NDBs. There are automatic weather advisories like AWOS which can give you the winds and you can figure out the runway. You can even relay a message from another aircraft should your destination have a weak Unicom station or none at all. How many of these have you tried or even considered?

I remember one cross-country when I was a commercial student. I was trying to find this mountain airport. I knew I was close; I just couldn't see it. I heard this other airplane just below and ahead of me report going to the same airport. I called him and asked if I could follow him in. No problem. He took me right to it. I'm not recommending this for regular use, but in this case it worked.

Upon realizing you are too high for the airport, most of you cut the power and try to dive the airplane. Since you can't see below the nose, a steep descent presents a real safety problem where other airplanes might be operating, like near an airport. At least do some S-turns to check your path. Students feel that once they start on in, come hell or high water, they are committed to make this landing happen. Remember, you can sit up there considering options as long as you have fuel. Forcing an approach is like forcing a landing. Most of you know about go-arounds and can do them when asked. Your instructor, though, wants to know if you will do it without asking. The approach works the same way. Will you try again if the circumstances dictate?

Your options are only as limited as your imagination. Unfortunately, you are your own greatest limitation. It starts when you give up your responsibility for your training and have to be spoonfed. It continues because your instructor is totally responsible for you and must control all that you see so he will know what you will do. You learn by rote procedures. You fear disapproval from incorrect, independent actions. You are not encouraged to think. You see the same things every lesson. You cannot transfer information. Therefore, you will not consider options that you have not been specifically shown. We are progressively linking the chains in a system desperately in need of an overhaul.

During your training, you will have too many flights where you go to one airport and beat touch and go's to death working on the mechanics of the landing flare. This gets you one approach at an airport where you are pretty familiar and then a very familiar approach when you go home. You never get practice in the weakest area for new pilots: approaches and landings at strange, new airports. I have an idea. Go visiting. Spend a whole lesson going from airport to airport with just one touch and go. Try as many different types of airports as you can. The best lesson would be where you could include airports you have never seen.

The other thing you could do is to do a series of single touch and go's at your home airport. After each one, go off in a different direction and altitude and fly out about five miles. Keep trying different combinations of headings and altitudes until you can approach your airport from anywhere. This will build flexibility and judgment.

Finding the correct runway can be a pain. My field has two sets of parallel runways; the ones used most are set about 130 degrees apart. That makes for eight possible patterns and touchdown spots.

Getting into busy airports can also be a pain. I used to teach at this airport just over the hills with only one set of parallel runways, but it has an ARSA (Class C) and lots of traffic. People from my current airport don't like to go over the hills because they are intimidated by the traffic and radio work. People from over the hills won't come here because airplanes come in from all angles and report their position over unknown local landmarks like "Bob's Diner." We also have airplanes doing combinations like right base entries to left runways, left downwinds from right runways, and helicopters going to and from everywhere. Intimidation is related to familiarity. Once you get used to new experiences, nothing is all that intimidating.

Back to finding a runway. You have to know way ahead of your approach which is the runway in use, because the numbers will only be visible from very close. People forget how the runways are set up. On the ground you know they are compass headings, but in the air you draw a blank when you need to find a runway. With a little imagination, you could just look at your directional gyro and see where your runway is going. Imagination, unfortunately, doesn't rank high on the list of student pilot virtues, not that you have time for imagination anyway while you are approaching an airport, so let's try something foolproof.

Say, for example, you are coming from the south, heading due north to an airport directly ahead, and you have to land on Runway 30. Put your outstretched index finger, either hand, across the directional gyro with the tip of your finger on 300 degrees. Leave your finger still and look at the airport to find the runway at the same angle. Touch the base of your finger with your thumb and that is the end you will land on. Your finger points the way you will be going.

Keep in mind that the numbers on the directional gyro and the runway are at opposite ends. Once you have done this a few times, you will just look at the gyro and match the runway without using your finger. However, your finger is there if you ever get confused in the future. You may require a different finger depending on the instructor.

Wake turbulence, that nasty unseen vortex from big jets is just waiting to roll you. The theory is taught pretty well; it is the practice of avoidance I find so interesting. One problem is overcaution. The tower calls your takeoff after the obligatory three minutes have past since a Gulfstream departed. You want another several minutes just to be sure. You have that right, however, except for certain wind and runway situations it is unnecessary. This is usually when the Piper Cub next in line gets your clearance and nothing happens as they fly away.

Another problem is putting faith in the tower to judge the wake. If a tower clears you to takeoff, you assume the wake is gone, when the controller may have forgotten to have you wait. You are pilot in command — ask for a clearance in three minutes. Some students don't know enough to be afraid of wake turbulence because they feel their instructors were put on this earth to save them from themselves, so they need take no responsibility. They are the ones who carry on as if it weren't there, while knowing all the time it might be there.

The most common and bizarre manifestation of this invisible hazard is that you really wish wake turbulence would just go away because it gets in the way of your normal routine. Rather than vary your flightpath for a takeoff behind a departing jet, you would rather sit until you can make your normal departure. You would feel silly making an early turnout over your flight school because you don't want your fellow students to think you are flying like a student. Even if the tower would appreciate the departure to help with traffic and your instructor insists it is safe, you still feel funny. People might be watching you. Investigate those feelings.

How about a long landing to overfly a wake? Will you feel strange landing halfway down the runway? In your presolo days, you were lucky to notice where you touched down. With increased skills comes a change in expectations and landing halfway down the runway now looks funny. Even though you know that is what you are supposed to do, you might just take a gamble on the wake. You need approval. If your instructor directs you to land down the runway and informs the tower for you, then it is okay because everyone now approves.

So what are you going to do if you have never seen a wake turbulence avoidance with your instructor, you are at a big, pilot-controlled field where no one is around to approve of your decisions ahead of time, and you are following right behind a three-engined Falcon on final? Will you land long, come back later for a normal approach, or gamble on the dissipation of the wake before you get there because you don't want to do anything out of the ordinary? Two of these will work; one might roll you into the ground. If you find you are willing to jeopardize your safety because you don't want to use a procedure even when you know it is the correct action because you think it might look strange, sit down for a long chat with your instructor.

Sometimes you can use the correct procedure, whether you have seen it or not, and you surprise everyone because no one expects it. You can feel just as strange for doing the right thing. The difference is that you will be infinitely safer. My first encounter with real wake turbulence came when I was to depart after a Boeing 737. I was in a Cessna 152. After three minutes I was cleared to go. With a very healthy right crosswind, my concern was a vortex remaining on the runway and on my flightpath. I climbed steeply and initiated a turn to offset to the right as soon as practical.

"What are you doing off the centerline?" blasted the tower.

"Avoiding the wake turbulence, sir, by offsetting into the wind." Always stand your ground when you are correct — just do it politely.

After a noticeable pause, they came back. "Oh — good idea; carry on."

Sometimes students are surprised to find they are sticking to the procedures far better than more experienced pilots. This is because you folks don't know enough not to deviate from the teachings. You repeat only what you are shown; so if you know only the correct procedures, that is what you will do. In the above example, I wondered what all the other light aircraft did when following a 737? As you get further from instruction, there is a tendency to let all those procedures gradually slip. You may hear towers warn you of wake turbulence for years and never hit any. This will build complacency as you pay less heed to the warnings and come closer to dangerous wakes. You will have no problem until your luck runs out.

Students who observe pilots neglecting procedures may assume they are being coddled where real pilots just plunge on ahead. With experience, you can use judgment and shade procedures to a certain degree. However, experience does not connote wisdom, and the student who follows the procedure shows more wisdom than the experienced pilot who does not. In the convoluted logic of safety, when you follow the wake turbulence procedures and find that you don't feel anything in the airplane, you wonder what was all the fuss about? Yet that is the idea — by using the procedure, you should never feel a wake. By not feeling the wake that you successfully avoided, you get the mistaken impression that it might not have been there to avoid. How close will you have to get to believe it is really there?

Before you can go off exploring the countryside on your own, you have to learn two new types of takeoff and landing, the "short-field" and the "soft-field." Practicing them will drive you crazy. The procedures are polar opposites so, of course, you have to learn them both at the same time. You have to be introduced to them but won't perfect them until the checkride. You might wonder why you are being subjected to these at all because your instructor is not going to send you to an airport requiring a short-field procedure. With rare exception, no insurance company will allow you to land on anything but a hard-surfaced runway, so you may never land on a real soft field. You don't need them, can't use them, but have to be good at them. You figure it out. I watch this big academy all the time send out students in their 152's to practice short- and soft-field touch and go's at their home base, which happens to be a major terminal in a large city where the runway is 200 feet wide and 7000 feet long. Now there is realistic training!

I learned to fly at that academy and pleaded for training on real soft fields. It was against the rules. After becoming an instructor at another school, I got permission to operate off this gravel strip with a big ditch beside it. All my previous soft-field landings had been on the big runways. Under the false assumption that my old procedures will work just as well in the real stuff as they do in a poorly simulated training environment, I felt confident bringing a student along for my first real soft-field landing. I don't make that assumption anymore.

Skidding toward a ditch with a student who had not learned to land straight woke me up. Gravel doesn't have the traction to straighten crooked wheels like a runway.

This indelibly etched in my mind the critical need for realistic training and the lunacy of learning techniques that you never see in their intended environment. This was the second red flag that my training had been woefully inadequate. The first came when I was training for my instrument instructor rating while flying in my first real weather. More on that later.

Back to the gravel. Hanging out all the flaps and leaving a little power to land as slowly as possible as I was taught nearly put us in the ditch because we had virtually no directional control when we dropped in. On the next attempt, we used half flaps, slightly more speed for control, and a lower landing attitude so we could see the strip. We ever so gently rolled onto the gravel, eased the stick back for aerodynamic drag, and let the gravel slow us down. This worked great. So much for my training.

Understand that learning a soft-field landing on a hard-surfaced runway will teach you only how to do a soft-field landing on a hard-surfaced runway. If you come in so slowly that the rudder is at the low end of its effectiveness, and you are holding the nosewheel off the ground, how do you maintain directional control on slippery gravel? By maintaining the speed necessary for rudder control, you can pick your time to lower the nose as the rudder is losing effectiveness. Conventional technique requires that the nose stay nice and high. This is so you don't stick in the imaginary mud. This makes sense for mud and for rough fields with holes that nosewheels can catch. However, this strip was very smooth and there was a greater danger from blocking forward vision because of the ditch.

You are taught one soft-field technique when there are many different types of soft field: gravel, mud, turf, grass, snow, dirt, or any combination of these. As long as the nose is clear of the ground, how can you nose over? Your attitude therefore depends on your particular circumstances. Until you experience real soft fields, you won't appreciate the natural drag; so all that "nose pointed skyward stuff" is misleading. The speed of the airplane while the prop turns will throw gravel and turf backward so the nose need not be overly high on touchdown. If possible, the second half of the roll is where you might want the nose higher. You may have to work hard for realistic training, but it will be worth it. If you ever have to put down on a real field in an emergency and you have only previously landed on big runways, you may be in for a surprise.

Soft-field takeoffs from hard-surfaced runways create their own delusions. Back at the academy, they pull those sticks back, pour on the power, and try to drag their tails down the runway. Eventually the airplane lurches into the air on the edge of a stall. They level off at about 100 feet searching for the ground effect. When the best rate of climb speed catches up with the airplane, off they go. This must be okay with the FAA because I have seen the same thing for years and pilots keep getting certificates. Had they actually trained in soft fields, they would see that surface drag is the biggest impediment to performance. Therefore, once the nose is clear of the ground, you aren't likely to flip over.

Do they think somehow that the wonderful aerodynamic drag created by holding the nose high on landing would suddenly disappear on takeoff? You won't perceive that on a runway because it has no where near the surface drag of a soft field. But if you combine that aerodynamic drag with real surface drag, you won't be moving very far very fast. You need the nose in the air for ground clearance, but putting it as high as possible delays your takeoff considerably.

You need to learn the attitude which allows you to leave the ground in the shortest distance. The window of ground effect is at most one wingspan. The closer you get to the ideal liftoff pitch, the sooner you break ground. The closer to the ground you level off while still remaining clear, the sooner you can accelerate to climb speed. These objectives are lost in the conventional technique because who needs it on a hard runway. Most pilots blow through the ground effect because of the excessively high attitude, especially in the high-performance aircraft. You don't have to.

Here is a secret. If you wait until you see the airplane

climb with your eyes, you will blow right through the ground effect before you can lower the nose, because it takes too long for the response to go from your eyes, through your brain, then to your hands. I think sight is the slowest response we have. Touch is faster. If you burn your hand, don't you pull it away before you see what was burning it? Flying by feel works the same way. You are told that the objective of the soft-field takeoff is to transfer the weight of the aircraft from the wheels to the wings as rapidly as possible. This is true.

Take the next step. You can feel the transfer of weight in your butt long before you will ever see the airplane leave the ground. As soon as the airplane "feels" light, lower the nose. With some practice you will be able to level off six inches from the ground. Sight is a bad way to judge flight close to the ground, especially if your nose is blocking all your forward vision.

Not only will you be taught to takeoff and land from imaginary mud, you will learn short-field procedures for how to operate over imaginary trees. These are regulation trees and always grow to exactly 50 feet. It must be genetic engineering. You will simulate landings with those trees sitting right at the threshold of the runway. Have you ever seen a tree at the threshold of a runway? Just outside the fence, sure, but not where you are told they will be.

There are many airports out there with short runways and interesting terrain. However, if you use good normal technique, you will be just fine in most cases. Considering that you can land a plane in fields too short for takeoff, you might be stuck if your field requires a real short-field technique. I have never seen a tree that wasn't outside the airport boundary. So you may wonder as I do, why students are taught to limp in about 75 feet over the threshold, chop the power, drop in, and flare just before impact. Must be those imaginary trees in the way. So much for the stabilized approach.

When you chop the power, you chop the airspeed and lift that comes with it. The only way to avoid the approaching stall is to pitch down — a lot. The pitch will build airspeed which must be dissipated as you float away from your short-field landing. All this for a tree that won't be there. Anyway, a few of us maverick instructors got together and figured out that if we just came in high enough on final to make a straight path to the runway and clear any realistic obstacles, we could fly a very precise, stabilized approach to the touchdown. We also found that by carrying a little power to fly at the slower approach speed, and by easing off the power to idle and flaring just enough to round out to a smooth but firm landing at the minimum speed, we could make some very short landings.

Short-field takeoffs are generally taught quite well. This is a refreshing change. Maybe this is because the length of the runway is irrelevant to the technique. You still want to be up and climbing steeply at maximum performance. Short-field technique becomes rather dry, however, if you never know the exhilaration and joy of barely clearing real trees. For that you need a genuine short field. Short fields are invariably narrow fields with obstacles and uneven terrain.

Good technique is essential for safety. There is no substitute for having to maneuver around trees on landing, or pitching for the best angle of climb and waiting those painful seconds for outstretched branches, power lines, windmills, or other hazards to pass you by. This kind of situation forces you to use the procedures you have learned. You can transfer much knowledge to various situations from short-field work. The short-field approach is a great way to lose altitude if you are too high, the flaps are already out, and you either can't or don't want to slip. The short-field approach also teaches you about the effect of pitch, angle of attack, the back side of the power curve, and induced drag. You must have this pointed out to you, however.

Frustration is built up in the minds of students whose instructors don't show them a realistic application for new skills, because it is stupid to learn something that has no apparent use. It is unsafe to let students practice maximum performance-maneuvers on their own in real situations so they won't see the logic then either. The danger of not seeing the proper application comes sometime after you have been out of instruction and you need these procedures to get in and out of that cute mountain airport by the lake. By this time, pilots have either forgotten the procedures, never seen them used properly, or thought because they passed the checkride from training acquired on big runways that the same techniques apply. We are all student pilots and you can always go back for recurrent training.

When your instructor thinks you have the degree of skills in the required areas, you get to go off on your own. To wing across the friendly skies with all the confidence you can have at this point is one of the great joys of flight training. The cockpit will be strangely quiet except for echoes of your instructor when you talk to yourself. In this new environment free from criticism, you will start to feel like a pilot. You begin to make decisions, at first because you have to and gradually because you enjoy the satisfaction of being pilot in command. There will be moments of uncertainty, like when you can't locate the checkpoint that is directly below your airplane. Your confidence will build, tempered by the increasing knowledge of your own limitations.

As a student, you will stick rigidly to procedures and shouldn't create any serious problems for yourself. Your first solo cross-country will likely be some place you have seen before. This is for confidence. Then will come the new places. Should you run into trouble, your training has been so recent that solutions are readily at hand. Much of your training covers the worst cases and emergency situations, so that a normal flight seems boring by comparison. You get to fly at your own pace, not that of the instructor. You will be as relaxed as the circumstances allow. After a few trips, you won't want the instructor back to pressure you again and ruin your newfound sanctuary.

You will be convinced that you fly worse with the instructor on board. His status changes from wise deity to necessary evil. There is a reason for this. When you fly by yourself, you do things when *you* want to do them. When the instructor comes back, so does the need for approval, so you try to do things when you think *he* wants you to do them. This constant second-guessing builds stress and causes hesitation on your part. You now wait for prompting where on your own you would just act. You find yourself after such prompting saying, "I was just about to do that." A skeptical look will be shot back. The instructor thinks you have digressed and insists on more lessons.

The only way around this is to try and fly as if the instructor were not there and make your normal decisions. This is extremely difficult. Most of us, no matter how much time we have at the controls, feel like a student when an instructor watches us fly.

Here are a few hints for your solo cross-countries. Plan to arrive at the airport at least two hours before your planned departure. By the time you have done that extra-careful preflight, gotten a couple of weather briefings, paced awhile considering all the variables, reviewed what you did on that overcomplicated flight log, tried to figure out where you put how to get to the destination (Did you use true heading?), waited for fuel and oil (that is somehow elusive when you need it) and gotten a mechanic to tighten something that may or may not have been loose, you might be ready to go. You may feel like you have already flown the trip before leaving the ground.

You make too much work for yourself. Many of these first trips are barely over the minimum 50 miles. On a regular lesson, you might cover 150 miles or more without giving it a thought. Keep things simple. Ask your instructor for guidance on what you really need. Part of flying is learning what not to do. Also, be sure to reserve the airplane at least two hours longer than you could possibly imagine needing it. That first cross-country causes students to tell everyone at the destination what they have just done.

You may also want to replan the entire trip home all over again before you return. Your instructor may have told you always to depart with full tanks, in which case you will top off tanks that are practically full. This takes time. You will find that you spend maybe a quarter of the entire experience actually off the ground.

During your solo flights, you are going to resolve all the problems yourself. The more experience you have under your belt, the more information with which to solve problems. Many students miss a chance to pick up creative solutions to problems when they bolt from the school after each lesson. Should you have a bunch of pilots gather in a flight school and start talking, you can build a storehouse of interesting information once you sift through the exaggerations.

I have a couple of my own memories that might be useful. I plodded around New England during my solo cross-countries at the tender age of 17. I had just flown this route with my instructor the previous week. Entering the pattern on my first solo cross-country, I made my first real mistake. I misjudged the speed and distance of an airplane on crosswind and turned downwind ahead of it when I should have turned away and come around behind it. As I turned, I realized the mistake, saw how fast the airplanes were closing, advanced the power, and called the other airplane to insure our separation. For all the world to hear, this crusty voice came over the speaker (we didn't use headphones in those days) and told the whole world exactly what he thought of my flying abilities.

After landing and sheepishly pulling up to the only FBO on the field, I tiptoed up the ramp to find a signatory for my logbook. Miserable and wishing I was anywhere else, I stood watching while a small man with a weatherbeaten face walked up. During his first few words, I recognized the radio voice that had so eloquently enumerated my pilot qualifications. After visibly losing what was left of my dignity, he seemed to sound conciliatory. Maybe it was my youth, my first cross-country, my babbling apologies with a promise to learn from the experience, I don't know. He signed my logbook and sent me on my way. With mixed emotions, I was stopped by a mechanic while moping back to the airplane.

"I saw the whole thing," he said, "and heard you on the radio. You goofed, but you didn't deserve that. He slowed when you sped up and everything was fine."

Okay, now who is telling the truth and who is being nice? I knew I had made a mistake, but I didn't want it lessened in severity just to feel better. A mistake is worth nothing unless you learn from it. We all make mistakes. On the other hand, why should I feel bad for something that was of little consequence? I hoped someday to be able to handle someone else's mistakes in the air and still be rational on the ground. Then again, I might use all the names I heard applied to me. You learn that if you do something wrong, it is a mistake; if someone else goofs, he are an idiot.

The next cross-country swirled me in mystery. I was on my way to Concord, New Hampshire. Along the route was the city of Manchester, which from the air looks very much like Concord. It has an airport that also looks like the one in Concord. Flying over Manchester, I thought I had arrived at my destination. But I had gotten there too fast. The arrival times didn't make sense. This was in the days before everyone used radar flight following. If it did exist in those days, I had never heard of it. As for the VOR, I had been shown it but didn't really understand it. That instrument was for instrument pilots. In those days, we all dashed around the countryside using deduced reckoning and pilotage.

Anyway, a decision had to be made. The sectional said I had arrived; the Estimated Time of Arrival (ETA) said I had not. Previous checkpoints did not reveal any noticeable increase in groundspeed. Fortunately, the towns are not that far apart so I could press on until my ETA and still keep the other town in sight. Soon another city loomed up that looked almost exactly the same. Only after descending did the landmarks confirming my destination appear. I arrived about when I should have arrived. While chatting with the folks on the ground, I learned that pilots regularly call in at one airport while approaching the other. Some even land at the wrong airport. This makes me wonder why towers sometimes broadcast "not in sight, cleared to land."

I learned that even if everything looks right, you must still verify.

If you are seeking night flying privileges on your student certificate, you need to have some night instruction before you go on a solo cross-country. I did not intend to fly solo at night until after my private, so my night training up to this point consisted only of some twilight training at the end of my regular lessons. Since the best of intentions are irrelevant to the reality in flying, it is a good idea to get night training before any extended solo flying whether you intend to fly at night or not.

I was returning home from my most beautiful flight to date, my long cross-country. A couple of choice locations along the New England coast provided spectacular scenery for my flight and unmatched serenity for many years. I felt like a 747 captain working my approaches and departures with the big towers. I was having so much fun that I completely lost track of time. The sun was setting by the time I was back near my home field of Bedford, Massachusetts. I had called the school from Cape Cod on my last stop to inform them I would be late.

The ground references faded as the sun lowered. I wasn't worried because I was following a familiar highway and knew my position exactly. It was all a fun, new challenge. It was also much lighter at cruise altitude than on the surface, a fact that became increasingly more obvious on the way in. A clear, summer night was overtaking the landscape as I made my approach. To me, there was nothing that seemed unusual. It was getting dark, big deal. I didn't know enough to be worried in my new environment or that night flying is a special skill.

I saw the airport, called the tower, and they were very matter-of-fact with me. The airport beacon and runway lights were really pretty, just like in the movies I thought, having never seen them this bright before. By the time I touched down for what became my first genuine night landing, it was dark. By the time I had tied down and walked to my flight school, the lights around the terminal made it seem really dark. My instructor, late for dinner at home, was there to greet me. He lectured to me for a while on the dangers of night flying, mentioning things I had never heard of, but he wasn't overly concerned as I was fine. He liked the part about following the highway when other references disappeared. My warning was to not do it again. Made sense. My first instructor is a credit to his profession and always treated me as a pilot first, not the 17-year-old kid I really was.

My father was also there to greet me. His tone was decidedly different when we got outside. "What the hell were you doing up there at night?" Dad is fond of yelling. How could I tell him I knew where I was, the visibility was great, I was working with the tower, and at my cruise altitude, it wasn't night. I just listened to him instead. Anyway, the experience was worth it to see the runway lights. I also made sure to get more night instruction.

You may think that the complaints and disagreements I have with the current system are simply results of my own mistakes. Hardly. Except for a few details of my personal accounts, nothing appears here that has not been confirmed by my observation of countless students, instructors, pilots, examiners, and FAA inspectors. The impressions I have came over time and formed clear patterns that need to be challenged. My stories are for your entertainment and possible enlightenment. I only hope that you have even better stories to tell.

Have fun on your solo cross-countries.

8 • WRITTEN, ORAL, AND FLIGHT TESTS

All right, you've had your fun — now it's time to get back to work. Gone are those carefree days when you wheeled over hill and dale. Vanished from the scene is the peace of cross-country solo flight without stalls, slow flight, steep turns, and other tedious maneuvers. You will notice that cross-country training is proportionally one of the smallest sections of your curriculum, but that is where you spend the most time flying after your rating. What the standard private syllabus teaches you is how to perform pretty yet unrealistic stalls that you are convinced will dazzle your mother on her first flight, how to fly very slowly with passengers who just want to get to the destination and back on the ground, and soft-field landings that no club or insurance company will allow.

Before you can demonstrate things to an examiner that you won't need for normal flight, that no passenger wants to see anyway, and forget techniques that no one will let you do, you have to pass a written test designed to measure your aeronautical knowledge.

What about this thing called "the written"? Well, before you can take the oral part of the flight test and demonstrate your knowledge to the FAA, you have to have the results of the written exam in your hot little hands — which demonstrated your knowledge to the FAA. The written test results you present to the examiner can be up to 24 calendar months old. You can legally walk in to a flight test just shy of two years after filling in the dots of your written and have satisfied the aeronautical knowledge requirement. How much can you remember after two years? How obsolete is the information? Considering the fact that regulations change every year and navigational charts contain new symbols and information every six months, do you think the examiner might want to test you on current information? If so, what good are the results of a two-year-old test?

Much of the test concerns information you could easily look up. No one remembers the exact wording of all the FARs. Why spend time in imprecise memorization when easy errors can be avoided in the real world with a little bookwork? No one can maintain all the fine shades of meaning in his head. Besides, the FAA inspectors look up stuff all the time when there is a question. Why then do we require student pilots to get into the habit of trying to memorize information that the people who wrote the rules are allowed to look up? Take the rules of the National Transportation Safety Board (NTSB) for example. After an accident, do you trust your memory from the written to perform all the tasks properly, considering your current emotional state?

Much of the written covers VFR navigation. Since most of you will take the written exam after having flown your solo cross-countries, doesn't that make this part of the test moot? In other words, if you couldn't navigate, you never would have been able to fly those trips. You are learning to fly, not to take written tests. So the fact that you were able to successfully navigate is a far more valid and accurate demonstration of your flight knowledge than being able to fill in dots from misleading questions. Granted, you may not be using pilotage, radio navigation, and deduced reckoning on every trip, but you will never get through the flight test if you cannot demonstrate proficiency in all these areas. All of this makes the written test irrelevant.

I really love it when written tests try to measure your capability for handling situations that only occur in flight. Do the people who invent the questions actually believe that they can test your ability to handle critical flight situations in the calm and controlled conditions of the examining room and expect that to have any relevance to your reactions under the stress of real emergencies in the cockpit? Take the recognition of critical weather situations, for example. The worst they can do is show you pictures of nasty weather or put charts with lots of dangerous looking lines in front of you. What are you going to do . . . tell the FAA on a test that you would proceed into weather resembling those nasty pictures? What this cannot test is the fact that private pilots who want to get home because they have other things to do will deny the existence of dangerous weather staring them in the face, ignorantly thinking they can squeeze through. On a written test, those same pilots will fill in the dot opting for the 180-degree turn.

A written test cannot evaluate all the possible experiences in the cockpit. The most amusing attempt to cover this on a test is the spin requirement. Following FAA logic, if you can fill in the correct dot, that must mean you can recover from an inadvertent spin. The spin requirement is only a knowledge test so you will never have to demonstrate spin recoveries until you go for your instructor rating. This came about because the FAA used to do as much to develop aviation as regulate it, and they didn't want to scare away potential pilots by requiring them to do spins. However, the gruesome statistics on stall/spin accidents demanded that some preventive action be taken. The inevitable compromise was to test for the knowledge and hope that it would translate into action at the appropriate time.

Well, you can talk about spins, you can take written tests on spins, you can even watch endless videos about spins; however, you will never know if you can recognize and recover from a spin until you have received training in them. Until that time spins will haunt you. Taking spin training out of the private curriculum was a mistake. In the interest of making the most of an inadequate requirement where the FAA wants only your knowledge to be tested, I have devised this question for the written exam.

When you find yourself in your first spin during solo

practice because you botched a departure stall you shouldn't
have been practicing on your own, do you?

  A. Calmly think back to the pictures in the handbooks
     to verify that it is in fact a spin and then review the
     recovery procedure in your mind.
  B. Pull the stick full back, after releasing some arbi-
     trary amount of back pressure, praying the nose
     will rise just like it does in stall recoveries when
     your instructor is on board.
  C. Close your eyes in sheer panic, let go of everything,
     scream, and then open your eyes to find the air-
     craft has recovered on its own.

The written really isn't a test at all. A test is when you
have a wealth of information to study, you get lots of
questions you have never seen before, there are no answers
from which to guess, and you have to struggle to come up
with the correct solutions on your own. This is the way the
bulk of the world tests its pilots. What we have in the U.S.
is a memory exercise. The only thing being tested is your
capacity to memorize questions and the appropriate answers. You do this by getting one of those written test
guides that contains all the possible questions and answers. By keying in on a few critical words from each
question and the matching answer, then grouping the questions by general concepts in your mind, you can memorize a
phenomenal amount of questions. You don't even have to
have any idea what many of the questions are asking.

When I took the Airline Transport Pilot (ATP) test, I
studied 1500 questions in the written test guide. The exam
would take 100 out of those 1500 for the actual test. By
using the above method, I knew the correct answers to
about 80 percent of the questions on my test. My score was
90 percent which means my memory probably failed for
some questions and some I figured out the old-fashioned
way. This is all legal; the book from which I studied was
published by the FAA. They may think that a pool of 1500
questions is daunting; however, anyone with a slightly

better memory than me would have gotten them all. When the results of this test are used for job placement, the preference goes to the pilots who can best fill in memorized dots, perhaps over pilots with better flight skills.

Before you can take the written test, you have to have an endorsement from your instructor. Before you can get that endorsement, you have to demonstrate that you have the knowledge to take the test. To demonstrate that, you have to prove to your instructor that you have already memorized the answers. If the endorsement is proof that you already know all the answers, why do you have to take the test?

There is a shortcut to plodding through written test guides on your own. You can take a weekend ground school, a concentrated program designed to cram as much information into your head in the shortest possible time to get you the highest score. Just like in a regular school, you will rapidly forget what you learned after taking the test. You will most likely pay more than ten times the cost of a written test guide for these classes because the people hosting the class know exactly what is going to be on the test. Through post-test interviews, they can greatly narrow the field of possible questions so that you need not learn anything that will not be asked. These courses are big business owing to large profits and low overhead. The competition is hot — just check the advertising.

To call these programs ground schools is fraud because they are not classes that will prepare you for anything to do with flying. It is only to get you through the written test, which is fine because the written hasn't anything to do with flying either. People are attracted to the programs with the highest pass rates and test scores. The incentive for the flight schools then is to get the closest tap on the exact questions on the test. What you pay for is not aeronautical knowledge; you pay for the best inside information.

If you want to build your storehouse of aeronautical knowledge, there are many effective ways to do so. You can take a comprehensive ground school from your instructor either individually or in a small group. You can supplement your learning with various video courses and individual tapes. You can dig into the handbooks and manuals for all the things your instructor has not told you. All of this will help you fly better and none of it will be of any use on a written test, which makes the written a farce. With no real purpose, the written has become a rite of passage that has long since lost its meaning.

To really find out if the written has any relation to pilot skill, the FAA should commission a study of accidents and compare the written test scores of the pilots. If the folks I have flown with can offer any valid generalizations, then here is how I categorize pilots by test score. Those who don't pass just don't care and should stay on the ground. Those who score from 70-79 percent are lazy, overconfident, and will make mistakes from careless and arrogant operation of aircraft and should be carefully watched. Those who score 80-93 percent are thinkers who don't have all the memorized answers but will strive for better performance, ever increasing knowledge, and will be the best pilots. Those who score 94-100 percent are inflexible nerds who are fine in controlled situations when everything goes according to plan, but they fall apart whenever the unexpected occurs and are just as dangerous as the lazy, overconfident pilots.

The FAA has two choices as I see it. They can abolish written tests entirely (my preferred choice), or they can make them real tests. If it is to become a real test, then any preview of the questions from either a test guide book or any prep course must be outlawed permanently and immediately. Since post-test interviews can still be used for insider trading of information, then any multiple-guess test should have a pool of 25,000 potential questions. Better still, dump the fill-in-the-dots test in favor of something written out, since strategies and the ability to take fill-in-the-dots tests are much greater factors in a good score than any ability to learn the material.

The FAA computers that grade all the dot tests aren't geared up to read essay questions. Therefore, why not have the person administering the exam also grade them and send in the results to the FAA? Spot-checking the examiners should keep them honest. We already have designated flight examiners; why not do the same for written examiners? These people would then be certified to give you a written test, grade it and give you the certificate or endorsement satisfying the written requirement. The private written could be reduced to three simple questions that would allow you to demonstrate what you know rather than what you can memorize.

1. Define and describe all airspace including weather minimums, equipment required, rules and procedures:
2. Plan a complete cross-country of 300 miles in your training aircraft given the following conditions:
3. Describe everything you would do following an engine failure seven miles from your home airport while cruising at 6500 feet:

Some of the questions do not even make any sense to ask because you would never find that situation in flight. Take the VOR questions. You are asked to identify your position in space; sometimes with full needle deflections, sometimes with *To* indications while heading away from the station. The logic behind this is to test your understanding of the VOR through trick questions and scenarios no instructor would permit. What does it prove to identify where you might be in space maybe, based on inadequate information and no logical orientation, navigational goal, or sequence of events leading from the problem to a meaningful solution? If I ever had a student try to identify his position with a full deflection instead of identifying the correct radial with a centered needle, we would be in for a long chat on the ground.

To answer written test questions that have no meaning in flight, you need a strategy to play the game. There is a foolproof way to attack any VOR question. Look at the VOR in the picture. Imagine that the station is in the doughnut. Draw the three triangle lines of the VOR symbol around the doughnut if it helps your visualization. Split the instrument face from top to bottom along the horizontal white dots of the degrees of deflection. The top 180 radials are where you will be in space when the indication is *From*. The bottom 180 radials are where you will be with a *To* indication. The same side safe rule tells you where to go, so reversing it tells you where you are. You are in the radials of the quadrant away from the needle. If you have a *To* indication with a left needle, you are in the direction of the radials of the lower right quadrant.

Why are there questions on the automatic direction finder (ADF) on the written? Most trainers do not have an ADF. There is no specific ADF requirement in Part 61 of the FARs, only vague references to learning radio navigation. Most instructors do not want to teach student pilots the ADF because their understanding of it has been so hopelessly and needlessly complicated from their instrument training that they are reluctant to teach something for which they are not sure, or that they feel is too much for a student pilot to handle. The written, however, will contain at least one ADF question. You will be told by most instructors not to worry about the ADF questions — just skip them. There goes your potential 100 percent score.

You will wonder why the FAA requires knowledge for an instrument you have never used and your instructor says isn't worth the hassle to explain. The standard method of ADF instruction is so hopelessly deficient that the only way your instructor can explain it to you is the way he learned it, which is totally confusing to the average student pilot. The solution for the instructors is to find a new way to explain the ADF so it is not confusing. To be fair on the written, the FAA should either insist on ADF training in the curriculum or drop the questions from the exam.

The elusive truth about the ADF is that for VFR navigation it is incredibly simple. It has none of the twisting and limited deflection problems of the VOR, making it easier to use. If ADF navigation were taught to you before the VOR, it would accelerate and demystify your understanding of radio navigation. Despite this logic, standard practice dictates that you skip the ADF and concentrate on the more difficult VOR, where your knowledge is tested with questions about scenarios that would never occur in flight. Understanding the ADF in VFR navigation would make its use in instrument training a snap.

Following the standard practices again, flight schools leave teaching the ADF until you are under a hood and have no visual back up with which to understand your use of the ADF. This creates a situation where you never know if you are using it correctly. This is why instrument students hate the ADF. Try this. Fly an airplane with an ADF and bring along a competent instructor. Tune in an A.M. radio station or a nondirectional beacon (NDB) and see that the needle points right to the broadcasting station. You may wonder why something nondirectional can be used to find your direction. Just rename the station by calling it a beacon and drop the nondirectional part.

Remember, the only directional instrument in your airplane is the magnetic compass. The ADF is not a direction finder; it is an angle finder. If you have one of those movable compass cards, put "0" on the nose and leave it there forever. Some ADFs have cardinal directions like an "N" for north. This you must ignore as well for it is not a compass card but an angle card. The ADF measures angles of a circle identified by 360 degrees. The "0" point is your nose and every arrow indication from 0 is the angle to the beacon from your nose.

You may have heard the term "relative bearing." Bearings are absolutes; they are not relative to anything. A bearing is a direction, and you can't get directions from instruments that measure angles, so banish the term relative bearing from your mind. Replace it with "relative angle." The relative angle is the number of degrees the arrow is pointing from 0, or the nose of the aircraft, relative to the station. Wherever the arrow points is where you will find the beacon. If the arrow is on 0, then the beacon is directly ahead. If the arrow points 90 degrees over to the right, then you will find the beacon off the right wing. If you want to fly to that beacon, then turn 90 degrees until the arrow is on the nose. Keep the arrow there and you will get to the station.

Once you have it firmly planted in your head that the ADF is really an automatic angle finder, we can now put in the directional part. Mentally superimpose the arrow of the ADF on to the directional gyro (reset to your compass), and it will give you the heading, or bearing, to your beacon. It matters not where the airplane is heading; the arrow will always point to the station because the instrument can automatically measure all 360 degrees. When you can visualize this, you will understand radio navigation and compensate better for the intricacies and limitations of the VOR.

Suppose you tune in a station and the arrow points 30 degrees to the right. If you want to get technical, you can look at your directional gyro, observe your heading, say 270 degrees, 270 + 30 = 300 degrees — that is your bearing to the beacon. Put more simply, if the arrow is 30 degrees to the right, then turn 30 degrees to the right. To make it easier still, turn to put the arrow on the nose. There may be some wind so you might have to turn periodically to keep the arrow on the nose. If wind is a problem, once you know approximately where the station is, then look out and find two landmarks directly in front of you and keep them lined up. Practice with an instructor to refine this procedure.

Every instrument pilot who reads this is cringing because I advocate that private pilots be taught to home on an ADF, something instrument pilots were warned never to do. Think about it: Private pilots fly in VFR conditions where they can see any hazards out the window. It's not like they have to fly an NDB airway in the clouds, so the only thing that matters is that private pilots be able to get from where they are to a known position.

The most important reason students should know how to use the ADF is that the majority of cities and towns have A.M. radio stations that broadcast signals which can sometimes be picked up for hundreds of miles. Wouldn't it be nice for students in trouble if all they had to do to get home was to tune in their local station on the radio and keep the arrow on the nose? They can't, of course, because trainers don't have ADFs, instructors won't teach them even if the airplanes are so equipped, and the sectional charts publish most of the A.M. broadcasting towers only as obstructions.

In order to apply all this ADF knowledge to written questions, just remember that the angle formed by the arrow from 0 is the angle from your heading to the beacon. Superimpose the arrow on the directional gyro, and you will get the bearing.

Let's say that now you have passed the written. You cursed yourself for forgetting some of the answers. Well, it doesn't matter. Anything over 70 percent is bragging material. A nice score might convince your instructor that you are ready for the oral part of the checkride and might lead the examiner to believe that he is in for a quick exam. However, there is no basis for either of these as many an oral has shown. You can score 100 percent and still not know what you are talking about.

One problem with the results is that you never know what questions you missed because the FAA refuses to tell you. They will inform you by code which subject areas you missed. Since you had to know those subject areas before taking the written in order to get your instructor's endorsement, it seems a little silly for your instructor to have to again certify that you relearned the subject areas that your instructor has already certified you knew, while still not knowing which questions and therefore which bit of knowledge you missed on the test because the FAA won't tell you. Once your logbook has been endorsed by your instructor certifying that you have received further instruction in subject areas that you already knew, you will have satisfied the aeronautical knowledge requirement for the flight test.

One of the most important yet final things checked before a flight test is your aeronautical experience. Strange, because these are the hour requirements that have guided your training since the beginning. Just another huge detail neglected because you are never concerned with matters you think your instructor should do for you. Consequently, when your instructor, who has better things to do than check and update the breakdown of your flight time, asks if you have all the hours you need for the upcoming flight test, only then will you discover any shortfall. This is the first time many students learn they actually have strict requirements to fulfill. Unless you are in an extremely well-organized flight program, or are faithfully following a 141 syllabus, chances are you will have a shortfall in some requirement. Just when you are preparing for your flight test, you have to shell out more funds to make up the flight time.

The breakdown of hours for the private hasn't changed in anyone's memory; however, everything else about aviation has changed. You need to have a total of 40 hours flight time to get a private certificate. This breaks down into 20 hours of solo and 20 dual instruction. Those 20 hours of dual are subdivided further into 3 hours of cross-country, 3 hours of night, and 3 hours of preparation for the flight test. You may not have noticed, but that leaves only 11 hours of dual to learn everything else. You won't do it, so any school that bases a price on the minimum flight time is deceiving you. Chances are you had at least 15 hours of dual before you soloed. There are also the 20 hours of solo time, 10 of which is for cross-countries. What about the other 10?

I do not think it is a good idea to have students practice all the maximum performance maneuvers right after soloing and before the cross-countries, which is when a lot of this time is acquired. Right before the checkride, you will have more of the experience and judgment to handle the stalls, steep, and slow stuff. Cross-country is where you need the practice because that is where you will spend most of your flight time after the rating. A better balance would be 15 hours cross-country and 5 hours solo practice building confidence. Better still, let the instructor determine where the balance should be for each student.

Students are sometimes surprised at how strictly they must adhere to the requirements. The long cross-country is a common trap. You must go at least 100 miles in a straight line from home and fly over 300 miles total, or you will have to repeat the flight. I know a student who went for a flight test with a 298 mile cross-country thinking it would be good enough. The examiner measured it on the chart. The checkride was halted and the cross-country had to be flown again — all for two miles.

The way these things usually work is that when you go to check your hours, you will have lots of dual time, probably enough solo time, and little to no night time. Somehow the total requirement for 3 hours and 10 landings just kind of falls by the wayside.

What is night anyway? That's the time after the instructor has gone home because most of you learn to fly in the summer when the days are longer. It is also the time when insurance companies and instructors do not want you to fly solo because you aren't a pilot yet. To fly at night in most of the rest of the world, you need either a separate night rating or to be flying on an instrument flight plan; night is that different from the daytime. I know of no school that permits student pilots to solo at night even though it is legal. We don't want to lose you. So you have to snare an instructor at night who is already making himself available for lessons up to seven days a week. Asking for the night time is an added strain on him even though it is required for you.

You could get a certificate with a night restriction, but sooner or later you will end up flying at night without the proper training, so that is no option. If your training is to be spread out over the year, then plan on getting your night time when the days will be the shortest. That will make it easy on everybody. You will only need 3 hours of night experience for the rating, but that won't make you an expert at night flying. Since you will have spent more money than whatever arbitrary figure you allocated for the certificate by now, you will most likely pressure your instructor to give you exactly 3 night hours and 10 landings and not one second more.

The hardest part about night flying is learning how to

land — no surprise. For those of you who insist on blocking out your forward vision with the nose when landing in the daytime, you will have quite an adventure learning to land at night. Your only reference out the side is the occasional white runway light, and that only helps if you happen to look to the side. Okay, now forget all that garbage about full stall landings and just do the same landing technique you learned in this book for the daytime. Put the red lights at the end of the runway just above the nose, keep them there, and everything will be fine. Use the white lights that you can now see because your nose is lower to judge where the ground begins. The lights are a foot or so above the ground, so just work your way down slowly. By keeping the nose lower, the landing light will actually light up the runway rather than blind pilots flying overhead.

If you want to gain something valuable from night training, try to go on a VFR night cross-country with your instructor. Night VFR is the bridge to IFR. You can get lost very easily when the ground references hide in the dark. When the horizon disappears, your instruments become very important. Towns and cities lose their distinction at night and radio navigation with flight following is virtually mandatory. Your safety level changes because you can't see emergency landing fields or the hazards around them. The highways (which are not necessarily the best choice in the daytime) light up very nicely at night. Despite hazards like overpasses, compared to a dark field at least you can see what is in your way. You can also light up all the airports with pilot-controlled lighting along the route for both checkpoints and safety.

The big dangers at night are the unseen clouds. You won't inadvertently fly into a cloud in the daytime, but at night you can waltz right into a cloud you never suspected, even when a brightly lit city is in clear view up ahead. Clouds hide at night. I have watched the world fuzz out on an IFR flight when I never suspected clouds. The record for night VFR is not good, so when you get your certificate after your minimum 3 hours, build your experience slowly and carefully.

Start at dusk on clear nights in familiar areas and work up to greater challenges. Give yourself high personal weather minimums. The best thing you can do is get an instrument rating to use at night. If you lose currency, you can always get an instructor for recurrent training. You will probably love night flying. The sky is usually less crowded, the airplanes are easy to see, and the radio is less formal. Night flying is for poets.

What happened to our romantic journey of flight? Except for some night fun and solo trips, flight training has been hard work. There is now a danger point to cross over. Some students get very comfortable in that time between the solo work and the checkride and don't relish taking up any more challenges. I call them "flounders." Flying without an instructor in familiar airspace takes all the incentive out of sweating out a future checkride. For those of you who are driven to clear goals, you will plow right passed this delay. Professional candidates especially have no problem.

But if you are flying just for fun, have no time limit, no clear objective, or no problem with money, you may well flounder before the flight test. I have known people to languish for years working up to a checkride and then continually floundering. It takes intricate timing to pass a checkride. You have to go for it at the peak of your knowledge and flight skills. Priming for a checkride only to flounder and vanish for several weeks is immensely frustrating to instructors. They work harder for you at this point than at any other. A two-week break before a flight test can set you back a month in getting you to the same point again.

If you keep floundering, you may find many different people in the right seat, depending on individual instructor patience. You cannot be babysat by instructors with endless supervision and solo endorsements forever. It is time to leave the nest or leave flying. This does nobody any good. Self-imposed delays will have you blaming the instructor who may have since moved to a better job, the school for not doing its job, the FAA for anything you can think of; you blame anything but the real cause — the flounder.

For you to pull off a checkride, everything has to fall into place. It's kind of like making huge pictures with falling dominoes. The chain can easily be broken, but it is most satisfying when it works. Line up all your dominoes. The weather has to be good enough for both you and the examiner. He will ask your opinion because you are supposed to be pilot in command, but many examiners will cancel a flight on a whim when you think it is safe to go. Other examiners will work with you to find the time and place even in marginal VFR to get the job done. Some examiners cancel the flight for you if they think there might be clouds in the forecast despite a clear sky out the window because they are petrified of the liability of breaking a regulation and losing their obscene profits. It all depends on the examiner. That is why your choice of examiner is important. When you find an examiner you are comfortable with (and you won't know that until after a checkride), use him for all your future ratings if you can.

The airplane presents another hurdle. Not only does it have to be working, it also has to be legal. There are very few airplanes in this country that are in absolute compliance with every technicality of the FARs. All the logbooks have to be checked for maintenance compliance and correct entries. Theoretically, you have to check the logs for the required maintenance and inspections before every flight. This is such a pain that at best you will find some status board in your school that lists the hours when inspections are due. This means the only time you will see the actual logs is just before and during the checkride. This is not for your benefit. The logs are checked thoroughly so the examiner does not get caught in an airplane that isn't legal.

One critical factor usually overlooked is the fact that all the cars have to start. Now, the owner of the school and the examiner make a great living so they can afford cars that start. However, people who work, teach, and learn to fly in flight schools all drive clunkers. You can't have a checkride if the person who opens the school can't get there. Don't laugh, I had a phase check canceled because of this once. Anyway, if the weather doesn't fall apart, the cars and airplanes don't fall apart, and you don't fall apart, you should have a checkride.

Who is that designated examiner who will test your mettle to the limit, all for the piece of paper that allows you to act as pilot in command? Contrary to student beliefs, the examiner is not the fire-breathing beast from hell, with huge horns and big teeth, just waiting to quickly dispose of yet another inferior student pilot. The FAA inspectors are like that; the designated examiners are much more human — although there are always exceptions. Pilots get to be examiners by hanging around aviation for years and successfully kissing the butts of other examiners and FAA officials. All this for the privilege of making $200 or more per checkride, pass or fail. At two to three checkrides a day, it adds up. How the FAA determines when a person can judge when a candidate is due a certificate is still a mystery to me. These positions are jealousy guarded as they are one of the few pilot jobs that make a reasonable living. The price for an examiner may seem steep, however.

Unless you are supremely confident of your skills or have an untouchable ego, I would not recommend taking your checkride with an FAA inspector. You only have to do that for your flight instructor certificate, so why rush the inevitable? Inspectors are assigned at random, so you will have no idea as to the temperament of your taskmaster. Inspectors are paid a salary by the government which means no matter what the duty or how hard they work, they make the same pay, and there are a lot of duties that are far easier than giving a flight check to a student pilot. They have no incentive to pass you because more pilots in the air just create more work. Consequently, anything about your performance, knowledge, or anything not absolutely perfect about the airplane will be grounds for termination of the test. You have to go through the inconvenience of rescheduling whereas the inspector can now go to lunch.

A designated examiner, in order to attract business, has to be fair. Inspectors have incredible job security so they can fail anyone for whatever capricious reason with no economic consequence. The designated examiner is much more willing, for example, to haul in a mechanic to correct a maintenance log entry so you can proceed with the flight check. The inspector will tell you to come back another day. Both the examiner and the inspector live up to the same standards; the difference is that the designated examiner is much more likely to work with you to accomplish your goal.

Most examiners get their positions after years of instructing so they understand the strain. They know that this isn't your best effort and that you are nervous. They have the judgment to see through problems and sometimes give you the benefit of the doubt or a second chance to prove yourself. The inspector is more likely to have never taught a lesson and go strictly by the *Standards*, looking for a reason to fail you. Like I said, they have better things to do. Designated examiners always have market forces at work. The word quickly spreads if they are giving away undeserved certificates, or are putting pilots through the Spanish Inquisition.

The single most important factor in your passing a checkride is your willpower. All the studying and preparation in the world won't help you unless you consciously decide to pass. When you wake up the morning of the test, say to yourself, "Today I will become a pilot." Look at it this way, you have put in hours of study and flight time in preparation and practice. Your instructor and most likely a phase-check pilot will have reviewed your abilities and knowledge and have endorsed your logbook accordingly. Everyone else knows you can do it, so the only deciding factor in your success is how much you want to succeed.

Do you have what it takes to reach for a logical answer when the correct one escapes you under the pressure of an examiner staring you in the face? Can you reach down deep in your character to make a maneuver happen after blowing it the first time? Do you have the confidence to keep going no matter what you think of your performance to complete the checkride? These are the things that are really being tested. The examiner needs to know if you can hold together under the most strenuous conditions, using all your faculties to analyze situations and find solutions. The examiner has to be able to project your ability today to a future situation with passengers and have reasonable confidence in your decisions and judgment.

An examiner once told me he hates to see a perfect checkride because it does not show anything except rote mechanics. However, if a problem arises such as a traffic conflict at a congested airport, or if the candidate makes a mistake, how the situation is handled tells everything about his ability to accept the responsibility of pilot in command. Making mistakes won't automatically jeopardize a checkride because everyone makes mistakes. It is the chain of events that follows a series of consecutive mistakes that leads to fatal accidents. Therefore, it is imperative to immediately correct the mistakes that you do make.

That is where your willpower comes in. That is also how the examiner judges how you will fly years from now. In all the pilots I have talked to after a flight check, the one thing they all have in common was that going in they believed they would pass the test.

Your objectives for any rating are clearly spelled out in what are called the *Practical Test Standards*. This is the FAA booklet that clearly delineates your tasks in knowledge and flight skill, and the standards to which they will be judged. This is a shining accomplishment for the FAA because it has gone a long way to remove the subjectivity and capricious decisions by various examiners. FAA inspectors, in my experience, still make up their own rules in spite of the *Standards*; kind of like when Congressmen exempt themselves from their own legislation. The purpose of the *Standards* is to protect you from unreasonable tolerances, tasks you never heard of, methods you were never taught, and examiners who just don't like your looks.

All tasks are to be judged by the same standard so theoretically if you meet that standard, all examiners should pass you. The variations in interpretation come because humans are involved. Fortunately, there is leeway built in so examiners can use their own judgment when you don't exactly match the letter of a particular task but are still qualified for the rating. When you go for a flight check, it is just you and the examiner up there. No one else will ever know what goes on.

Before the test standards, there was no control over exactly how a certificate would be awarded. There are many correct ways to perform procedures, and you could fail simply because your correct method was different than the examiner's correct method. The *Practical Test Standards* require that you meet the standard, not fly as the examiner would have taught you to fly. This is the grey area where judgment comes in because no two people fly the same way. Even with the *Standards*, there are still abuses of power. I had a student nearly fail a commercial checkride because the examiner didn't like the way he performed lazy-8's. Since I had passed many checkrides myself and had many students get through without complaint using the same technique, the argument with this examiner was simply a matter of style.

"It should be like a dance, like an aerial ballet," came the command of the examiner.

You won't find such nebulous tolerances in the *Standards* fortunately, so I had grounds to dispute our difference in method. Besides, unless I flew with the examiner, I would never know what exactly was his idea of an aerial dance. Try using those words on a student and see what kind of lazy-8 emerges. Thanks to the *Standards* and some delicate negotiating, I was able to prove that the student had in fact met the tolerances of the maneuver and was therefore justified in getting the certificate, even though the examiner would rather have seen his own version of lazy-8. This saved me from having to complain to the FAA, although I never used that examiner again. While the large academies may have certain examiners they either put on staff or rely on, instructors at smaller schools are more free to select examiners who closely match their teaching style, so such conflicts should not occur. Your examiner will expect to see the same methods your instructor has taught you. Such consistency is greatly in your favor.

The most critical section of the *Standards*, the Introduction, is probably the most overlooked. The bulk of the booklet just lists in rote fashion all the stuff you have to do. In the Introduction you will find the concept of the test. Examiners have the responsibility to conduct tests in strict compliance with these standards and determine that you meet the tolerances. Your instructor is responsible for training you to an acceptable standard to perform each task and certifying by endorsement that you are competent to do so.

There is no reason to fail a checkride because you should have demonstrated that you can do everything before you get to take the test. The examiner checks the phase-check instructor who checks your instructor. That makes three people certifying your competence before you become a rated pilot. The only thing that can fail is your own willpower.

The *Standards* delineate the examiner's responsibility and the flight instructor's responsibility. Conspicuously absent, however, is any mention of the applicant's responsibility. This is because the fundamental flaw in our training is that the FAA does not require any responsibility of the person being trained. Take a look up through the commercial and instrument standards; there is no section on the applicant's responsibility. It is true that you have to demonstrate that you meet all the standards; however, only the instructor is responsible for training you to meet the objectives. Since you have no responsibility for insuring that you meet the objectives, your instructor will receive all the blame from both you and the examiner should you fail. I would add the following statement to all *Practical Test Standards* Introductions:

"Applicant's responsibility: All candidates for a certificate or rating are responsible for insuring that they have acquired the training necessary and have the knowledge and skill to meet or exceed the tolerances of all objectives and tasks in these *Standards*, and they certify that they understand and accept the privileges and limitations of the certificate or rating sought."

Keep reading the Introduction. You will note that the examiner gets to use realistic distractions throughout the test. They love to do that. The reason is that pilots have flown into the ground when distracted by something as simple as a seat belt dangling out the door, slapping the side of the airplane. A seat belt dangling in flight sounds like the airplane is falling apart, which can be very distracting. Your examiner won't sneak a belt out the door just to make your life miserable; that isn't fair. However, should you be taxiing with the passenger door open because it is a hot day, and you fail to tell the examiner to close it before takeoff, he may either point it out to you at the last minute, or let you take off with the extra ventilation.

Either way, this may be enough to rattle you for the whole flight. This is how such a simple thing can build up and pile on the stress. The examiner is within his rights to act just like a baby passenger, treat you like a god, and ask stupid questions at the most inopportune times. You are pilot in command, tell the examiner to shut up — nicely.

Many distractions take the form of technical questions that appear to be testing your knowledge, when they are really testing your judgment. You may be asked what the altimeter setting is while you are on short final. With the ground plainly in sight, who cares. Lesser mortals, however, will attempt to get the setting from the ATIS or tower and blow the landing. You are not there to impress the examiner. You are there to show you have the stuff to pilot the airplane with safety and judgment. You should politely tell the examiner you will recheck the altimeter when it is safe to do so. Try this line, "For the safety of this aircraft, I must request that you not talk to me until after the landing." Who can argue with that?

Here are some hints for the big day. Don't get up and put on fancy dress clothes that you have never worn in an airplane. This is the one day to keep your routine as consistent as possible, so please limit anything unfamiliar to you. Guys, keep the ties loose; you will be uncomfortable enough without being strangled. A good meal the night before, no alcohol, and a good night's sleep are far better than last-minute cramming. Studying at the eleventh hour just highlights how much material you do not know or for which you are not completely sure. You can't know everything, so why build up anxiety for something you can't help.

When you wake up make a promise to yourself that come hell or high water you are walking away with a pilot certificate. Sometimes there are factors beyond your control: hurricanes, locusts, oil embargoes, and dead car batteries. But if you use some ingenuity, you might still get through. Say, for example, that the weather forecast calls for deteriorating weather in the afternoon. Even though it is traditional to have the oral exam first, why not fly first and chat during the bad weather? Most students are more concerned with the flight anyway which distracts students during the oral. The oral takes lots of your energy, leaving you little for the flight.

For these reasons, I recommend flying first so you can relax during the oral. Examiners, if you fly well, are more likely to give you a shorter oral anyway because they tire after the flight. Most of the oral centers on your cross-country planning (you can still fly first and defend it later), so this is no time to craft an elaborate plan that you can't use or explain. You may know what you are doing, but the examiner has to be convinced you know what you are doing, so keep it simple.

Things will be very quiet when you enter the room and first meet your examiner. This is most unnerving. The first duty and responsibility of the examiner is to make sure both you and the airplane are legal to take the test. The certificate you seek is a lifetime binding contract unless suspended, revoked, or returned. This is why you sit quietly while the examiner has his head buried in your logbook.

This is where your attention to detail will pay off. You have to have your 87-10 form filled out to perfection. Forms are an art. See if you can meet with the examiner ahead of the test with a draft of this form to insure compliance on the test. Nothing is more frustrating than to hold up a checkride because the form isn't correct, except to have a test terminated because you don't yet meet the requirements. It is imperative that you take the responsibility to insure your eligibility. There are a surprising number of checkrides that never get beyond this point because an applicant hasn't met the qualifications.

Students try to slide by all the time and it almost always blows up in their faces. Here is how it happens. Your instructor is a busy person with lots of students and can't keep track of all your hours and will not know your situation as intimately as you. Because of the fundamental flaw, you know you are not the one who is responsible. You also want the certificate so whatever it takes to get there is fine with you, including shading the requirements you don't have to check. For example, say you have logged 2.9 hours of night time with 9 landings and you refuse to have a certificate with any night restrictions. Your instructor has flown with you at night and probably figures you have the required time.

Your instructor will usually just ask if you have met all the qualifications. Most of you will automatically say yes whether you know you have the time or not to please your instructor, avoid any embarrassment, and get on with the checkride. Your instructor wants to treat you like an adult so he will trust your answer and not add up your hours himself. If he did check your log after asking you if you have the time, it would show that he doesn't believe you. You, of course, don't want anything to stand in the way of your certificate, least of all going through the hassle and expense of another night flight, so you hope no one will catch the discrepancy.

You may get away with it and get your certificate,

though it is highly unlikely; it is also fraud. When a discrepancy is discovered by the examiner, the test will be stopped. You will have to schedule the night flight. Your instructor will feel betrayed because he trusted you. Although you won't be held responsible, you will have some explaining to do, and this is where you will blame your instructor. This is a waste of the examiner's time, and he will also blame the instructor. You will be embarrassed as you have to explain to your peers why the test didn't get done. All this because you wouldn't take a little time and responsibility to follow the rules.

This happens all the time and there is no reason for it. Flight checks are commonly suspended for insufficient night, solo, and cross-country hours and distances; incorrect, neglected, or expired endorsements; overdue 100-hour, annual inspections, airworthiness directives, and missing or incorrect maintenance log entries; inoperative equipment that could have been easily fixed; missing paperwork such as the current weight and balance, airworthiness certificates, registrations, operating limitations and placards, checklists and required pilot operating handbooks. I know an aircraft that flew for months without an aircraft identification plate which was removed for painting and never replaced. The examiner found it on a checkride. Anyway, when the examiner is satisfied that both you and the airplane are legal for the checkride so *he* won't get in trouble, the exam will proceed.

Up until the time you get the certificate and for a short while afterwards, the people who will get the blame for your mistakes are your instructor and your examiner. You can do something stupid and survivable in training and cause your instructor to lose his certificate while you go on to get yours one day. This is why you are babysat by instructors during training. Student pilots lie, forget things, don't know or care about the consequences to others of their actions, and take no responsibility for training because none is required. Now all of a sudden on a checkride, you are pilot in command and expected to be absolute ruler.

You have had no preparation in developing your thinking in this way during training, so this isn't fair to you.

The examiner has the responsibility, like your instructor, to make sure you don't get him in trouble. Whereas the instructor has to make sure you don't get him in trouble during training, the examiner has to project that he won't get in trouble from your flying after your training. They test this by completely leaving you alone and simply passing final judgment. The fundamental flaw creates split and conflicting roles for instructor and examiner because you, the applicant, are left out of the equation. When you are required to guide and supervise your training and develop your thinking, the checkride will become no big deal.

In a small, dark room, you sit transfixed by an incandescent bulb glaring in your face. A small, double helix of cigarette smoke slowly curls towards the ceiling. In the backlight glow, a faceless entity covered by a fedora hovers in gleeful inquisition.

A gravelly voice shatters the stony silence, "What is basic VFR in a control zone (Class D), hmmmm?"

Numbers swirl in your head immediately followed by nervous questions. What will you say now? Is basic VFR different from regular VFR? Which VFR is being asked? What about special VFR? How can you remember all the types of VFR?

Welcome to your first oral exam. A soul-wrenching experience designed to make you feel small, really small. That lump in the pit of your stomach grows with each painful second you deliberate how to answer the question. The examiner's patience withers quickly as that initial smile degenerates to a scowl combined with that look of "here we go again." The tick of the clock reverberates in your brain. You are briefly reminded that all you wanted to do was fly an airplane, why should you be put through this?

In typical fashion you have let the environment control you rather than the reverse. It happens to everybody. In flight training you got used to the way your instructor asked questions. You quickly learn all the answers because instructors use the same stock questions asked in the same fashion. What you acquire is not the information, but how to answer your instructor's questions with stock answers. All your information is lost like a crashing computer disk when a new questioner appears.

The information being tested doesn't change, but the style of asking it does. When you acquire your knowledge in a particular rote method, memorize it, and repeat it back the same way all the time, you won't recognize it when confronted with exactly the same questions differently phrased. You appear to have lost all your knowledge, when in fact you know it but can't access the information because an unfamiliar code was used in the question. You feel stupid, embarrassed, and frustrated because you thought you knew this stuff. You become anxious because you want the certificate. These all combine to block your ability to reason.

There are two ways to correct this situation. One is to develop a flexible working knowledge of all materials as you go through the training so no matter how the question is asked you can discuss the general subject. This won't happen though until the FAA changes the way we educate pilots. The other way is to develop a strategy to match wits with your tormentor and retain your sanity.

Back to our question. When an examiner mentions something like basic VFR in a control zone (Class D, you will usually blank out. Here is why. A review of the FARs shows a chart under the heading "Basic VFR weather minimums." This is where you get all your cloud clearances. A subpart of basic VFR provides that in a control zone you can't operate with less than a 1000-foot ceiling. So when basic VFR talks about cloud clearances and VFR in a control zone talks about ceilings, you can easily become tongue-tied with an examiner staring at you when you really aren't that sure how they combine. When you are asked an imprecise question that lumps concepts you are trying desperately to keep separate, you will try to read the mind of the examiner rather than deal with the question. Such a simple phrasing misunderstanding can leave you dumbfounded.

The examiner is wondering why you don't know what VFR is all about and why you are here for a flight test. He has no clue you are weighing several possible responses and the various consequences of each, hoping to hit the one the examiner wants to hear. When an examiner asks a question he has posed thousands of times, he concentrates solely on your responding with the words he expects to hear. He doesn't listen to himself asking the question because he is preoccupied with your answer, so any vague form of what he thought he was asking could come out. The examiner will only perceive what he meant to ask to get his expected answer, not what he actually said. You can't know what he meant to ask so you have to do your best with what you are given.

The big problem is that examiners come up with an answer first and then mentally backtrack to form questions they think should lead you to their answers. However, the examiner is the only one who knows the progression of connected thoughts that lead from question to answer. When you get a general question and respond with a perfectly logical and complete answer, you can still disappoint the examiner because you didn't follow the same mental sequence of steps he used. You can be wrong even though you give a correct answer to a question the examiner didn't even know he asked. He can't understand why you don't follow the same mental progression to come up with his answer, so he won't give you credit for your thoughts because it is not what he expected to hear and won't admit that maybe he didn't give you the correct information to lead you to his answer. This is how you can know the material and still not give an answer satisfactory to an examiner. This is unfair and hardly a demonstration of your knowledge, but if you want the certificate you will have to play the game.

Back to our strategy. You can't leave gaps of silence after the questions are asked because it will be interpreted as ignorance. Your strategy is to discover what the examiner expects to hear. Start by reciting what you do know.

You know that basic VFR in controlled airspace has weather minimums that include visibilities and cloud clearances. Control zones (Class D) are controlled airspace. You can't operate in a control zone with a ceiling below 1000 feet. So basic VFR in a control zone would include the ceilings, visibilities, and cloud clearances. Now rattle off all the numbers so you can show you know them.

Somewhere in all this should lie the words the examiner expects to hear. If not, at least you have demonstrated you know where the answer can be found. Failing that you have laid the groundwork for further discussion where more questions will follow. This establishes your credibility and from here you can discuss matters pilot to pilot, instead of master to silent student. You have just taken control of your oral exam. If you really want to prepare for an oral, go out and take a bunch of job interviews.

After an hour or two of abuse, after you feel totally ignorant and weary of questions, after you pass the point where you want to scream, the examiner looks at you and says, "Okay, that will do. Meet me at the airplane in 20 minutes." Wow. You did it. You aren't quite sure how, but you said enough to convince the examiner that you have an idea what is happening. Chances are the examiner also taught you quite a bit without you realizing it.

The private is a learner's permit, as they say. It allows you to learn without killing yourself or others, or causing major catastrophes with an airplane. Experience is what makes a pilot. You don't know this yet because you feel all those shaky, incomplete answers make you a failure. Your examiner has seen hundreds of other pilots before you came along, so even though you weren't flawless, a flawless performance was never expected. It doesn't take all that long for an examiner to get a feel for what you know. If you keep your cool, struggle, and work hard for solutions, you should be all right. How you will react in the airplane is another matter; hence the flight check.

Your preflight will be observed by the examiner. You now get to engage in a ritual called "what's this?" This is where by magic the examiner points to the only parts of the airplane that you have no idea as to their identity and function. You figured your instructor would tell you everything you needed to know about the airplane, so anything else is extraneous. What is happening is that the examiner is pointing out many things that you either don't need to know, can't use in flight, or aren't certified to do any work on. Why bother?

You will probably be asked something like, "What are all the antennas for?" This is because sometime earlier an FAA inspector told all the designated examiners in your area to ask that particular question. Pilots don't have to know that. All we have to know is when one is missing. The people at the avionics shop have to know what they are for. Most of the time even they can't tell you until they remove the cover. The examiners don't know what they are all for either, so as long as you sound like you know what you are talking about, the examiner can then go back to the FAA inspector and report that students know what the antennas are for. Playing "what's this?" is great for finding out if you know where the cabin vent outlets are, how the oleo strut works, where the fuel tanks hide, where the frize is on an aileron, and where all the paperwork hides that you have never thought to read even though you so dutifully memorized the ARROW acronym.

When you hop into the airplane, make a concerted effort to slow down. Nervous energy breeds rapid actions which lead to mistakes. The perception that real pilots speak fast on the radio and act fast in the cockpit usually comes back to haunt you on a checkride. Remember that real pilots know 90 percent of the radio calls in advance and appear to move quickly only after years of repetition and practice. If you want to look professional, carefully and faithfully follow the checklist.

One hard part of the test is treating the examiner like a passenger. Since student pilots are not allowed to carry passengers, this isn't easy. However, you are assumed to be a private pilot on the flight test. You can log pilot-in-command time on the flight check. This is possible because it is actually the endorsement of your instructor certifying that you are safe and competent to operate as a private pilot that gives you the privilege. The examiner is there to see if you will get the certificate to keep the privilege. He acts like a passenger to see if you have accepted the responsibility of the certificate you desire. That is why you have to brief the examiner, fasten his seat belt, close the door, and give him a safety chat. You know he knows how to do all that stuff, but your future passengers do not, and it is that future passenger for which the test is based. So even though you feel silly, brief him, and buckle him in.

It is eerily quiet flying with an examiner. He never tells you when all is well, so the less he says, the better things are going. Don't expect any feedback and no matter what happens, keep on flying. This is in sharp contrast to the endless drivel of your babbling instructor, so it will feel strange. All the teaching has been done; the examiner is only there to evaluate.

Checkrides are famous for the "I've never done that before" phenomenon. This is where you do something wrong on a checkride that you have never done before in training, and there is no rational reason for it. Or it is something that you have always been able to do without even thinking that now suddenly escapes you. Many checkrides are busted because of this. The problem is that it sounds completely phony to tell the examiner that this has never happened to you before. How can he believe you? This is his first flight with you.

Your instructor can't correct a problem like this because it only happens on checkrides, and you never know what form it will take in advance. Should you get remedial training after the checkride, the flaw will disappear causing the instructor to doubt the examiner. This is confusing for everyone, especially the instructor, who would not have sent you for the checkride had he not seen you perform everything up to standard.

I had a student bust a checkride because in three tries

he could not glide the airplane with the engine idling, from abeam the runway numbers to anywhere even close to the runway. The fact that he had just done the same thing perfectly on the previous lesson was irrelevant. On this flight he could not have landed on a superhighway between major cities. Why this happens nobody knows. I have talked to many students who have not been able to demonstrate something on a checkride they were perfectly capable of doing any other time. There is no cure because we don't even know why it happens. The best you can do if something doesn't work is try to figure out what went wrong and do something different the next time. You should also try to have your procedures as clearly, orderly, and simply organized in your brain as you can.

You may fall victim to the self-fulfilling prophecy. This is where you are so worried about a particular technique that you make sure you blow it completely. This is a will-power issue. If you have decided to pass, then you won't sabotage yourself this way. If you are really worried about passing, then whatever maneuver gives you the most concern can only be done badly on the checkflight. If you are convinced that you are not ready for the checkride, then you will use this maneuver to bust it.

One student was so worried about his soft-field landings that everything else went well because all the worry went into the soft-field landings. Consequently, those landings were anything but soft. The flight wasn't busted because everything else went well and because of some fancy talking to the examiner explaining the proper technique. After a barely passable second effort, he made it through.

What you are saying by undue concern for a particular task is "I am a failure" even before you go for the flight. You give yourself a way out of succeeding. The examiner wants to pass you. I have never met an examiner who didn't have the best interest of the candidate at heart. However, you have to make it as easy as possible for him to pass you. If you have to have that certificate, you should be able to put your fear of any particular procedure in perspective.

When you train to fly, you gain so much more, for flight training changes the way you live. We have created a society which progressively distances the individual from responsibility and the consequences of his actions or lack of actions. Passing the buck has become a virtue. However, the authority of the pilot in command in the air is the closest thing to absolute power and responsibility this society has to offer. Because of your background, education, the fundamental flaw, and standard rote training methods, you are not prepared and adjusted mentally to take that responsibility until it is thrust upon you during the checkride.

This book is a call to change our system so that we start right from the first day teaching you the proper attitude that prepares you by the time you get to the checkride to accept your legal responsibility as pilot in command. It is up to you, the reader, to demand that change. You now earn the privilege of commanding an airplane when you can demonstrate through personal responsibility, decisions, and judgment that you are worthy of being pilot in command — despite your training.

The oral and flight tests are not to see if you can regurgitate rote factoids and repeat reflex mechanical actions in an airplane, which is how our current system teaches you to fly. They are to gauge how well you will handle the responsibilities and privileges of your certificate long after it is granted. In the oral, you will get scenarios to solve where many rules and procedures must be weighed and applied. This will demonstrate your flexible and comprehensive understanding of the whole system.

You have to prove that you understand the material up to the level of your certificate, which for the private is quite marginal. Even so, like a doctoral thesis, you will be held accountable for what you say and have to defend your statements. You may talk a good story so your words will be tested in flight. You must prove your skill in the environment for which you seek to operate. You are alone with the examiner; one applicant, one judge. Perhaps for the first time in your life, you are being tested to the limit, where the outcome is solely and totally in your hands. Nothing in your life has adequately prepared you for this experience because we do not value the power of individual responsibility and decision.

Look at our society. Throughout your education, you were permitted to return shoddy work that was not your best and still slide through. The information you crammed into your head for a test could be dispensed with shortly after the course ended. In sports you learn that you can succeed or fail without any real-life consequences. On the job you learn to avoid initiative, pass decisions on to superiors, spread the liability over a committee, and hold endless meetings without arriving at conclusions. Nobody is held accountable for what he says anymore. Politicians are expected to violate their pledges. Salespeople are expected to lie. Lawyers are hired to pass blame to the deepest pocket. We sue for any reason to avoid the personal consequences of our decisions. We promote and advance through political correctness and hierarchical conformity rather than merit and ability. The individual is persecuted, shunned, and ground down.

With this background, we test you on a checkride for all the qualities this society has tried so deliberately to breed out. You are tested on your capacity to exercise total authority and accept total responsibility for your decisions and actions. That is what flying is all about. Should you prove your worth, you will become pilot in command. Should you succeed, you will be changed forever. With a challenge successfully met comes incredible confidence and satisfaction from your accomplishment. You set out to become a pilot and you made it happen. It is yours forever. Flying an aircraft is difficult and humbling. Not everyone can do it. You have to have special talents in the right combination. You succeed because your willpower has made you succeed. Congratulations . . .

9 • HIGH PERFORMANCE

You passed the checkride. Nice going, bucko! You survived the abuse of your instructor, the frustration of unpredictable weather, the peculiarities of airplane maintenance, the intricacies of the air traffic control system, and the idiosyncrasies of checkrides and FAA examiners. You had a goal to learn to fly and you made it happen. Of the millions who dream of flight and the thousands who try, only a small portion actually make it through to complete the checkride. People drop out all along the way for various reasons: They don't understand all the material, their money runs out, flying an airplane is just too tough. Oh well, there is always bowling. Some people just disappear without a trace right in the middle of training.

But not you. You stuck it out, beat the frustration, bureaucracy, paperwork, study, practice, exams, and now you are a private pilot. With gleeful exuberance, you will abuse your phone bill. This took quite an investment of time and money, and all your friends and relatives should know it. You can now thank those who believed in you and snub your doubters and detractors. Tell them all they are welcome to fly with you as long as they bring some gas money. You can revel in the wicked pleasure of watching your first passengers squirm ever so slightly through their first 30-degree bank. "It's nothing," you calmly relate, feeling just a little superior, "you ought to see the world from 60 degrees."

Shame on you. You know full well that you don't want to kick the plane over that much with passengers, especially since your attempts at 45-degree turns require most of your skill, and your 60-degree attempts usually end up pulling lots of "G's," or in a spiral. You also know your passengers will refuse the steep stuff because 30 degrees of bank is quite enough for most people, thank you. Some pilot in command you are turning into, what with the ink on your certificate still wet.

It may take weeks, it may take months, it may even take years, but sooner or later you are going to get bored. You will be bored with low-performance aircraft that can't get out of their own way, bored with flying in the same local area, bored with flying by yourself because all of your friends have had their one and only ride, and they don't want to play in airplanes anymore, bored with the tired and limited fare offered at the local airport restaurants, bored because all your flights are so routine, bored because with rare exception there are no real emergencies to cope with like you simulated in training, bored because you have no instructor to torment you and build up tension with endless maneuvers and procedures, and bored because all the challenge of flying appears to have gone. You are bored as you never thought you would be bored.

The simple truth is that the challenge and reward of acquiring a pilot certificate is far more exciting than what you can do once you get it. Welcome to the big lie. The big lie will never come out of the mouths of instructors or flight school management. It would be bad for business. The big lie is that unless you are a paid professional pilot or flight instructor, or have the resources to make long exciting trips across the land, your little local flights will eventually get boring.

When you were training, you were ready for anything, especially simulated emergencies, because you never knew exactly what your instructor would throw at you. You subconsciously built the impression that emergencies are just waiting to happen so you better have the adrenaline ready. After your checkride, you quickly find out that emergencies are the exception, if they happen to you at all. Over time, you will wonder what all the fuss was about. Complacency will replace your lightning student reflexes such that you could be totally unprepared when one actually does strike. Then you will rediscover how exciting flying can be.

You will miss the rush of training; that heart-pounding, sweaty-browed gauntlet thrown at you by your instructor. You may seek that rush again. You may fly in ever higher wind velocities. You may experiment with increasingly marginal weather. As you seek the rush, your confidence will tempt you beyond your ability, and you will make a mistake. After much reflection, you will conservatively pull back until you get bored again. You may stay bored in your flying or repeat the rush cycle. Either way you will be in a rut. You may find other things to do and flying becomes progressively less a part of your life. The average private pilot flies only 50-100 hours a year. That isn't much. Which is why we professionals worry about you sunshine warriors. You lose your edge while you think you are getting bored.

The further you get from your flight check, the more you will cut corners, slip from the perfection you sought for the checkride, and give away the qualities and demands you made of yourself to get your certificate in the first place. Your boredom leads to complacency. That is why there are biennial flight reviews and currency requirements. You learn to fly in the world of low-performance aircraft, and many of you convince yourself that you will be safe and happy remaining in that world. You could be happy for years.

Eventually most of you will find that the Cessna 172 that used to seem like such a handful is now way too slow. You tire of seeing the gear just hang out there holding up your progress. The engine controls are just too simple and you want more fussing. You observe your compatriots dashing forth in sleek, high-performance singles. You jealously observe the gear tucking neatly away with a corresponding increase in speed and climb. The purpose of airplanes, except for you hopeless romantics, is to move people and stuff around very fast; and low-performance aircraft just are not fast.

Your boredom now moves on to frustration, where the only way to raise you out of the ashes of tedium and restore your spirit is to take on a new challenge. The beauty of aviation is that there is an infinite variety of new airplanes that are bigger, better, and faster. Go ahead and call your instructor and tell him you want more: more speed, more horsepower, more complexity, more challenge, more frustration . . . more, more, more! The bug has bitten again. Your desire to fly has been restored with newfound vigor. Off you go into the unknown again just like when you first started flying.

You folks who desire a career in aviation have felt all along that low-performance was a stepping stone to better things, so you have drooled over fast airplanes ever since you started flying. You people are never satisfied where you are, which is what drives you to succeed in this crazy business. Your continued impatience is clearly evident. You can't wait to solo after your first few lessons . . . can't wait to fly cross-country as soon as you solo . . . can't wait to try high-performance before the checkride . . . can't wait to fly in the clouds, in a multi-engined aircraft, at night, with an engine out, on your first lesson, with the instructor incapacitated by appendicitis, the airplane has a blown tire, the controllers are all on strike, during an earthquake, thunderstorms on final approach, Air Force One is behind you, TV crews waiting on the ground, and you come in and make the perfect landing. This is what pilots are made of — or so you think. Suffice it to say that if you are bored with flying, you will either quit or move up to satisfy your craving for adventure. That means advancing to high-performance.

What is high-performance? It is when you have so many new knobs and handles to play with that you couldn't possibly be bored. Officially, however, high-performance is where you have an engine that is 200-horsepower or more, or is an aircraft with a controllable pitch propeller, retractable gear, and flaps. You will still hear the term "complex aircraft" bandied about, like on the commercial standards where you need to bring a complex aircraft for the flight test. Well, by definition all complex aircraft are high-performance because of the prop, gear, and flaps, so the term really doesn't amount to much anymore. Since you need a high-performance endorsement to fly anything over 200-horsepower, the complex aircraft definition lingers on simply to confuse you. The Cessna 172 RG Cutlass has a 180-horsepower engine and the variable prop, retractable gear, and flaps, so it is high-performance. The Cessna 182 with fixed gear is also high-performance because it has a 230-horsepower engine.

The FARs say you can't act as pilot in command of a high-performance aircraft until your instructor has endorsed your logbook certifying your competence. You get to fly a better airplane and your instructor gets more money — what a deal. One question, then, is whether you can log pilot in command (PIC) time before you get the endorsement. The FARs are not specific on this point. They do say that you can log PIC time anytime you are sole manipulator of the controls for an aircraft you are rated to fly.

Unless you are rich enough to get your private in a twin, you will have a single engine class rating. This includes high-performance singles because no distinction is made between high- and low-performance in the rating. However, you cannot act as pilot in command of high-performance aircraft without the endorsement. The key then becomes the word "act." You cannot act as pilot in command, but you can log PIC time. Figure that one out.

You will notice that there is nothing in the FARs to prevent you from learning to fly in a high-performance aircraft as long as you have the appropriate endorsements. Japan (Airlines) in Napa, California, uses the A-36 Beechcraft Bonanza as their primary trainer. Maybe you professional students (if you can afford it) might consider taking part or all of your training in high-performance aircraft. One option might be to move up to high-performance after you solo, flying your cross-countries in fast airplanes. This would save you some hours later on as you would already have the endorsement when you get your private, and you could save the extra high-performance time required for the commercial.

The secret to professional training is to cover as many logbook categories on each flight as you can. If you can take a multi-engine, actual instrument, night, cross-country, PIC flight, you will be ahead of the game. Investing in higher cost training initially can sometimes save you money in the long run. This is not in the interest of the flight school, so they will not point out such options. You may have to be insistent. You will also stand out from your peers if you do something different, which brings pressure on you. In professional training, especially at the big academies, conformity is the rule, not what is in the best interest of the individual student. No one knows any better, unfortunately, until after his training is over. Your instructor may know, but he cannot go against flight school policy if he wants to keep his job. Therefore, it is up to you to be open to seeking and trying new and different options that will best suit your needs, not the needs of the flight school. After all, who's paying the bill?

There is nothing drastically different about high-performance aircraft. In many ways they are easier to fly when you get used to them because of their increased power, weight, and stability. The propeller maintains its RPM despite attitude and airspeed changes. You have the power for greater altitudes. Greater speed control is available by virtue of various gear and flap combinations. What you have to learn to do is change your thinking to take advantage of your new capabilities.

You will learn to set the engine power with one control and the propeller with another. In low-performance the throttle does everything. This screws up your thinking for high-performance. You will also learn to monitor engine instruments more carefully, manage power more exactly, lean the mixture more finely, and learn to think much further ahead because everything happens faster. Thinking ahead will help your instrument training later on. All knowledge is cumulative.

The workings of the throttle and propeller control are different from low-performance aircraft, so before you get all hung up on trying to memorize which one goes first in various configurations and flight situations, take some time to understand their workings. The throttle regulates the power developed in your engine. Power is the measure of force available to turn the crankshaft. Power in the engine is controlled by regulating by the amount of air going into the engine.

Combustion takes place when that air coming in is mixed with the appropriate amount of fuel at a ratio of approximately 14 parts to one and ignited by the spark plugs. The more air going into the engine, the more fuel can be mixed with that air, the greater the combustion, and the more power is available. The same atmospheric pressure that allows air to flow into your lungs so you can breathe also forces air into the engine so it can breathe; hence the term "air-breathing engine." Therefore, the limit on available power is the limit of the atmospheric pressure around the engine.

Air flows into the engine through an intake manifold. We measure that pressure the same way as a barometer measures air pressure, by the force exerted to move mercury up a vacuum tube a certain distance measured in inches. Hence the term "inches of mercury." Hence from that, "inches of manifold pressure." Your throttle controls the flow of air into the manifold and that regulates your power.

Your manifold pressure gauge is your engine barometer. Atmospheric pressure comes from the force of gravity and the weight of the atmosphere above you. As you climb higher, there is less atmosphere above you exerting less weight and therefore less pressure with which to flow into your engine. That is why you lose power as you climb. You will get the most power from the engine at sea level, or below, with high air pressure. If the atmosphere exerts 30 inches of pressure, some 27-28 inches will be available for the engine. Nothing is 100 percent efficient.

When that air is mixed with fuel and burned, the resulting combustion develops power which turns the crankshaft, which is attached to the propeller. That power is measured in horsepower, which is the ability to move 550 pounds, one foot, in one second; hence 550 ft/lbs per second. A 300-horsepower engine has lots of power to turn a crankshaft. To avoid any further confusion, always think of the throttle as the control that regulates the flow of air and therefore the power developed by the engine.

The speed at which your propeller at the end of the crankshaft will actually turn is regulated by another system operated by the propeller control. There is a balance between the power developed by the engine to turn the crankshaft and the resistance to turning from the airload on the propeller. Your propeller speed is measured by revolutions per minute (RPM) indicated on a tachometer. This also requires a change in thinking. In low-performance aircraft, you get the idea that the power control regulates the speed of the propeller. Now you have a separate control for that purpose. The propeller control is no big deal as you can get by with most flights using only three settings; full prop for takeoffs and go-arounds; top of the green arc on the tachometer for climbs; and a cruise setting in the green arc for everything else. The separation of engine and propeller controls allows you great flexibility. You can regulate the engine power separately from the propeller allowing many different power settings for any given RPM, or different RPMs for a given power setting. The flexibility of high-performance means maximizing your performance for your conditions, circumstances, and needs.

To understand the constant speed propeller, we need some definitions. Since the propeller can vary the angle at which it strikes the air, which is called pitch, it is a "variable pitch propeller." Changing the pitch changes the angle of attack and the resistance of air on the propeller. Through an internal component called a governor, the propeller changes angles automatically to maintain the RPM that you set. That is why it is called a "constant speed propeller." You will be confused by the mixing of terms like high and low pitch, and high and low RPM, so use the British nomenclature and always refer to coarse and flat pitch.

When the propeller moves through the air, the force of the air because it is a fluid seeks to twist the blades to flat pitch. That is the "aerodynamic twisting force." To overcome that force and move to a coarse pitch, the engine pumps oil into the hub to twist the angle of the blades. The propeller is in balance when the air pressure load on the propeller is matched by the corresponding oil pressure in the hub.

Thrust is the measure of force generated by the propeller to move the airplane. Thrust will be maximized by changing the blade angle to achieve the most efficient air load, or angle of attack of the blades. So much for definitions.

Here is how to operate this system. The propeller control is set full in for takeoff, giving you the flattest pitch. The propeller can generate the most thrust on takeoff if it can absorb the most power from the engine. It does this when it turns at the fastest available RPM. This is possible from the pitch which offers the least air resistance. For an airplane accelerating from a standing start on the ground, the most efficient airload on the propeller is achieved from a flat pitch.

As you accelerate and climb out, the airload on the propeller changes because instead of spinning on a nonmoving aircraft on the ground, air is now rushing at the propeller at your airspeed. The angle of attack and the airload on the propeller has changed from ideal to inefficient. By twisting or pulling back on the prop control, you change the tension on a speeder spring, changing the setting of flyweights in the governor, which moves a pilot valve, which pumps oil into the propeller hub, which increases the coarseness of the pitch, which creates a greater airload on the propeller, which slows down the RPM to the climb setting. Sound familiar? By pulling back on the propeller control for cruise, you do exactly the same thing for the same reason. You get better performance because you have set the most efficient airload on the propeller to generate the most thrust for a given percentage of power from the engine.

The only other setting you make is to put the control full in before you land in case you have to go around. Setting the RPM is the first operation of the system. The second operation is taken care of automatically by the governor. The flyweights that you set with tension on the speeder spring are spin-driven by centrifugal force by connections to the turning propeller. Should the propeller have a change in airload from any pitch, power, or airspeed change, the flyweights will restore within limitations the original tension of the speeder spring. The governor regulates the RPM by adjusting the blade angle to the airload that restores the flyweights to the speed which matches the speeder spring tension you set with the propeller control.

One of the myths that gets passed down the instructor/student chain is the idea that you should never have the manifold pressure exceed the RPM. Why? Good question. I have never seen an operating handbook that requires matching the manifold pressure and RPM. This is one of those myths that probably started decades ago when one instructor, trying to make things simple for a student being introduced to high-performance, required that student to synchronize the controls. Strange, because as you know, manifold pressure and RPM are apples and oranges. That instructor probably told the student that if the manifold pressure ever exceeded the RPM, the engine would overboost and instantly disintegrate. Pass this down through enough pilots, and like the 100th monkey theory, you reach a critical mass and everyone believes it.

Normally aspirated engines aren't even boosted, so how can they be overboosted? I just read through a Cessna Centurion manual and never even found the term overboost. Engines that are normally aspirated can only use the ambient atmospheric pressure to bring air into the engine. The only way to increase the air pressure in the intake manifold above what is normally found in the atmosphere is to boost the air pressure with a turbocharger. Should you greatly increase the pressure going in to the engine combined with a coarse pitch propeller, the engine in attempting to turn a propeller with too much air resistance will cause the massive internal pressures and temperatures you know as overboosting. The extra power developed in the engine can't be absorbed because the propeller can't turn fast enough to absorb it when there is too much air resistance.

Any engine, either turbocharged or normally aspirated, is going to have normal operating ranges of manifold pressure and RPM for various conditions set by the manufacturer. There is an infinite variety of combinations within those ranges that is available to you. However, because you never question your instructor, same old problem, you lock yourself into a myth of matching the manifold pressure and RPM. There is even a term for it now. It is called "squaring" your settings.

Back to our Centurion manual. At 4000 feet in standard conditions, you can cruise at 63 percent power using 25 inches of manifold pressure and 2200 RPM. Oh horrors! None of you out there who buy into the myth would even consider using such a combination because without any logical reason you have decided it is overboosting. Even though the manufacturer clearly states many such combinations in the performance section, as far as you are concerned, they don't exist.

How can you justify flying high-performance aircraft if you aren't even willing to learn how to maximize the potential of these aircraft? Most of you are so restrictive that you will only fly 27 squared for takeoff, 25 squared for climb, and 23 squared for cruise regardless of the recommendations in the manuals. It is up to you to break this cycle, think for yourself, stop passing around myths, challenge your instructor, and demand to learn the full range of possibilities in these new aircraft.

Your relationship with instructors forever changed

shortly before your private checkride when you made the demand to fly only the minimum night time necessary for the certificate. Your instructor went from a god who streaks across the sky to just a means to an end. In high-performance it is the insurance companies that set the hours of instruction and experience required to operate various aircraft. Your next demand of an instructor will be to insist on a high-performance endorsement in exactly the minimum time set for your airplane regardless of your knowledge or capability. The rationalization will be that these airplanes are more expensive and therefore you need a break.

Garbage! If you are going to fly high-performance, you have to have the money, so skimping on training because of your newly inflated self-worth is pure arrogance. Now that you have a pilot certificate, there is a tendency to think that you know what is best from here on out. I have taught obviously dangerous students who, as soon as the minimum time had been acquired, twisted their brains to believe that they were now high-performance pilots and demanded their endorsements. Your instructor will bend over backwards to accommodate your wishes if possible, even succumbing to bullying and giving you an endorsement before you are quite ready in an effort not to lose your business, unless you are so dangerous he is more worried about losing his job.

People do not worry about getting the private in the minimum time. All they want to do is solo when everyone else does. However, after that any instruction time beyond the required minimum is regarded as a personal failure or milking by the instructor for more flight time. The dark side of training you to make decisions on your own is that you now feel comfortable making the wrong decisions. Now we have to train you to select accurate information as the basis for your decisions. Have the independence to ask the right questions, not to think that you have all the right answers — there is a big difference.

It is worth flying high-performance aircraft just to retract the gear. You will feel like an airline pilot. Listening to the machinery and the inevitable thump as the gear locks tight is really fun. You feel a new acceleration as the airplane reconfigures. Streamlined now, your climb rate increases and you feel cleaner in flight. It looks very strange without the gear the first time you look out the side window, so for a brief instant you will wonder if it will come back down again. You will get used to that.

Since the gear is only useful for rolling on the ground and as drag to steepen your descent, any airplane that hangs the gear out in flight will look ponderous to you now. Fixed gear limits the potential of any airplane. The smallest and lightest trainers leave the gear out because they don't go fast enough for the reduction in drag of retractable gear to justify the extra weight of the system. With today's technology though, most airplanes having four seats, 180 horsepower, and a constant speed prop can benefit from retractable gear.

Retractable gear is easy to use; bring it up after takeoff and put it down before landing. Since pilots occasionally forget the latter part, a large segment of your training will go to ensure that you have firmly imprinted in your mind enough places in flight to check that the gear is down and locked before you land. There is a saying out there that says there are two types of pilots: those who have forgotten to lower the gear and those who will. An old instructor friend of mine pointed out to me that many pilots fly their whole career without forgetting the gear, so the perpetrators of that saying are probably the ones who *have* forgotten the gear. If you do your checks faithfully, you will not forget the gear.

People have funny ideas about when to raise the gear. I like to raise the gear as soon as the airplane is established in a climb, also called a positive rate of climb, just like the multi-engined airplanes. For single engine aircraft, the general rule for most pilots is to raise the gear when you are out of usable runway. No one ever defines usable runway the same way, however. Pilots have translated it to mean when you have completely flown past the runway.

For the sake of argument, my definition of usable runway is the amount of runway that would be required, if your engine failed, to glide to a landing and roll to a stop. What many of you folks are doing is waiting until you fly over the departure end of the runway to raise the gear. Well, unless you can stop all forward motion, descend vertically, and instantly brake to a stop, you have long since run out of usable runway.

Let's investigate the Centurion manual again and play with some numbers. You will find that on a standard day, at sea level, maximum weight and maximum performance takeoff, it takes a little over 2000 feet to clear a 50-foot obstacle. To land over a 50-foot obstacle and roll to a stop using maximum short-field technique under the same conditions, it takes about 1500 feet. If you lose your engine at exactly 50 feet after takeoff, what will it take to stop? Let's assume normal takeoff technique, so add an extra 500 feet to get going. Add another 500 feet for the transition from the momentum and reconfiguration of the airplane from takeoff through engine failure to landing. Total this up and you will need a runway of at least 4500 feet to have any usable runway at all after climbing to a height of 50 feet and instantly losing your engine. So anytime the gear is left down after achieving a positive rate of climb on a runway of less than 4500 feet in standard conditions, you are maintaining unnecessary drag, squandering altitude, and wasting performance.

Most high-performance pilots seem to raise the gear at a height of 200-300 feet as the last vestige of runway slides beneath the airplane. This is just a guess, but let's say that a Centurion descends at an average of 1000 feet per minute after an engine failure with the gear down while transitioning from the best rate of climb speed and configuration to the best glide. If you lose the engine at 250 feet, you will be on the ground in 15 seconds. The best glide for a Centurion is 85 knots. That translates to roughly 144 feet per second, so in 15 seconds that is 2160 feet. If you add the standard 800 feet for roll out, you would need almost 3000 feet of runway remaining after reaching a height of 250 feet to come to a stop following an engine failure. This would require a runway of at least 6000 feet in length before it would make sense to leave the gear down while climbing to a height of 250 feet off the ground. Interesting.

Why do you folks leave the gear down so far past any hope of usable runway? Why don't the manufacturers publish numbers on gliding distances following engine failures at various heights above the ground? Okay, what is the big deal of leaving the gear down? Well, extended gear gives about a 400-feet-per-minute penalty in rate of climb. While an airplane is laboring through 250 feet while climbing at 600 feet per minute with the gear down, the same airplane would have been at 416 feet raising the gear up at a positive rate of climb and climbing 1000 feet per minute. You might need that altitude one day.

Should you lose the engine at 416 feet, you will have more time, altitude, and distance you can travel before contacting the ground. You will have 20-25 seconds before you land and since gear can extend usually in 8-10 seconds, you can then decide to land with the gear up or down depending on what lies ahead. You have definite advantages over the folks who leave the gear down too long. All of my numbers are at best just guesses and not intended to be accurate performance figures.

What I am trying to illustrate is that you always have options to consider and that accepted practice may have no basis in logic. On short runways you would want to get the gear up immediately. On long runways you could leave the gear down to a predetermined height off the ground. If the manufacturers of high-performance singles could publish balanced field lengths for their airplanes like they do for more advanced aircraft, you would have accurate performance figures to tell you how much usable runway you would need should you lose your engine at say 50 feet, 250 feet gear down, and 500 feet gear up. That way you could plan each takeoff and decide when is the best time to raise the gear.

Upon first inspection, your high-performance airplane is crowded with lots of new things with which to fiddle. If you go flying without sorting out in your mind where each control is and what it does, you will spend your early lessons just learning your way around the cockpit. That will cost you lots of money. Why not sit in the airplane for nothing and practice until you can touch every device within reach when blindfolded and describe its normal and emergency purpose and function. Give yourself two hours at least for a single. When you get to twins, figure on three to four hours per model.

When you are flying faster than you are used to and there are more things to adjust, you can become confused very easily. As a result, your first high-performance climb-out might go something like this:

"Okay, bring the power back, ah oh, not the prop, ooh, wrong one, mixture, not that, throttle, oh well, eek, oops."

Our current system does not encourage you to properly prepare for your first high-performance flight, preferring instead to have you pay the big bucks to learn in the air what you could have learned on the ground. I had a student who kept shutting off the engine with the mixture while aiming for the RPM because he refused to sit on the ground and learn the controls. New controls are bad enough, but you have to use them in the right order. Now was that power first or prop first when you speed up? Is it flaps and then gear, and then flaps again to slow down? What a dilemma.

Most books on flying aren't much help. They write out a complex formula that only confuses you more, followed by some conciliatory, qualifying excuse for their inadequate information about how everyone gets it backwards at first anyway. Thus programmed to fail, you are relieved from using the proper technique which your flight manuals have already relieved themselves of giving you. That isn't good enough. It gets worse.

Take the Cessna Cutlass, a typical first high-performance aircraft, and look at the descent and prelanding checklists. From the fuel selector on the floor, back up to the throttle, across to the carb heat, back over to the mixture, down again to the cowl flaps, up to the seat belts, back down again to the fuel selector, up to the landing gear, recheck the mixture, recheck the carb heat, push in the prop control, and turn off the autopilot. This is disgusting.

There is a better way. Use the flow, Luke! You can't forget anything if you move methodically from one side of the panel to the other doing the appropriate thing with each control, which in most cases is nothing. Whenever you want to "power down," always start at the far left side of the panel, move all the way across to the center, and then "down" to the fuel selector. That should include every possible thing you can adjust in the airplane. Whenever you want to "power up," start at the fuel selector, move "up" to the engine controls, and then over to the switches on the far left. As long as you always flow using power up and power down, you will never use the throttle and prop control in the wrong order, unless of course you fly a Baron.

Power up will be used for things like takeoffs, goarounds, initiating climbs from level flight, setting up for stalls, minimum controllable airspeed flight and recovery, maximum performance maneuvers, and a final landing check. Power down will be used, for example, in setting climb power, setting cruise power, prelanding and downwind checks, and slow flight. Here is how power down applies to a prelanding check in the Cutlass. Start on the far left:

Primer - locked,
Master - on,
Mags - both,
Circuit breakers - in,
Gear - up,
Carb heat - in,
Power - 21",
Prop - cruise,
Mixture - set,
Flaps - 0 to 10 degrees,
Cowl flaps - closed,

Trim - set,
Fuel - both.

Now you can use the checklist to verify you got everything. However, the beauty of this system is that you can't forget anything, nor can you do it in the wrong order. You can stop trying to memorize the order of complicated procedures and theories of engine operation to determine what goes where, how much, and when.

The next flow for landing will come on downwind. Please don't go checking all the stuff you just checked for prelanding when you had lots of time and airplanes weren't screaming around the pattern with you. I refer to the master, mags, and fuel selector stuff. Downwind is no time to go fumbling around the cockpit of a high-performance airplane. Look at it this way: If the radios are still on, so is the master switch; if the engine sounds the same and develops the same power, the mags are still on both; and if you are not a glider, the fuel is on. Limit yourself to the engine controls, gear, and flaps from here to the ground.

My Cutlass downwind check, power down:

Gear - down,
Carb heat - on,
Power - 21",
Prop - cruise,
Mixture - rich,
Flaps - 10 degrees.

Once you are set up this way, you can now do almost the identical landing you learned from this book for low-performance airplanes. Most of you, unfortunately, are taught by your instructors how to land all over again using completely new procedures. This really extends your training.

Anyway, back to our landing. Once you have done the downwind stuff, don't touch a thing until the 45-degree key point past the runway numbers. Just before you turn base, bring the power to 15", just like 1500 RPM in the low-performers; prop - full; mixture - rich; flaps - 20 degrees; and you are done. It's that easy.

How about if you want to power up and climb? Easy

again, just do everything in reverse starting with the fuel selector:

Fuel - both,

Cowl flaps - open,

Flaps - up (depending on your situation you may have to wait until you get the power in),

Mixture - rich,

Prop - climb,

Throttle - climb,

Carb heat - in,

Gear - up.

When you do your power up check before landing or stall practice, you are all set for adding power because the power up check sets the mixture and prop before you get to the throttle. Anyone can learn this system and with a little practice use it with ease.

Familiarity breeds comfort, so you will find instructors gravitate to what they teach most often; those good old presolo maneuvers, stalls, slow flight, and steep turns. It is true that you have to show competence in these private repeats for the commercial certificate, but there is no reason to perfect them with an instructor for a high-performance endorsement. To get the feel of your airplane, do a couple of imminent stalls, a little slow flight, some steep turns, and them leave them alone. You can practice to your heart's content after your training.

Most instructors and flight schools are in a low-performance rut where you will greatly delay learning to think like a high-performance pilot because you are practicing presolo maneuvers at low altitudes. What you need to do is learn to fly fast at 8000 feet and then plan visual descents to various types of airports, because that is where you are going to have the most trouble adapting to high-performance.

The low-performance rut is evident whenever the airplane is to be slowed down, especially for slow flight. You will probably be taught to bring the manifold pressure from cruise power back to 15", just like your low-performers which go to 1500 RPM. First of all, big engines don't like big power changes, so if you can find another way to slow down, the engine will thank you. Secondly, as the airspeed and cylinder head temperatures fall because of the low power setting, you will be putting on the carb heat, enriching the mixture, dropping some flap, going back to the gear, then dropping some more flap, opening the cowl flaps, trimming, and checking the fuel selector to make sure you are not a glider. After all this is done, you will have lost several hundred feet and be in an imminent stall because 15" won't hold you up with all that stuff hanging out. Now you have to put in lots of power and try your best to recover and stabilize the airplane in slow flight.

If you wanted to fly it like a low-performer, you should have stayed in one. To avoid the low-performance rut, change your thinking. Low-performers have little power; they get lots of lift from big wings, and because of their low speed — some drag. High-performance aircraft have lots of power; they get their lift from lots of airspeed, and because of the increased airflow and variable combinations of gear and flaps — lots of drag. In a low-performer, when cruising above the white arc, the only way to slow down is to reduce the power.

Enter the world of high-performance. At normal cruise in a Cutlass, you will be indicating about 130 knots. The gear can come down at 140 knots. You won't think to go to the maximum gear speed because most instructors only let you drop the gear when you are slowed for descent or downwind because that is all they have ever done. Try this: Drop the gear with everything else set for cruise. That will put you well below 130 knots where you can now extend the first 10 degrees of flap. This will then put you in the white arc where the rest of the flaps can come down. You will be stabilized in slow flight having moved only two controls, three if you add in trim. To resume cruise, just reverse the process. To set up for stalls or minimum controllable airspeed, do a power-up flow check. Options, always think of options. Find out how you can expand your knowledge and learn to fly high-performance airplanes better.

Those laundry-list procedures adapted from low-performance ruts that are in use in many flight schools are simply bad procedures. You will get hung up on the rote order of events rather than learning how to think about which control makes sense to use given your circumstances. The problem with most students and instructors is that you continually try to make the best of bad procedures rather than consider new options. If you know the limitations from the operating handbooks, you also know the range of freedom available to you. Whether you use that freedom or not is up to you.

Many aircraft manuals request that you leave the gear down when practicing touch and go's so that you don't forget the gear in the short time between takeoff and landing. Well, how do you learn how and when to use the gear if you can't use the gear? How do you mark the checkpoints in your mind to verify that the gear is down if it always stays down? More interesting than the manufacturers attempt to limit their liability is the fact that this recommendation is universally ignored. No student has ever pointed this out to me and asked why we do not follow it. When I ask students if they read the manuals, they respond, "Oh yes!" Mention this tidbit and that blank look common to students on their early flights returns.

Everyone turns into a baby student with an instructor on board. You can read about keeping the gear down in the pattern and override any thought of it simply because the instructor ignores it; therefore, it must be irrelevant. Granted this recommendation was probably written by lawyers preoccupied with avoiding potential lawsuits from gear-up landings, but what never ceases to amaze me is the blind cult following that students of any experience level give to their instructors, such that myths like matching manifold pressure and RPM are carried as fact, flaps are only extended in the middle of base and final when they will most destabilize your landing approach, and recommendations clearly printed in manuals are never even considered. Will you take enough initiative to question your instructor when your aircraft manual and your teachings differ? It is one thing to raise the gear while remaining in the pattern because you have made a conscious decision in order to get the best training. It is quite another to imitate the rote actions of an instructor who may not have ever considered the manufacturer's recommendation.

The way to improve on the basic touch-and-go pattern is to do the old one-touch-and-go airport tour mentioned earlier. With the superior speed of high-performance aircraft, you can visit more airports. This exercise will get you the flexibility, agility, and familiarity you need for more advanced aircraft.

When you concentrate on relearning your presolo maneuvers, you miss acquiring the full potential and capability of your new aircraft. Not only do you miss new options for use in the air, you also miss planning for full utilization from your ground preparations. Most newly endorsed high-performance pilots I have taken for aircraft checkouts flew at low altitudes and knew only one cruise setting. What a waste of training. Why do you think the manufacturer went to the trouble of publishing all those possible combinations of engine settings for various altitudes? I won't get into turbocharged engines in this chapter because you won't find them in most training environments. Flight schools that don't teach you to fly properly won't trust you with expensive turbocharged aircraft.

Back to the performance charts of our favorite Cutlass. If you want some power setting generalizations, use 55 percent for economy, 75 percent for speed, and 65 percent for the best of both. Now you have some options. Looking at the 6000-foot chart, you will find 65 percent, + or - one percent, at three different settings: 21"/2500 RPM, 22"/2300 RPM, and 23"/2100 RPM. Which one to use? The fuel consumption is the same for all of them because the percent of power used is the same, so that is not a factor.

The first option spins the prop fast without a lot of power behind it. You will get the performance at the cost of the most noise and engine wear. This is the least desirable to me.

The second option is what most of you will select because it is the closest to your ingrained square settings, and the manifold pressure is below the RPM so you won't worry about overboosting your unboosted engine.

I would choose the third option because it is the quietest and keeps the wear and tear to a minimum. You will note that the 23 squared setting you are used to only gives you 65 percent power at 2000 feet, down near where you practice your presolo maneuvers — what a coincidence.

You have to think about the needs of your flight when selecting your power combinations. I flew a Cutlass non-stop from Phoenix, Arizona, to Concord, California, at 8500 feet with a power setting of 21"/2100 RPM and landed with over an hour's fuel, remaining. By being able to control the engine and propeller separately, you open up a range of possibilities including long distances, better speed, quieter cabins, and higher altitudes for better true airspeeds and maybe better weather. Dig into your performance charts and see how you can get the most from your airplane.

There is a trick to selecting the most favorable altitudes after considering the forecast winds aloft, which are invariably wrong. The thing to remember is that the true airspeed of your aircraft will increase with altitude because the progressively thinner air creates progressively less drag. The catch is that you still have to be able to maintain a certain amount of power to take advantage of the reduced drag because in a normally aspirated engine the power also drops with altitude. The altitude with the highest true airspeed will be the highest altitude with full throttle and RPM where you can maintain the percentage of power you want.

For example, you could hum along in your Cutlass at 74 percent power at 6000 feet where full throttle would get you 24" and you could set a reasonably quiet 2300 RPM. Your true airspeed would be 135 knots and fuel burn 9.8 gallons per hour (gph). Your range, though, with 62 gallons usable is only about 710 miles. For 65 percent at full throttle, you could cruise at 8000 feet with 22" and an RPM of 2200.

Your speed would drop to 129 knots and your fuel burn would drop to 8.8 gph. Please note that for a 2000-foot increase in altitude, a nine-percent drop in power results in only a five-percent drop in airspeed. Your range would increase to 780 miles. For those who want distance, you could get 57-percent power at full throttle from 20" and 2100 RPM at 10,000 feet. The speed drops to 121 knots and the fuel drops to 7.8 gph. The range goes up to 830 miles. This altitude and power setting would only help if you wanted more range because the loss of airspeed is much closer to the loss of power than the difference between 6000 feet and 8000 feet.

Play with all the numbers of your own plane and see what interesting combinations of power, airspeed, fuel burn, range, and endurance you can find. Chances are that for the typical high-performance single, 8000 feet will be your best all-around altitude. You will find that altitudes lower than 8000 feet allow more power, airspeed, and shorter travel times, at the cost of much higher engine settings and more fuel burned. It will be better to cruise higher than 8000 feet only if distance is your goal because the drop in power is very significant to your operation above that altitude.

To take advantage of the extra performance, you learn to fly high and fast. You will also have to learn how to efficiently come down from those high altitudes so as not to blow past the airport. The procedure is the same as we covered back in solo cross-countries. Plan on losing 500 feet per minute; so if you have to lose 6000 feet, it is going to take you 12 minutes. Now, the low-performers are fairly close in airspeed, so one speed and distance kind of fits all. Not so with high-performers, so to make your mental calculations easy, round off the descent airspeed to the closest of 120, 150, or 180 knots, which will give you miles per minute of 2, 2.5, or 3 miles respectively. In our example depending on your speed, you will cover 24, 30, or 36 miles over those 12 minutes.

You wouldn't think that such a simple formula could cause people to screw up their descents. Well, the formula works fine, but pilots start mixing in their personal feelings. They can't believe that the descent has to be started so far out, so they don't descend. Those who try to eyeball descents from 10,000 feet will soon be embarrassed enough into following the formula. You could save lots of training time if this concept were introduced early in training; sadly it's not. The faster the airplane, the harder it is to believe that you have to descend as early as you do.

If you have to descend 10,000 feet and are traveling three miles per minute, that's 60 miles. You may be reluctant to do this in unfamiliar territory if you have only ten miles visibility. However, if you try to eyeball the approach in limited visibility, your decisions are based on false perceptions, which is why it doesn't work. We have instruments and a brain to make calculations because our senses aren't geared for measuring anything like the speeds and distances possible from our airplanes.

You would think that having made the same mistake in private training you would have learned by now. However, the nature of our education system is that you can forget everything after you learn it after a test, and our flight training system allows you to shirk any responsibility for remembering your training. The result is that you have to redo your mistakes and relearn how to fly every time you try something new. Take responsibility for your training and break this cycle.

Back to our descent problem. High-performance airplanes travel half again to more than twice as fast as low-performers, yet for passenger comfort and your comfort, you are still stuck with the same 500-feet-per-minute rate of descent. Your planning, therefore, has to be better. If you have to eyeball the approach, set up a descent airspeed and constant pitch and use power to maintain your reference in the window just like you do on final. You can also stair-step your way down. Pick checkpoints and target altitudes along the approach. That way you will take one imposing task and make it a bunch of small, easy steps. You can monitor your descent profile using the checkpoints as well.

When you come down from high altitudes and you are not used to it, your new sight pictures will screw you up. Even though your speed remains constant, the lower you get, the faster you will think you are going. If you subconsciously perceive that the airplane is continually accelerating, you will get anxious and feel progressively more rushed as the ground gets closer. You will feel further and further behind when in reality, nothing has changed. Stick to the formulas, checkpoints, and checklists, and your transition to high-performance will be smoother.

You may be wondering how do the big guys handle this problem? If they were limited to 500 feet per minute, a jet descending from 30,000 feet flying at 480 knots down to 10,000 feet and then 240 knots below 10,000 feet would need 400 miles to get to sea level. That is longer than some airline flights. The difference is that they are not limited to the same rates of descent because they have pressurized cabins. In a pressurized cabin, the pilots and passengers only feel the rate of descent of the cabin, so if you are flying at 30,000 feet with a cabin pressurized to 8000 feet, you can bring the cabin down at 500 feet per minute and have a sea level cabin in 16 minutes.

The worst combination for descent planning is a turbocharged airplane that isn't pressurized. Now you have the power to get up high, the cabin that is right up there with you, and the engines like small, incremental adjustments in power. One advantage you can take is to file an instrument flight plan when going anywhere new. The controllers who are familiar with the area usually have reasonable descent profiles established for you. You are still pilot in command, and they may want you to come down too fast, which may require some negotiating. But many times your planning can be greatly simplified by the removal of VFR guesswork and replacing it with published routes, altitudes, arrivals, and approaches.

Now that you have entered the world of high-performance, if you really want to enhance your safety, fly every airplane as if it were high-performance; preflight with a checklist, always use the descent formula, and use performance charts to optimize every flight. You will learn that your low-performers can only get the same 65 percent power at higher altitudes by increasing the RPM, and therefore the noise. With constant speed propellers, you never hear the increase in manifold pressure, only the change in RPM.

The most important reason to always act like you are flying a high-performer is to always check the gear. Whether fixed or retractable, this is the best way to avoid a future gear-up landing. Don't wait until you fly retractables after your checkout to start putting back the gear checks. There is no high-performance currency requirement, so if you wait six months between high-performance flights, you will have six months of not checking the gear to overcome.

You will be competent in high-performance aircraft after your initial endorsement for about a week. This endorsement has to be used to solidify the procedures in your mind. After a week you will start to forget the new thinking you worked so hard to develop, and you will become just another low-performance pilot trying to fly high-performance aircraft. Airports on the weekends are full of them. You should also do complete flow checks all the time no matter what you are flying, so that when you have the extra controls of high-performance airplanes, the hands will know where to go because the brain still has the habits. My students make great jokes of flow checks in low-performers:

> Fuel - by gravity,
> Gear - welded,
> Mixture - rich,
> Prop - fixed,
> Systems - boring.

You might feel silly at first, but you won't forget anything when the critical time comes. There can be great delays in between the blocks of high-performance work during your training. If you are on a budget that requires maximizing your time, you may fly high-performance in short bursts followed by long delays. It is in the interest of the flight schools to separate your high-performance training as much as possible so that you forget the maximum amount and have to spend the most money getting back your knowledge and skill. Treating every airplane from here on out as high-performance will cut back on the repeat time.

One might wonder why we still use such a peculiar acronym as "GUMPS" for a final landing check. This is the most bizarre and ludicrous procedure in all of flight training. There you are in a brand new aircraft (to you), with lots of new things to do, going much faster than you are used to flying, coming in for a landing (which is one of the most critical phases of flight), and you are suddenly expected to use new names for the controls as you go groping around the cockpit, when you should be looking out the window and flying the airplane. Does this make sense to anyone, or am I the only one who finds this practice not only strange, but dangerous?

"G" stands for gas. We never call it gas — we call it fuel. You check the fuel, drain the fuel, check for water in the fuel, and switch tanks with the fuel selector. But on base leg in a high-performance aircraft, it suddenly becomes "gas." The G could stand for gear. That would make sense as it is often the first control in a power down flow check. No, I'm sorry, the G could not stand for gear, that would be logical and consistent. Of course, whenever you do a GUMPS check, the first word out of your mouth is guaranteed to be "gear."

"U" stands for undercarriage. This just fractures me. Do we all suddenly take out British citizenship when we fly our first base leg in a high-performance aircraft? Honestly, undercarriage? That word is three syllables too long for use on base.

Of course, now being British, we would have to call the fuel (excuse me, gas), "petrol." I suppose we could change the acronym to PUMPS for consistency. The only places the British have good names are when they describe prop pitch.

"M" is for mixture. I recognize that word.

"P" is for prop. I recognize that one, too.

"S" is for systems.

Having a lengthy systems check right before landing is nuts because we now know that any pilot who doesn't want to grope when he should be flying does all that stuff way out on the prelanding check during descent. Besides, after the prop is checked, no one has the time or inclination to go through the systems. Most students just say "systems" to humor the instructor and hope that it covers the check.

You already checked the fuel, so toss out this nasty word "gas" that we never use. "Undercarriage" — be serious. What we have left are the essentials of "mixture" and "propeller." What about the gear? Following the crazy order of GUMPS sends your hands in a star pattern around the controls just like the published prelanding checks. You could very well forget the gear as you try to pronounce "undercarriage." If you are that distracted, you could very well not hear the gear warning horn (or is that undercarriage warning horn?), no matter how loud it is. The stuff you land on should always be called gear, period.

Well for all you folks who learned the GUMPS check, we can modify it to Gear, Mixture, and Prop — "GMP." You can still pronounce it "gump" and keep the familiar sound of your check with familiar names for the controls and use them in a familiar order, without unnecessary groping. Even on airplanes that change the classic positions for the controls a GMP check will hit the biggies.

Anyway, I'm sure students will be taught the GUMPS check for some time because, like child abuse, these things just get passed down the aviation family. There will come a time when logic, initiative, simplicity, and making the system better will triumph over rote memorization, trying to make the best of bad procedures, and training without any conscious thought or responsibility, but we have to make it happen. I still think the GUMPS check came to us from some English mole instructor trying to sabotage our flight training system.

The problem with dumping a useless check like GUMPS is that it is so imprinted on so many pilots' minds that there will be tremendous resistance to change. Pilots and instructors go through such incredible efforts to learn these inadequate and inferior procedures that even if something better comes along, instructors still want students to suffer as they did. It's like some archaic fraternity ritual. All us new and improved GMP types will conflict with the old GUMPS instructors who will insist their way is preferable, simply because after doing all the work to learn it, they are used to it. When will we stop adapting ourselves to the procedures and start adapting the procedures to ourselves? For the rest of your life when on base and final, say "gear, mixture and prop" and end your groping. If you forget that, then go back to the old power-up/power-down flow checks. Either way, you will get all the important stuff.

Chances are that you have been thinking about the move up to high-performance aircraft long before you actually take the plunge. You were probably caught in a bind of your own making between the low cost yet boredom of low-performance and the high cost and excitement of high-performance. Is that necessarily the choice? Do you know any commercial operators who have fleets of low-performance aircraft? They all use high-performance. Why? Because they are cheaper to operate, that's why. Commercial operators do not use low-performers because they can't carry enough, fast enough, far enough, out of enough airports, to make a living. Even with the extra maintenance of retractable gear or a constant speed propeller, high-performers are still more profitable. How can that be?

The problem goes back to your thinking, so you won't see the proper comparison. As a student renting airplanes, you always think in terms of cost per hour because that is how your training is set up. Using that criteria, high-performance aircraft are always more expensive to operate. Now start thinking in terms of time, distance, and useful load; criteria that come under the umbrella of utility.

Think about the distances to the exciting places you want to visit. Think about how fast you can get there. Think about how many people and how much stuff you want to take with you. The reality is that depending on the model, once you pass a certain number of miles, high-performance airplanes become cheaper than low-performance.

Let's take three different airplanes and fly them over a 500-mile course. We'll use the Cessna 152 Commuter, 177RG Cardinal, and the 210 Centurion. First the 152. It has two seats and a 110-horsepower engine — definitely low-performance. You can rent a 152 for about $40 an hour, depending on the place. You will need a fuel stop with the 152 because the furthest you would ever try to fly would be 300 miles. Figuring in two starts, two climbs, and a cruise of 100 knots True Air Speed (TAS), 67-percent power at 6000 feet, and 5.4 gph, you will use 31 gallons of fuel. With two starts and stops, figure about six hours on the Hobbs meter for the trip. Add another two hours for all the inevitable delays that come with a fuel stop, and you will need seven to eight hours total travel time to make this trip. At $40 an hour, six hours on the Hobbs will cost you $240. With that much travel time, plan on arriving tired and not doing much else that day.

Okay, how about the Cardinal RG? The Cardinal has four seats and a 200-horsepower engine. It can probably be rented for about $70 an hour. At 8000 feet using 65-percent power, you will get 140 knots TAS and 9.3 gph. Allowing for start, taxi, climb, and cruise, you will use 36 gallons for the trip, and it will take you three hours and 40 minutes. At $70 per hour, that's $255. If you get an early start, you can be there by lunchtime and have the rest of the day to play.

Our Centurion has six seats and a 300-horsepower engine and can be had for about $100 an hour. Using the same basic conditions of 8000 feet and 65-percent power, you get 164 knots TAS and burn 13.5 gph. This will cost you three hours and ten minutes of your time, 43 gallons of fuel, and $318.

This comparison was taken right from the operating handbooks and based on standard conditions, gross weight, and, when possible, normal rates of climb. The one most likely to be at gross is the 152, so the numbers for the high-performers should be better.

Let's see what we can get from all the numbers. The most dramatic statistic that leaps out at you is that for an extra five gallons or $15 you can take a Cardinal and save yourself many hours of travel time and the aggravation of a fuel stop. How much is that worth to you? If you only think in terms of cost per hour, the Cardinal is almost twice as expensive; but if you think of utility, you will save much more with the Cardinal. The fastest of the bunch is the Centurion. It will get there just half an hour quicker than the Cardinal, but it burns more fuel with that extra horsepower, so it will cost you more. For a time and fuel comparison, the Centurion doesn't offer much advantage for the cost.

For strategy and further comparison, see if you rent by the hour, or whether you pay for the fuel separately with what are called "dry rates." If you pay strictly by the Hobbs meter, go for the best speed. If you pay for fuel separately, find the most economical cruise.

Since these figures were calculated using gross weight, let's fill the seats and recheck our utility. This is where the Centurion pays off. Dividing the total cost of the rental by the number of available seats, the Centurion will only cost you $53 per seat, the Cardinal will run you $64 per seat, and your "low cost by the hour" 152 is going to set you back $120 per seat. The extra capacity of the Centurion gives you the most utility if you have six people to move, despite the fact that it has the highest operating cost per hour.

To carry the same six people in three Cessna 152's, it will cost you $720 and take you more than twice as long as one 210 to get everyone there. Even if you only put four people in the Centurion and three in the Cardinal, that will still only cost you $78 and $85 per seat, respectively. Still not convinced high-performance is cheaper? With only two people in the Cardinal, you will pay $128 per seat, verses $120 per seat for the 152. Big deal. The Cardinal is still twice as fast over that distance.

The key is utility. Find the airplane that most closely matches the speed, load, and distance you want to fly. Do your own comparisons at your school with the available airplanes. Once again, your options are only as limited as your imagination and your ability to analyze and challenge the conventional wisdom. By any comparison, except when you are a student building time, the Cessna 152 is the worst overall cross-country airplane. Did you ever wonder why the commercial operators who have to minimize their operating costs use fast, high-performance aircraft, and your flight schools have fleets of low-performance aircraft available for you to rent for trips?

High-performance aircraft are exciting and challenging. They will add to your prestige as a pilot among your peers. If you do decide to move up, a whole new world will be opened up to you. For your new elevated status, you are required to have superior judgment and responsibility now that you are in faster, more complicated, and more expensive craft. The equipment will give you far better performance, but it is much more easily abused, so your care of the aircraft and pilot skills have to be better. Remember that every time you take a step up in aircraft capability, you need a corresponding increase in pilot capability.

10 • INSTRUMENT TRAINING

Ah, the instrument rating: that elusive notation on your pilot certificate that separates the babies from the real pilots. Of all the pilots out there, only a small percentage of you are active instrument pilots, with most of this group being professionals. Instrument flying is the standard of the airlines and many commercial operations. The instrument rating is required for any pilot who expects to work in this industry beyond a few specialized operations like banner-towing and crop-dusting. This is where we leave the "fun in the sun" amateur pilot behind. From here on out it is serious study.

There are private pilots who get the rating for various reasons, like executives who make frequent business trips and need the increased capability of the rating. Many of these pilots, though, get the rating only to let it slip away through lack of use. Instrument skills are the most fleeting, passing quickly unless frequently practiced. The currency minimums in the FARs are the barest minimum to maintain some measure of competency.

You must think and plan carefully if you decide to get the rating and you are not intending to make a career out of aviation. How much are you really going to use it? Will you get recurrency training when you need it? Will you analyze and strictly adhere to your own personal limitations? If you decide to get it, then you have to keep using it.

What is a rating anyway? A rating is a special new condition that adds new privileges and limitations to your certificate. An instrument rating gets added to either a private or commercial pilot certificate. Pilot certificates can get lots of ratings; for instance, you have an aircraft rating that probably says single engine land. Anyway, you are about to go through a lot of trouble just to have the word "instrument" typed on your certificate. The good news is that to an even greater extent than high-performance, a totally new world of flying will open up to you.

You have crossed a bridge. You are entering a strange new world like nothing you have ever known. You will navigate and fly when you can't even see. In the clouds there is no up, no down, no left or right, no outside reference at all, only the instruments. Should you succeed in your endeavor to acquire the rating, you will be envied by those who failed to meet the challenge or those who dared not try. To be able to fly and navigate without visual reference is an accomplishment of which to be proud. When we leave the visual world behind and enter the nonvisual world, we enter the world of imagination. Welcome to the other half of flying. Welcome to the world of IFR.

Did you like that hype? It's there for a reason. The perception that instrument flying is somehow magical and mysterious is nothing more than an excuse for not having the willpower to learn some new techniques. If you really learned to fly well during your private training, you already have 90 percent of the skills you need to be a good instrument pilot, especially if you learned to fly from a busy terminal with frequent IFR operations. IFR stands for Instrument Flight Rules. You knew that.

There is a misconception out there among those about to embark on instrument training that because it is mysterious, you have to learn to fly all over again. The similarities, however, to what you already know far outweigh the differences. Break down IFR into its individual letter parts. Most of the "instruments" you already know, especially the flight instruments. As for "flight," you are still in an airplane so all the physical laws you already know still apply. That leaves only the "rules." The biggest thing you have to add to your pilot repertoire is a new set of rules.

Almost all of the problems associated with getting an instrument rating boil down to the inability to change or maintain headings; altitudes; and airspeeds. These basics go back to your earliest flight lessons. You believe you have to learn to fly all over again when the sad truth is that you never really learned properly the first time. The average recreational amateur compared to a sharp instrument pilot is at best sloppy. This is not a derogation of sunshine pilots, because with good visual reference to the horizon, you have such a wide safety margin that you can have a sloppy technique and still continue a long and happy relationship with aviation. Try the same thing under instrument meteorological conditions (IMC) and if you are lucky, you may only violate the rules.

When you are flying VFR, the air traffic control (ATC) system and the designated controlled airspace are both designed to keep you separated from IFR traffic. ATC does it through radar where available and position reports the rest of the time. Controlled airspace mandates separation through adequate weather minimums, so you will see and avoid the IFR traffic. All of your flight experience up to this point has been outside "the system." By learning to fly under IFR, you get to work inside the system. You will find that it is far easier to be a part of the system than to have to comply with all the restrictions to remain outside it.

The IFR system is designed for you to fly and navigate without hitting another aircraft, them hitting you, or you running into anything. That is it. For you to be successful in this system, you absolutely must have a solid grounding in basic flight skills. Before you ever start instrument training, go out and practice in your airplane until you can hold a heading within five degrees, maintain a standard rate turn through a variety of airspeeds, hold a constant rate of climb and descent within 100 feet per minute, hold airspeeds in any normal attitude within five knots, and be able to do any combination of them using the same tolerances. When you can do that without any problem, you will be ready to begin your instrument training.

Well, what about the rules of IFR? Considering all the fuss and consternation surrounding them, you would think they take up their own volume in the FARs. The truth is that when you leave out the CAT II and III stuff there are only 11 instrument flight rules. They cover:

1. fuel requirements
2. flight plan information
3. VOR checks
4. clearances and flight plans
5. takeoff and landing
6. minimum altitudes
7. cruising altitudes
8. course to be flown
9. radio communications
10. lost communications
11. malfunction reports.

All the rest of Part 91 you already know. What makes these rules so difficult to grasp is that our current training methods hardly ever show you how the rules are applied, leaving you high and dry after the rating to figure out for yourself how to follow the rules and utilize the system. How you train for the checkride and how you will use the rating bear little resemblance to each other. Standard instrument training is only designed to please FAA examiners in VFR conditions. If you combine together inappropriate training, misguided objectives, and inadequate flight skills, it is no wonder instrument training has the reputation it does.

The instrument rating has the singular distinction of being the only rating whose standard curriculum and training do everything possible to keep you out of the environment for which the rating is designed. The whole idea of instrument flying is to provide a system for which you can safely operate in weather below VFR minimums, yet you can get an instrument rating never having flown in weather below VFR minimums. If instrument flying is so critical and has such strict requirements, why does the FAA allow brand-new instrument pilots to engage in the equivalent of on-the-job training?

There are three categories of instrument time: actual, where you are flying in real clouds; simulated, when you are wearing a hood or other visually restrictive device; and flight simulator, where you are sitting on the ground. Under Part 61 of the FARs, you need 40 hours of instrument time. It breaks down to 20 hours maximum in a simulator, 15 hours in flight, and five hours where I don't know what you have to do. Of your 15 hours that you actually have to spend off the ground, only five hours have to be in an airplane or helicopter. Fascinating. This means you can spend 20 hours playing in a machine on the ground, five hours in an airplane or helicopter as appropriate, ten hours off the ground in something with instruments like a sailplane (for example), and five hours that are not defined except that they can't be spent in a simulator. With this minimum time, you could go for the flight test. Unbelievable!

The Part 141 folks get an even better deal as they can go for the rating with only 35 hours of instruction. The theory is that ground trainers and hooded time in the right proportion are as good as actual instrument experience. Therefore, according to this theory, instrument students who have never seen the inside of a cloud can transfer the training from a completely different environment and be able to safely operate in a hard IFR environment. Horsefeathers.

We know that student pilots take no responsibility for their training because the FAA takes it away from them and places it all on the instructor. Therefore, students only remember what they have learned by rote repetition and have no ability to transfer skills from other areas. This applies to every single certificate and rating a pilot seeks. The proof that there is absolutely no foundation for the belief that simulated conditions are as good as actual is the safety record of all instrument pilots. Just as the spin requirement that only requires a demonstration of your knowledge while sheltered on the ground has not prevented stall/spin fatalities, so the simulation of instrument conditions will not stop pilots flying into the ground in real weather.

I am not saying that training solely in actual conditions is either practical or desirable; I just want you to recognize that you cannot equate simulated training with actual instrument flight. According to the *American Heritage Dictionary*, "simulate" means: to have or take on the appearance, form, or sound of; imitate; to make a pretense of, feign; pretend. The fundamental problem is that anything simulated is not real. You cannot expect students to simulate instrument flight and have any competence or confidence in real conditions. We do not simulate like this in any other area of training. You cannot simulate touch and go's 3000 feet above the ground and be signed off to solo. You cannot simulate a cross-country by sitting in an airplane on the ground and talking through the flight log. You cannot simulate night by flying in the daytime wearing really dark sunglasses.

Why? Because none of it is the same as the real thing. The FAA recognizes this and makes you fly touch and go's from real airports, has you fly cross-countries over real terrain, and has you fly at night when it is dark. However, when it comes to flying in the clouds, the FAA says you can get the equivalent experience on the ground in a comfy, air-conditioned simulator. This is like the difference between watching a love scene in a movie and doing it yourself. Which seems more real to you?

Let's take a look at our pretend instrument conditions. First the hood. Ah, the hood; a medieval torture device that restricts your vision as long as your head remains perfectly still. This causes massive neck pain and headaches after a while, which is so unnatural that you will have to move your head. You can't help peeking out the window which ruins the whole effect the hood is trying to create. You just may peek on your own anyway. You can also watch the sun move across the flight panel as you turn the airplane, giving you a great visual clue unavailable in the clouds. The hood restricts your vision to the instruments like it is designed to do, so you might wonder why all the fuss? By not knowing the answer, you prove that the hood cannot substitute for at least some actual weather training.

By narrowing your definition of instrument training to simply flying by instrument reference, you miss the whole point of the rating. This limitation comes from our current system of training, which is why it is not good enough. The hood does what it is designed to do, well sort of; the real problem with the hood is what it cannot do. It cannot duplicate the feeling you get psychologically when you fly in the clouds. With the hood there is always a way out — all you have to do is peek or take off the hood. There is no way out of the clouds until you are out of the clouds.

The hood also distorts your perception of instrument flight such that you think IFR means you fly from just off the deck to the minimums on the approach without visual reference. In reality you can fly IFR in brilliant sunshine, in and out of broken clouds, in, above, or between solid overcast, and any combination you can imagine. The hood lets you believe that IFR is always the same. You miss learning that no IFR flight is purely IMC, and that to be effective in the IFR system, you have to be not only a good VFR pilot, but be able to easily transition between IFR and VFR conditions while still following IFR rules. This is one way how you miss properly implementing the IFR rules in your training. If you have to use a hood, try training at night as much as possible. Even if you do peek at night, it probably won't give you any useful clues, and the sun won't tell you where you are.

The hood is bad enough, but training on the ground for instrument flight is worse. Ground trainers stay on the ground; it is a completely artificial environment. Ground trainers can't pick up ice, they don't bounce, and you can't run into anything. What kind of training is that? Once again the fundamental problem of limiting instrument training to simply reference to instruments rears its ugly head.

Having said that, it is true that the airlines have developed very sophisticated simulators that are extremely close to the real thing. This makes infinite sense because it costs too much to fire up an empty 747 for a few blasts down the old Instrument Landing System (ILS). You might figure that if the airlines do it, then it must be good enough for you. Hardly. The reason that the airlines can simulate instrument conditions is that pilots of that level have experience, know, and understand flight in actual weather. It works for them because they already know what is real, so they can make use of a simulation despite its limitations and still derive a great benefit.

You cannot use a simulator properly because you don't yet know what is real. This is aggravated because current wisdom starts your training in the simulator, which gives you completely false perceptions of IFR flight, and then expects you to perform well in the real thing. Once again, everything is backwards. You can only simulate what is real once you know what is real. That is why you should never use a simulator until the end of your training after you have cloud experience.

Our current system of training starts you off with two strikes against you because before you even get into an airplane, you have already been subjected to the wrong training at the wrong time. It is no wonder then that you will have trouble with the rating. Think of our movie love scene example again. Didn't those movie scenes have a much different effect on you once you had done it yourself?

Simulators cloak you in this safe shroud where your training resembles a video game. From there you go flying in the brilliant sunshine with an instructor to watch over and protect you. You get the idea that the flight part is just another game. There are no real-life consequences to your actions. There are no other airplanes because the instructor is looking for them. There is no ground because the instructor will keep you clear. Without these consequences, there is no need for your personal involvement beyond moving the needles. Training is reduced to a video game mentality where students on the simulator and under a hood learn to move the instrument needles in the way that pleases the instructor, without any need for overall comprehension. Many students spend hours moving needles around, having absolutely no purpose in their minds for their actions; they are just happy to be instrument training because that is what they perceive real pilots do.

You will notice that flight schools make it a point of prestige in their sales stuff when they have a simulator, as if it is somehow a great thing not to train in an airplane. I would like to know the profit margin of operating a simulator compared to an airplane. An airplane has a purchase cost, insurance, wear and tear, parts, mandatory inspections, fuel, labor, and maintenance costs. A simulator, which can cost you as much as many airplane rentals, has a purchase cost or rent payment and an electric bill.

There are those of you out there who buy into the belief that ground trainers are as good as airplanes so much that you actually log your simulator time in your total flight time. Tell me, if you use your simulator at night, do you log night time? Do you log actual if your flight school building is surrounded by fog?

What is the essence of IFR? Why do we have this system? The entire, intricate system of weather service, traffic control, airspace, navigation, and regulations are designed for one purpose only: to allow you to fly from here to there without hitting anything. If you keep that concept in your mind, then everything else you do in training will make sense. When you train only by hood and simulator, you never learn this concept, so all your training becomes a new collection of disjointed and disconnected procedures without any purpose. By learning incorrectly that all these procedures are somehow new, you learn to forget what you already know.

This is why instrument flying is perceived as being so different, when it is really so similar. VFR, for the most part, is geared for cross-country flying by visual reference. IFR is the same thing only without visual reference. The rules of IFR are designed to get you there safely. You are still flying an airplane just like you have always done. The airplane does not know it is in the clouds, you do. All the same flight skills still apply; you just have to be much better at them. The essence of training, therefore, is to get you mentally ready to fly in the clouds so that the airplane can get you safely where you are going.

IFR, like every other step in aviation requires a change in thinking. Here again, it is the weakest area of current training. You may learn to fly a simulator very well. This is a useless skill when your personal security and confidence go out the window the first time you hit the clouds. You now have a mental conflict because up until now you thought you had been instrument training when in reality all you have been doing is wiggling the controls of general aviation's answer to video games.

How will you handle the clouds in different models of airplane? Most people train in low-performers in order to save money. You will never learn flexibility this way. If all your training is in low-performers, how will you know when to incorporate all the extra things you have to do in high-performance airplanes? You also get the feeling when you train on the ground or under the hood during VFR that you are the only aircraft in the system. This is a false impression because in neither case are you in any system at all. You will only learn nice, neat procedures with unchanging airspeeds or varying altitudes this way. This is hardly an accurate representation of the constant changes and deviations from your nice, neat procedures required by ATC in a hard IFR environment filled with other traffic.

The most important thing you can learn in instrument training is to learn about yourself and for what you are capable. Flying in clouds challenges all your emotions. You have to learn your individual reactions, how to control your emotions, and deal purely with logic. No simulator can duplicate this feeling. No hood can create this anxiety. To become a good instrument pilot, you have to train your brain to handle your emotions.

You cannot be a good instrument pilot without excellent basic flying skills. The fact that most instrument students spend the bulk of their time trying to hold headings and altitudes bears this out. Wasting time learning to fly a simulator well just delays the process because you still have to be able to control an airplane. Most of you aren't properly informed as to the critical nature of good basic flying skills before you start your training, which really sets you back and costs lots of money; but it doesn't have to be that way.

The FAA has an excellent requirement for the rating that you should accomplish before you begin instrument training, and most of you waste the opportunity. You need 50 hours of PIC cross-country time with flights over 50 nautical miles from home before you can get your rating. Most students blast through this time, considering it only an obstacle rather than constructive training time. Although your curriculum does everything possible to convince you otherwise, IFR flying is really just cross-country flying.

Commercial operators and airlines have people and stuff to move and schedules to keep. That means flying cross-country in all but the worst weather. Your job then is to learn the system to be a good IFR cross-country pilot. Therefore, it is during the VFR cross-countries when you should be preparing for IFR training. Take this time to master the basics and your control of the aircraft. Since IFR in the real world is cross-country flying, you may wonder why the FAA requires only one long cross-country flight for the instrument rating.

What is so similar about a VFR cross-country that makes it such good training for IFR? Take a look at the similarities. To fly VFR, you prepare a navigational log (keep it simple), check the weather, file a flight plan, do a preflight, takeoff, fly a departure route, cruise, navigate with radio navigation and deduced reckoning, descend, make an approach, and land. Guess what? You do the same thing for an IFR cross-country. The difference is that the flight plan is mandatory in order to obtain the IFR clearance to enter the system and land at your destination. You won't find much pilotage in IFR unless you have good VFR conditions. Then again how much do you use pilotage after you have had your private for a while?

IFR is in many ways easier than VFR. All your magnetic courses, distances, departures, arrivals, approaches, airport diagrams, and minimum altitudes are published for you. In VFR you have to request flight following; in IFR you are guaranteed positive control. In VFR you have procedures to strictly follow when you desire to operate in ARSAs and TCAs (Classes C and B); in IFR your clearance just carries you through. In IFR you get traffic separation in military operations areas (MOAs). Once you get the rating, why would you ever fly VFR in unfamiliar airspace again?

There is much made of the great secret of successful instrument flying: the development and implementation of a good instrument scan. This usually takes the form of hub-and-spoke rapid eye movement where you center on the attitude indicator and then whip your eyes as rapidly as possible to all the other instruments, returning each time to the hub. What an eye strain. Even if you try it, you won't be able to keep it up for long because your concentration moves from flying the airplane to moving your eyes. You may see the instruments, but you certainly won't be reading them.

Like the manifold pressure/propeller control explanations mentioned earlier, the handbooks give you some inadequate justification for learning this scan followed by the qualifier that it really doesn't work. Knowing this, why hasn't anyone come up with an alternative? Our current training still encourages the teaching of this method of eye movement for aircraft control which everyone acknowledges is impossible to keep up, doesn't work, isn't used by professionals, and therefore will not be used by students. Amazing.

You are also told as you begin your instrument training that there are certain primary instruments to concentrate on for all your various flight conditions. Okay, which is it? Are there certain instruments that are primary for every maneuver, or should you move your eyes equally over all the instruments? If the scan centers on the attitude indicator, isn't that the primary instrument for everything? If not, why does the scan center on it? If the primary instrument is other than the attitude indicator, why doesn't the focus of the scan keep changing to center on the primary instrument?

Dare I make it worse? You will be told that one of your major faults during training is fixating on certain instruments. At this point, you will go back to the simulator to correct your problems, using a scan that does not work, on a machine that does not accurately depict actual flight. This is the point where you should be ready to throw in the towel because you can't possibly win. The reason that you have no idea where to look and why you can't control the airplane well is because your training is so contradictory that you don't believe any of it, so you try to figure out where to look on your own. This can be costly and will greatly delay your training.

When you are learning to control an airplane with the instruments, try to develop a sense for which group of instruments to emphasize for any particular situation. This is what you will teach yourself to do eventually anyway, so why not start out that way? My idea of a scan is to take in and process the information from all the instruments to see the total flight profile in your mind. Instruments are like pieces of a puzzle, and it is up to you to complete the picture. There is no primary instrument for any situation, rather there is a small group of instruments that provide more immediate information, backed up by trends relayed to you by the other instruments.

You actually started developing an instrument scan on your first private lesson. You developed that scan to be able to fly relatively well in VFR conditions. How else could you hold or change your altitude, airspeed, or heading? Your scan continued to build when you incorporated navigational instruments, primarily the VOR. You also had to be able to control your airplane solely by reference to instruments before you could get a private certificate, so you have to have a scan already developed. Why throw that all away with some new and bizarre technique that doesn't work?

For the instrument rating, you are asked only to learn

some new procedures, fly to higher standards, and learn to use some new navigational aids. This will take lots of practice and is not easy; but there is nothing really all that new, so there is no reason to learn a new scan. What you really should be doing in instrument training is learning to improve the instrument scan you already have. When you can analyze the flight instruments, interpret the navigational instruments, and control the airplane, you will have a good scan.

You come to instrument training with a VFR scan where the instruments are backed up by reference to the horizon. What you now face is the challenge of flying pretty much the same basic things as before and learning some new procedures, all this while flying without a reference to the horizon. To do this, you need your imagination. The reason instrument flying is more art than science is because of the huge impact that your imagination has on your success. You have to see what the plane is doing in the air through the instruments, what some flight schools call "vision through instruments."

Your method of scanning is not to blame for lousy airplane control, even though it will be blamed, for that is far too simplistic an explanation. When you hear talk of scanning, think of visualization. What you have to develop on your own (because no one can teach you this) is the ability to take in the instrument readings and, through your imagination, visualize what your airplane is doing through the air, and through your own analysis of this information, make the airplane do what you want it to do. That is the real essence of instrument flying.

The link between scanning and visualization is never properly made in the mind of the new instrument student, and that is the fault of our training system. You therefore must make that link yourself. I worked at a school that prided itself on talking about vision through instruments, even though they could not make the link in the minds of the students who still moved the needles around to please the instructors. That philosophy is on the right track, but we have to take it one step further for it to make sense.

Try to think now of "purpose through instruments." What is the goal of each procedure? What parameters will allow you to reach that goal? What planning is required to set up the parameters? What instrument readings are required to match your planning? What is required of you to get those instrument readings? By working the plan backwards from the goal, you know in advance what you have to do to get there.

Since even the backwards is backwards, in our current system you will be trained to start with your current instrument readings and without any purpose try to accomplish your goals. This won't work. You end up practicing procedures without purpose in nonrepresentative environments of actual flight, thus taking your training progressively further from the objectives of the rating.

If we could get out of the simulators and out of some of the flight time rehearsing bad procedures without any purpose behind them, and just sit around discussing why we do things in IFR flight, students would gain far more insight into the real IFR system than they can through our current peripheral training. Before you can understand instrument flying, you have to understand the instruments. Dive deeply into how each flight and navigation instrument works, how that relates to controlling your airplane, and how that relates to what you are trying to accomplish. Your purpose will then have direction.

I have known many instrument students to shoot pretty decent approaches without the slightest clue as to why they did the things they did. This can work for a while; eventually however, a new situation would arise that would cause them to fall apart. I had no idea of the reason or logic for any procedure when I got my instrument rating. I passed the flight test because I had been through all the motions of those same approaches so many times, and I got through the oral because none of the questions were new to me. I still had no understanding or purpose. My early students had no purpose either. It was then we stopped flying and sat down together to make sense out of the system. Doesn't it seem just a little silly now to blame sloppy aircraft control on a bad scan?

If you take a look at the FAA *Instrument Flying Handbook*, you will find precious little attention paid to the thing you think is the most important and can't wait to try: the approaches. What you will find is a great deal of attention paid to the basic control of the aircraft, IFR regulations and procedures, and a detailed description of the instruments. By skipping over most of this information that you are not interested in learning, and going right for the fun stuff, you miss the entire development of how and why things like approaches come about and how they are used. It's like trying to understand the ending of a mystery novel, having skipped most of the middle chapters. This happens when you get a bunch of baby instrument students together and all they talk about are approaches. Thus glorified in your mind, you put way too much importance on them. Students always brag about what a mess they are making of their approaches, which gives you permission to screw them up and still be able to brag to your peers.

An approach is nothing more than a few flight skills thrown together. You will float past the basics without any retention because you are not impressed with the importance of basics so your approaches when you get to them will be a mess.

The basics are called "attitude instrument flying." This term is meant to refer to the attitude of the airplane. However, what is really needed is a massive correction in the attitude of the student. It may be too much to ask, but it would be really nice to have you come back from a lesson and brag about how well you controlled the airplane. Then when you did get to approaches, you wouldn't have a reason to complain like everyone else. Take the time to learn all you can about the instruments and the IFR system from the training manuals and then do everything you can to master the basics in flight.

Once you have a reasonable idea how to fly the basic

attitude stuff consciously, you have to move up another level and learn to fly by instinct. When there is nothing else to think about, most people can fly the basics pretty well. That isn't good enough because there is so much going on in a busy IFR environment that you don't have the space in your brain to spend all your energy concentrating on pitch and power settings, headings, and altitudes while neglecting everything else you have to do. Everyone has a threshold of overload. For instrument students, that seems to be reached during holding patterns and approaches.

There are only so many thoughts you can hold in your brain at one time; therefore, the only way to add more information to the conscious mind is to move some of your responsibilities from there into your realm of instinct. Your attitude instrument flying has to become sharper without having to think about it. If you have to think about how you are going to make the airplane descend on the approach, you will not have the ability to think about when the aircraft should descend on the approach, and you will blow it.

Errors like this will haunt you throughout your training, they will never go away, nor will you be able to master any complex procedure until you develop your instinct. The first level of errors common to students without instinct is the inability to hold or change headings, altitudes, and airspeeds. That is easy for the instructor to correct, or so it seems. For example, by telling you to correct your heading, the instructor has dealt only with the symptom, not the problem.

When you have to concentrate on basic flying and it occupies a significant portion of your brain, and you have to fly complex procedures at the same time, the resulting overload will cause a brain blockage which will bring about the second level of errors. This second level includes: the inability to hear or understand the directions of controllers; the inability to set up the approach on the navigation instruments; leaving vital instruments out of the procedure; the inability to initiate an approach; missing any step in the approach sequence; the inability to figure out and enter a holding pattern; wandering from the navigational track; wandering through decision heights and minimum descent altitudes; forgetting to descend at all; a complete loss of position on any part of an approach or hold; the inability to determine the strength and direction of the wind; and an inability to communicate clearly on the radio. This is by no means a complete list.

Most instructors will only treat the symptoms of the second level of errors because that is all they can see. They are not taught to treat the real causes which are in the first level of errors. The second-level errors will never go away though until the underlying first level of errors has been dealt with by the student by making basic attitude flying instinctive.

Okay, how can you develop this instinct so you will just know what to do with the airplane? Remember way back in your first few hours when we talked about flying and reconfiguring the aircraft by specific pitch and power settings? Remember also that if you were consistent in technique and always used the same settings, you could vastly shorten your training time because you didn't have to relearn how to fly every single flight? Remember that once you learned the basics it freed you to master all your other techniques? Well, the same exact thing applies to instrument flying. This proves once again that all knowledge in flight training is cumulative, you can't forget anything, and that the system continually falls apart because it treats each new rating as brand new knowledge. Nothing ever changes.

Those pitch and power settings I use for the private are the exact same ones I use for the instrument rating. All my private students are learning their instrument pitch and power settings from day one. I use the hood time during the private to perfect basic attitude flying. When they are ready to start work on the instrument rating, my students are already halfway through the program. By using the same aircraft for private and instrument work, they save even more time by not having to learn a new aircraft while going for a new rating.

The only difference from the VFR settings is that the pitch will be a specific amount of degrees measured with the ball on the artificial horizon instead of fingers on the real horizon. As far as the aircraft is concerned, the pitch is identical. With a little foresight, any of you can plan a course of action like this, which may save you dozens of hours and thousands of dollars over the course of your professional training. All you have to do is be willing to question and challenge the system and find an instructor who will train you the way you want.

Here are the instrument pitch and power settings for the Cessna 172 that I use. Takeoff is full power and about eight degrees or two ball widths after rotation. A cruise climb puts the ball on top of the artificial horizon (the old and better name for the attitude indicator). Straight and level is 2400 RPM with the ball centered on the artificial horizon. All holdings, level flight between step downs and circling approaches are the same as the VFR downwind setting of 2100 RPM, where the ball is one-third above the artificial horizon. All descents will have the ball touching the artificial horizon line from below, with power settings of 2100, 1900, and 1500 RPM for cruise descents, ILS descents, and nonprecision descents, respectively. The attitude remains mostly unchanged for all descents; it is varying the power that regulates the airspeed and rate of descent. These are all the settings you need for all your instrument flying in a 172, except for unusual speed requirements from ATC.

If you can learn the settings for your airplane so well that you never have to think about how to configure it for any instrument situation, it will have become instinct and your brain will have been freed to concentrate on everything else that is happening. If you fail to learn your settings and have to struggle to remember them every time, you will never progress any further nor ever become an instrument pilot. It is just that simple. Just as instructors of private students leave you to rediscover how to fly each maneuver every lesson, many instrument instructors leave you to rediscover how to fly each instrument situation. It is up to you to insist on regular pitch and power settings.

Sometime after the rating, you will learn to quickly devise pitch and power settings for all the airplanes you fly. There is no reason to wait until you get the rating to figure out your settings, or to waste the extra time and money it takes to get there, when you could have memorized all your necessary pitch and power settings before your first instrument lesson. Put a copy of your settings on your bathroom mirror, on the dashboard of your car, hang them at work, repeat them at lunch — do whatever it takes so that you will have instant recall of the proper setting at the appropriate time in the airplane.

You can save about a third of the normal time it takes to get the rating if you have your settings memorized before you start training. Take the 50 hours of VFR PIC cross-country time to perfect them and you will really be ahead. Most of you are too lazy to memorize your settings and too anxious to get to approaches, so it usually takes a few embarrassing lessons of botched procedures to convince you to learn your settings. We then have to go back to basics to make them instinct. After that, you will be ready to begin your real training.

When you first learned your pitch and power settings for the private, you found that you really didn't need any instruments other than the altimeter, compass, and tachometer to fly the airplane. Your outside visual reference to the horizon and the feel of the airplane supplied all the other information you needed to fly. You can fly by the seat of your pants because that is where you make the best contact with the airplane.

To be a good instrument pilot, you have to throw most of that out, ignore everything you feel for interpretation of the airplane, and rely strictly on the instruments. Nothing new here. What the books don't tell you is that to be a really good pilot overall you have to be able to continually switch from IFR to VFR flight and back. Seldom will you fly an entire flight in IMC from directly after takeoff to the minimums on the approach. Yet if you train exclusively with the hood, that is exactly what you will think. Hood flying is just a game; cloud flying is real.

In VFR conditions you have the comfort of the horizon to assure you that all is well. However, when you are in the clouds, you engage in a mental wrestling match between your fears, your feelings of what you think the airplane is doing, and the reality of what the instruments are telling you. Fear brings out your primitive instincts, which is to try to figure out what the airplane is doing by its feel, just like you learned to do in VFR conditions. You learn from cloud flying that it takes positive rational action on your part to control your emotions to read and believe your instruments. Training in the clouds allows you to gradually dispel your fear and replace it with cool logic.

Whenever you go into the clouds, you have an initial adaptation time to both get used to flying exclusively on instruments and to accept that you are in the clouds. Once you go in and find that the airplane will continue flying, you get used to the clouds and do your job. When you are constantly switching from clouds to VFR, this whole mental adaptation process must be repeated. Every time you pop out of the clouds, you immediately regain your VFR horizon assurance and lose your emotional cloud adaptation. The more you fly in various cloud conditions, the shorter your adaptation time will become and the better an overall pilot you will be.

I do not distinguish between IFR and VFR as far as piloting goes, for it is all part of the world of flying. A good pilot should be equally comfortable in all environments.

There are illusions in flight that can only be experienced in the clouds: changes in acceleration, lift, up and down drafts, slips and skids, attitude and airspeed changes that can only be appreciated for their illusionary effect in an environment without reference or sensation of forward movement. You have to learn to recognize the deceptiveness of these illusions and fight their effects with rational and logical interpretation of the instruments. You won't get this training from a hood because sooner or later you will get a glimpse of the ground or the sun will move across the panel. Needless to say, sitting upright and still in a simulator is worthless for this training.

When you fly in the clouds, the level of light around the aircraft is all the same. There is no sensation of speed, movement, up or down, left or right. You are terribly alone the first time you enter the clouds, instructor notwithstanding. Over time you will learn to love the clouds. This will come after you have learned how to force your brain to control your emotions.

You also have to learn how to override your senses. We rely overwhelmingly on sight for our sensory input. Since seeing is believing in our culture, we have lost the specialized acuity available in our other senses. Once your sight is effectively lost (like when you fly in the clouds), the most amazing thing happens. All your other senses come rushing in to fill the sensory void, and you won't know how to process the information. You will hear the most minute change in the engine. You will feel every movement of the airplane. You will smell every smell and taste every taste in the airplane. Every dormant sense is now very much alive and lying to you. You have to learn to override this sensory input and rely on what you see from the instruments because the truth is that seeing really is true knowing.

A final word about simulators: They are fun to play with; they help if you just want to practice some procedures and build some skill; they sometimes aid your training because you can stop the action and discuss where you are and what you want to do; they are in reality nothing more than a supplement to learning because all they can teach you is how to move needles in a machine on the ground, and that is only a tiny portion of the knowledge you need for the rating. I do not believe that any time spent on a simulator should be able to be logged or considered adequate for meeting any instrument requirement for either training or currency. I place so much value on the ability to train in the clouds that the instrument rating requirements should be changed so that a minimum of three hours and five approaches, including, if possible, one missed approach, must be accomplished in actual instrument conditions before taking the flight test. If this is impossible, then substitute 15 hours of night training, although this is far less desirable.

Nothing is guaranteed to give you more frustration and aggravation than the automatic direction finder (ADF). Not because there is anything inherently difficult about the instrument (because we learned in the private that there isn't), but that the standard instrument training method for using the ADF so complicates this, the simplest of navigational instruments, that it is rendered effectively useless in the mind of instrument students. You learn to hate the ADF and all procedures associated with its use. A chorus of groans rises from any group of instrument students whenever the ADF is mentioned, such is the collective misery imposed on students by our current training system. How long will students have to suffer through the ADF?

The biggest problem is that you never know when you are using it correctly because in instrument training there is no visual back up, so you can't teach yourself how to use it. You should have learned by now that flight instructors only tell you when you are doing something wrong; the rest you have to teach yourself. If you don't know when you are correct, you can't do this, which is one reason the ADF appears impossible to master. The ideal way to get this reinforcement is to learn the ADF during your private. Failing that, spend some time without the hood in VFR conditions until you have the solid visual back up to understand the instrument.

We know the VOR is more complicated than the ADF to understand, so the only way the ADF could cause more trouble for students is to needlessly complicate the explanation of how to use the ADF beyond that for the VOR. The two best books on IFR flying are the FAA *Instrument Flying Handbook* and the *Airman's Information Manual*. However, neither of these have even a remotely plausible way of understanding how to use the ADF for IFR flight. Check out the explanation of the ADF from the *Instrument Flying Handbook*. Whoever wrote this method of understanding the ADF had to be doing some serious hallucinogens. Take a look at a sample of their ADF orientation:

"The ADF needle points *To* the station regardless of aircraft heading or position. The relative bearing indicated is thus the angular relationship between the aircraft heading and the station measured clockwise from the nose of the aircraft. A bearing is simply the direction of a straight line between the aircraft and station or vice-versa. A true, magnetic, or compass heading is measured from the appropriate north [I didn't know there were inappropriate north's], and a relative bearing is measured clockwise from the nose of the aircraft. Thus the true, magnetic, or compass bearing to the station is the sum of the true, magnetic, or compass heading, respectively, and the relative bearing. True bearing to the station equals true heading plus relative bearing. Magnetic bearing to the station equals magnetic heading plus relative bearing. Station-to-aircraft bearings are true, magnetic, or compass bearings plus or minus 180 degrees. The relative bearing shown on the ADF dial does not indicate aircraft position, the relative bearing must be related to the aircraft heading to determine direction to or from the station."

Yuk!

What a mouthful. It is all true of course, but how are you supposed to understand it and use it in an airplane? Try my way. To find a bearing *To* or *From* a station, superimpose the ADF arrow on the directional gyro.

There is only one situation I know of where you have to figure stuff out in the air using the ADF: When you have to identify an intersection or fix along an airway or approach using a magnetic bearing. Now, you could drag out a piece of paper and try to calculate your relative bearing (which you know doesn't exist because bearings aren't relative to anything), by using that formula where magnetic heading + relative bearing = magnetic bearing. This can be tough in a bouncy airplane in the clouds. Anyway, the formula (besides being vastly overcomplicated) is in the wrong order. Why would anyone have instrument students memorize a formula where the solution — the magnetic bearing — is published on the charts?

Well, there is a short cut. Always think of relative angles, for the only thing an ADF can measure is angles. Say, for example, you are flying an airway with a published course of 25 degrees. The published bearing to the station is 100 degrees. Your relative angle is the number of degrees in between them; in this case it is 75 degrees. As soon as the ADF arrow indicates a 75-degree angle from the nose, you are at the intersection. Sure beats that formula, doesn't it? This works fine for stations on your right.

For stations on your left, you just have one extra step. Instead of using the degrees published on the card, you have to count back from "0" on the left-hand side. In this example, you have a course of 300 degrees and a published bearing of 180 degrees to the station. The relative angle on the left is 120 degrees. Counting back 120 degrees from 0 on the left side, the arrow would indicate 240 degrees on the compass card when you were at the fix. Once the concept of relative angles catches on, we will have ADF cards written with 180 degrees on both sides with zeros at the nose and tail. When that happens, we won't need any extra steps for stations on the left. Try your own examples. Once you find the relative angle, the rest is easy . . .

Okay, you purists, bring on your questions. What if there is a wind? With a wind just use your current heading instead of the course. What about the needle wiggling? Or what if you have to keep changing your heading to maintain your track? Just take the average of the wiggle and your wind correction; no ADF intersection is all that critical. If you are worried, then use another way to identify the intersection, like having the ground controller call you when you get there.

Unless you are resistant to change, or really like playing with formulas that derive numbers too precise to use in flight, my method should make sense. If you develop a better way yourself, please let me know. I am always looking for better ways to fly and can use the help.

Tracking a nondirectional beacon (NDB) course, especially on an approach, is an absolute nightmare. Here is some of what the FAA handbook says:

"When a definite change in azimuth shows that the aircraft has drifted off course, turn in the direction of needle deflection to re-intercept the initial inbound heading. The angle of interception must always be greater than the number of degrees of drift. The magnitude of any intercepting turn depends upon the observed rate of bearing change, true airspeed, and how quickly you want to return to course. As you make the turn with the heading indicator, the azimuth needle rotates opposite the direction of the turn, and the interception angle is established, the needle points to the side of the zero position opposite the direction of the turn. When the needle deflection from zero equals the angle of interception, the aircraft is on the desired track. You remain on the desired track as long as the azimuth needle is deflected from zero opposite the direction of drift correction, an amount equal to the drift correction angle. If the needle moves further from the nose position, the drift correction is excessive. If the estimated drift correction is insufficient, then the needle will move toward the nose, requiring a further correction to regain track. When tracking outbound, a change of heading toward the desired track results in needle movement further away from the 180-degree position."

People who wish to retain their sanity should neither try to understand this nor ever try to implement this in an airplane. What you will do is play mental games trying to divide your attention between flying the airplane and figuring an interception angle from the magnetic bearing and the relative bearing, and then holding the needle opposite the magnetic bearing the amount of your correction while trying to calculate as the arrow wiggles when you will be back on track. This is provided, of course, that you can hold the correction constant, which you usually can't because too much is happening. You then have to recalculate what would be opposite your present heading to see if you have corrected enough and are on track. During all the doubling and subtracting of old and new relative and magnetic bearings, you should get lost.

When the needle moves further away as you add in more correction, you assume the wind has picked up. You then add even more correction which only aggravates the situation. You end up turning in circles and have absolutely no idea why because you thought you were following the procedure. Does this ring a bell with anybody?

To solve this problem, let's try something different. We are going to change our thinking and develop a new strategy rather than try to mold our brains to understand inadequate procedures. I'm kidding you. What you should do, though, is give up any idea of doing in-flight math whenever you track an ADF course because you don't ever have to and it will always confuse you if you do. Our new strategy is to always center your thinking around your "reference heading." Your reference heading is the published course you are trying to track. As long as you fly your reference heading, you will always be either on or parallel to your course; you will always be going in the right direction; and you will never have to do any math. Promise me that until you become an ADF expert, you will never look at the ADF while tracking a course unless you are riveted on your reference heading.

If the wind has blown you off the track and you are flying parallel, the arrow will always point your way back to your track because it always points into the wind, and you will never have to do any math. For example, suppose that while on your reference heading, the arrow points to the left. Your track lies to the left. The easiest way is just to correct by turning slightly to the left. Turn back every so often to the reference heading to check if you are back on track. When the arrow points to "0" while you are flying the reference heading, you are back on track, and you never have to do any math.

This is a great method if you are close to the station

and only minor corrections are needed, like on an NDB approach where the distances are so small that flying the course is the best way to handle the limited impact of wind and the inaccuracy of the instrument. The last thing you want is a big correction and in-flight math on approach, so if you can nibble at the drift and check your track with the reference heading, you can concentrate on the rest of the approach.

For those of you who want a more sophisticated method for use when you are miles from the station, there is a lot of wind, or you need a big correction, this is a method I took from the handbook but explain it here in English. The secret to bearing interception comes right from the words of the FAA: "When the needle deflection equals the angle of interception, the aircraft is on the desired track." Unfortunately, these words are hidden in confusing verbiage like buried treasure, so you could easily read right past them for years.

Anyway, phrased in my terminology, it translates: "When the relative angle equals your correction angle, you are on course." Say you want to fly to a beacon on a particular bearing. Turn to that bearing; it is also your reference heading. The arrow points, for example, to a relative angle of 20 degrees to the right. If you turned only 20 degrees, you would fly directly to the station on the wrong bearing, so that won't work. The usual practice is to double the relative angle, so turn 40 degrees to the right.

Anytime you turn past the bearing to the station, you put the station on the other side of the nose, so the arrow swings to the other side of "0." Your correction angle is 40 degrees. When the arrow indicates a new relative angle of 40 degrees on the same side as the station, you are on your desired bearing. This is the perfect way to intercept the final course of an NDB approach where your interception angle will be either 25 or 30 degrees from a vector, or 45 degrees if you are flying as published. What could be easier?

This method works for any interception either *To* or *From* a beacon. When you are flying *From* the station, the relative angle will be relative to the tail; everything else is the same. Yet another reason for dual-side, 180-degree ADF compass cards. Remember that you never have to do any math, just look at the angles. So much for the nasty ADF.

Back in your private, you were bored with weather until we made it live. The same applies here — the difference is that we can really make it live! If you sit down with your dry charts and prognostics that are reporting altitudes far above where your trainer can fly, you will drown in the tedium. We have to give the weather phenomena that you study meaning and significance. The best way to do that is to have you fly in it. Standard weather training prepares you to fill in dots on written tests describing charts that FSS people never let you see, and it prepares you to answer esoteric, pet questions by FAA examiners on checkrides. None of this will help you after your checkride as that knot builds in your stomach when you approach clouds where you have no idea what lies inside because you didn't receive any training in real weather before you got the rating. What you get from most training environments on weather, like the rest of the rating, is only about 25 percent of what you need.

Most of you will have knowledge about the weather when you get the rating. However, what is critical for you is to have experience with the weather; which is of course why this isn't even required by the FAA. Everything is backwards. No one can say with absolute certainty what you will find in that cloud up ahead. The problem lies in the unlimited number of variables that combine to form the weather out there. No computer can include every possible factor necessary to accurately predict the weather all the time. Your trainer certainly won't have radar. Even if it did, there would be limitations. Therefore, the only way you can predict what might lie in the clouds is to build a reservoir of practical knowledge from as much varied weather experience during training as you possibly can.

You need to develop personal boundaries of what you can safely handle. Those will change with greater experience. When you become a professional, the question changes from should you go to how and where can you go to still get the job done. During your training, you should try to continually raise your comfort level with weather.

Start with low stratus, particularly coastal stratus if you can. This stuff is light and smooth; it is safe and fun to fly in, and the tops aren't that high. You can move from there to fair weather cumulus, where you will bounce more and have to transition from visual to instrument flight. Please avoid any significant build-ups, however. Always be aware that many VFR pilots have a funny idea about how far 2000 feet is, so make sure you look for traffic as you pop out. Moving up to build even more confidence, try multiple stratus layers soaked with rain. You will be in the clouds for extended periods of time which will be very exciting.

You will also realize that the engine keeps going, usually, and the airplane keeps flying despite being poured on by driving rain. Flying from rain-soaked and windswept runways will give you new techniques for takeoff and landing. What you do not want is rainy cumulus that will bounce you severely and frighten you. You will not enjoy the experience and you certainly will not learn anything. Massive lines of thunderstorms mean an automatic ground school. However, with isolated thunderstorms you could still fly and gain valuable insight in how to work around them with the help of ATC. This is all designed to build your experience and your knowledge of weather, for you will need them both.

You need to learn your own limitations. Are you the stout-of-heart type who can calculate the perfect wind correction angle on the outbound leg of a holding pattern while circling in a big, dark cloud? That happened to me with a student once. Although this person was quite proficient with the procedures and flew very well under a hood, the pelting rain completely freaked him out. It is hard to discuss holding patterns with someone who can't remember his own name. Here is the perfect example of why endless practice of procedures on a simulator or under a hood doesn't do one iota of good if the first time you get in some real weather, you turn into a vegetable. This particular cloud was amazingly dark and rainy. The funny part was that it was very smooth to fly in, so it actually looked far worse than it was.

My student was a rated instrument pilot who had come back for some recurrent training. He had his own airplane and had received all the best procedural training the FARs required and my school had to offer. This was glaring evidence to me that procedures are meaningless if you do not receive the training in actual conditions so you are able to use them. Since cloud flying is not required, it is your responsibility to get this training from your instructor. If your current instructor is unwilling to train you properly to operate safely in the clouds, then it is incumbent on you to get another instructor.

The weather system itself has limitations. There are only two choices of information:

1) a report, which is already old news
2) a forecast, which is at best an educated guess of what might occur in the future, maybe.

From this information, you have to decide whether you want to take on the weather. Since you will never know exactly what you will find, especially once you are in the clouds, the best thing you can do is to look for trends, especially in long flights where extended travel times and changing terrain can make for vast changes in the weather. The next thing to do is constantly recheck the weather en route and at your destination. In weather flying you really need a plan B all the time, and then a plan C for plan B.

For most of you, the only realistic IFR training flight you will take is the long cross-country, so you have to learn all this stuff in just one flight. Do yourself a favor and take your instructor for a few long cross-countries before your flight test. While in the clouds on a long flight, get your instructor to challenge you with questions on various emergency scenarios. You will find that just thinking of losing your comms while you are flying in the clouds brings out a completely different feeling in your gut than dryly reciting the lost comms procedure in the FARs while sitting on the ground.

When I get a briefing, there are two things for which I pay particular attention: One is the satellite picture on radar (because that gives me the best overview of the weather and what I can expect over time), and the other is the forecaster with the most experience at that particular Flight Service Station. If you find some old curmudgeon hiding in the back briefing station, with ever so slightly bloodshot eyes and a telltale red nose from a long life filled with good whiskey, and he comes to the desk and looks you over slowly through round, gold-rimmed spectacles, if you are that fortunate, you are going to get a *briefing*! Computers are limited in their forecasts because they can only work within the finite number of known variables in their programs; therefore, they can only come up with what is predictable. If you want the whole story, you need to know what is unpredictable, and the only way to get that information is to draw on the longtime experience of a real person who knows the area very well.

We will forever lose a certain quality of service if the FAA ever takes away the personal briefings. It is far more important for instrument students to learn how to ask the right questions from briefers than it is to try to learn all the bits and pieces of the many complex charts that are jealously guarded by specialists who won't let you get any closer to them than a flash on a screen. We don't ask briefers to fly our airplanes for us; why should we be asked to read their charts for them? Your training time would be better spent gaining a general knowledge of the charts and the information they contain, and then going flying to see how the reality matches the predictions.

It is time to play with the FARs and give you some questions to ponder. You all know that you are required to file an alternate airport whenever your destination within plus or minus one hour of your estimated arrival is forecast to have a ceiling less than 2000 feet or visibility less than three miles. You also know that you are encouraged whenever possible to take advantage of the "tower en route control" (TEC) program, a service whereby you fly solely within the airspace of specified approach control areas. Towers have nothing to do with this, so why not rename it "approach en route control?"

Anyway, when you file a TEC plan, the FSS specialist will only take the information on the top half of the flight plan form. Now, the space to file an alternate is on the bottom half of the form. The question becomes: How do you file an alternate airport when you are on a TEC clearance and your destination has weather that requires an alternate? Also, how many of you go on lessons using a TEC clearance without even having an alternate planned? As pilot in command, you must follow the FARs despite any contradictions in the programs because you are still ultimately responsible. It is bad enough that the FAA will gladly violate you for any infraction of the FARs, but this situation is entrapment. They have actually created a program that when used as intended causes violations of the rules.

The only way to file an alternate is to file a full IFR flight plan, which causes the very congestion in the system that the TEC program is designed to alleviate. Because of this blatant contradiction, the only way you can be legal on a TEC clearance is to fly only during VFR . . . when you don't need it. You know the rule concerning alternates but completely ignore it because you and your instructor go happily off with the blessings of the FSS person who accepted your flight plan. This problem exists because of the same, old, tired, fundamental flaws everyone learns in their earliest hours of instruction. No one ever thinks to question his rote training, so now we have a system of blissfully ignorant pilots and briefers gladly violating the rules.

It gets worse. Suppose you have a complete radio failure in IMC conditions on a TEC clearance and your destination goes below minimums. After you miss the approach, where are you going to go? What altitude and route are you going to fly to get there? How is ATC going to separate you from other traffic when they don't know where you are going next? It is your responsibility to file an alternate when required, not for the briefer to baby-sit you and ask you for one. Learning the rules by rote is never enough; you have to understand the system to remain in compliance of the FARs, even when the system is stacked against you.

When should you cancel your IFR clearance? ATC controllers love for you to cancel as soon as possible because it relieves them of the higher responsibility they have for you while you are IFR. You will have a tendency to want to cancel your clearance early because you think that's what the pros do and you want to impress the controller. You still don't want to look like a student. However, once you cancel your clearance, all that airspace that you would have been cleared and routed through now pops back up, and the controller will tell you to remain clear of it. You may have lost your usual ground references under the clouds, so this could be a problem.

You also have to maintain your VFR cloud clearances and visibilities. Do you have a clear VFR path to your destination? Have you thought this through or were you just in a rush to cancel for absurd reasons of image?

You will have particular problems with your area so ask your instructor about them. I teach on the West Coast where the low stratus in my area tends to top out right at one base level of a TCA (Class B). The controller would say "report canceling on top." By trying to be nice to the controller and complying with the request as soon as we broke out, students would unknowingly want to cancel within 1000 feet of the clouds, in the TCA, or both. The controller will gladly cancel your IFR clearance as soon as you request it. If that gives you two violations of the FARs, then that is your problem. Remember who is pilot in command. Just because a controller wants you to cancel does not mean it is either safe or legal to do so. You can never be required to cancel IFR so stand your ground. Here is the sequence of how you can get in trouble.

"Piper 54321 report canceling on top."

"Roger 321 will cancel on top." (What you should say is that you will cancel when 1000 feet over the top with a TCA clearance from the controller.)

"Piper 321 would like to cancel my IFR clearance."

"Roger Piper 321 cancellation accepted, say altitude climbing through."

"321 is climbing through 3300 feet."

"Piper 321, could you please give us a tops report."

"Roger approach, they appear to be at 3000 feet."

Great, now it is on tape. Just wait until you get back from this lesson. Being stupid is bad enough, but you don't have to tell everyone. When a student beats me to the push to talk switch, I have been known to report to approach that my student has not yet learned cloud distance estimation and that in my expert opinion the cloud directly under us is actually 2300 feet. All the rest are about 3000 feet, however.

A big distinction in the FARs is made between precision and nonprecision approaches. In the civilian world, our precision approach is anything with an electronic glide slope. The precision comes from the ability to track a constant descent pathway to the runway. What we have then is the ILS approach for precision, and everything else gets lumped into the nonprecision category. Are they implying by calling these approaches nonprecision that you are not responsible for flying them as precisely? Naming certain approaches either precision or nonprecision is misleading and meaningless because the degree of precision on any approach depends on the accuracy of the navigational aids, your ability to use them, track them, and fly the airplane. The degree of precision between a localizer approach and an NDB approach varies greatly, yet they fall into the same category. You can be very precise in holding your altitudes and tracking despite this unfortunate name.

The language is what is not precise which leads to misunderstandings and misperceptions. For consistency, let's rename our categories with a concept you already know. When you first learned to land, you were told to hold your heading, airspeed, and rate of descent constant so that you could fly a "stabilized approach." Since the ILS requires the same thing, why not also call it a stabilized approach? Therefore, a nonstabilized approach would include anything that requires a stepping down from various altitudes.

To make a landing under IFR you need three things:
1) the required flight visibility
2) the runway environment in sight
3) the ability to use normal maneuvers and rates of descent to get there.

Consider each one of them separately. Determining the flight visibility can be a tricky thing. Your flight visibility is how far you can see from your cockpit. There is no gauge on board to measure your flight visibility, so it comes down to your best guess.

The transissometer that gives you a range of visibility on the runway based on light intensity still can't measure what you can see from the cockpit. The weather reporter in the tower at the other end of the airport who makes the ATIS broadcast may be sitting in the clouds while the runway environment is clear for you. This means you may exercise your authority to land when the reports may not indicate that the visibility exists. You may have to defend your actions, however, once you land. The flip side of this is that some pilots try to sneak in when it is obvious either from the transissometer or the tower folks that the visibility is well below minimums.

The IFR rules are absolutes and are designed for your safety. Within that framework, there is some grey area where you can exercise your own judgment. Your goal is to land if you legally and safely can do so. What happens if a procedure requires one mile of visibility and the ATIS is calling only three-fourths of a mile? The commercial flights cannot start the approach unless they have the required visibility, but under Part 91 you can take a look. It is amazing sometimes to listen to the reported weather miraculously improve for an inbound commercial operator, and then immediately deteriorate after they land.

Anyway, in controlled air your weather observers will report the weather every hour, and more frequently if there are big changes. Those reports are based on known distances from objects visible from the tower or weather station. This creates the grey area because they are not reporting from the runway. If the ATIS reports less than the visibility required, why not ask the weather folks for a visibility check right now. The last report could be up to an hour old and things can change fast in IMC weather.

In our example, the difference between being able to legally land and the reported visibility is only one-fourth of a mile. If you fly an approach when the reported visibility is just below minimums, how can you decide if you have the legal flight visibility to land? You take a distance you know to prove it; for example, the length of the runway. If you break out of the clouds and can see the far end of an approaching 5000-foot runway, you have at least a mile of visibility. If your runway is only 3000 feet in length and you can see a full runway length between you and the approach end, then you still have a mile visibility. This grey area will only be tricky when the weather is about the minimum. When reported way above or below the minimum, your decisions are easy. The system works for you because the final decision on visibility is left to the pilot in command.

I was shooting an approach with a student where we needed a mile to land but had only the three-fourths of a mile visibility from the ATIS. We informed ATC that we would most likely have to shoot the missed approach because of it. We broke out of the rain to find the end of a 5000-foot runway in clear view. (Now you know where all my examples come from.) The conditions had changed so recently that we were the first flight able to land. We called the tower and advised them of the new conditions and our intention to land instead of taking the missed. They then changed the ATIS to one-mile visibility. This is how the system is supposed to work.

The system is designed to allow a landing based on

your particular situation. You won't learn this with a hood because it is an artificial environment with the instructor determining all the variables. The interesting thing about that approach is that this was one instance where we really reviewed the missed approach procedure. We know we are supposed to review this every time; however, the truth is that you almost always know ahead of time if you are going to make it or not because of the reported weather. By overstressing the need for reviewing the missed approach in training and then never having to use it is crying wolf, which is why pilots become complacent.

The problem comes when you are in the grey area and have to make a decision. That is the time we should stress the missed approach procedure more. The system works both ways; there are times when you can land when the reported weather is below minimums, and there are times when you need the missed approach procedure when the reported weather is above minimums.

The rules concerning the runway environment are straightforward — you either see the required items or you do not. Any one of the items listed constitutes the runway environment which gives you permission to land. Confusion, however, arises around the approach lights. The thing to remember is that the approach lights are not part of the runway environment; they are part of the approach — that's why they are called approach lights. They are much brighter than the runway environment lights so that they may cut through the low visibility to guide you to the runway environment.

There are times when you will see only the approach lights, know that the runway is right beyond them, and still have to go missed. This is an experience well worth putting under your belt sometime in your training. You will be amazed at your desire to continue your descent in direct violation of the rules in an effort to catch a glimpse of the runway environment. This is how pilots become statistics. If you can't see the lights of the runway environment, you certainly aren't going to have the required flight visibility to safely land.

The FAR concerning approach lights is so poorly written that my old flight school created and believed one of the all-time great training myths. The rule says you are allowed to descend to 100 feet above the touchdown zone elevation using the approach lights as a reference. If you then see the runway environment, you can land; otherwise, you have to execute the missed approach. The exception is that you can continue descending when you can see the low-intensity, red lights of the terminating bars or side bars of the approach light system.

Now, only the most sophisticated systems even have red terminating or side bars; and when they do, the threshold lights are only a matter of feet ahead of you. This is so you can be led to the runway via the approach lights, the terminating bars, and then on to the runway environment lights so you can land. Many students and instructors never see terminating or side bars because none of their usual approaches have such a system. Since no one at my school (including me) took the time to find out what terminating and side bars were, we interpreted that FAR by substituting the only red bars we had seen, the ones at the far end of the runway. By not understanding the system, we could twist the meaning to actually believe that the approach lights could only be used to go below the decision height if we could see the far end of the runway. I always wondered after the truth dawned on me how many lessons at the school resulted in missed approaches, where aircraft flew away over threshold and runway lights that were plainly in view.

What are normal maneuvers and normal rates of descent? I don't know for sure. Neither does anyone else I know; however, we all have our own opinions. I also believe that what pilots tell examiners are normal maneuvers and what they will do to remain clear of the clouds on an actual circling approach may vary slightly. Having taught fighter pilots to fly general aviation airplanes, normal can be a rather broad definition. My personal ideas of normal are my ILS approach speed and below, my nonstabilized descent rate of 800 feet per minute or less, and no more than

30 degrees of bank on a circling approach.

People get hung up on what you think should be normal when what you are doing is not normal. When you are sliding down an ILS, you are not at a normal landing speed, yet somehow you expect to land like a normal VFR approach. To achieve this, upon reaching the decision height, you immediately chop the power and dump all the flaps in a futile effort to make your normal landing, thereby completely destabilizing the airplane. Does this sound like normal maneuvering? You will not make your usual runway turnoffs when you approach some 30 knots faster.

Since approaching faster is normal for an ILS, why not change your definition for the normal landing out of an ILS? Runways served by ILS approaches are invariably long, which gives you plenty of time to ease back on the power, gradually add flaps, and bleed off excessive speed until your normal approach speed is achieved and you can make a landing. The only difference is that you are farther down the runway. Big deal, that is what long runways are for. If the FAA didn't want you to use that much runway, they would have make it shorter.

One of the great sources of confusion concerns nonstabilized approaches where the "missed approach point" (MAP) ends right over the approach end of the runway. How are you supposed to make normal maneuvers to a runway several hundred feet directly below you? This is one of those questions that kind of hangs in the back of your mind and never gets answered. The reason is because instructors and examiners will pull your hood at a point where you can make a straight in or circling approach, giving the impression that you never get all the way to the real missed approach point.

What you fail to learn once again is that grey area which happens in the real world when the clouds are moving about the MAP. Whether you can make a landing or not in this case is dependent on luck and timing. Since you can't use normal maneuvers to drop straight in, you think there must be no point in putting the MAP directly over a runway.

Many instructors have decided on their own that the FAA is nuts and that you can't make a landing from the actual MAP, so what they have you do is to make up your own MAP one minute or two miles earlier on the approach. The logic is that the extra time and distance from the threshold will give you the opportunity to make normal maneuvers and still land straight in. This is a brilliant and imaginative solution that not only tries to solve a problem that doesn't exist, but also creates a whole new one.

Think about the rationality of teaching students to make up their own missed approach points. Shall we make up our own runway environment and "minimum descent altitudes" (MDAs) next? What about making up our own procedures entirely? Who in their right mind would ever consider giving students permission to make up their own IFR procedures? That is the new problem. The one that doesn't exist comes from not flying approaches all the way to the missed approach points.

Keep in mind your objective — to land. The FAA has done a wonderful job of providing the instrument pilot with every possible option in which to safely accomplish your objective. Why would you short-change yourself and break off an approach early in IMC conditions with an airport that could very well be in the clear? I have seen it happen. Here is the sequence of events. If you break out well before the MAP, you will be visual and there is no problem landing. If the runway in use is not the runway on the approach, you will have to circle. If you break out before reaching the circling minimum, once again you will have no problem.

Let's move into the grey area. One of the nice things about flying to an airport below VFR minimums is that they only have one airplane coming and going at a time, which means many times you can pretty much land on any runway you choose. This means that if you break out at the circling minimums at the MAP, you have a whole airport in which to maneuver to any runway, including the one you are now flying over. The FAA in their wisdom brings you right over the airport to circle if the weather permits. The goal of approaches is to bring you to a place where you can visually make a landing and not hit anything on the way in. Taking the approach up to the threshold of the runway and the open space of an airport accomplishes that goal.

Let's take the worst case. You are on a straight in approach at the MDA. You are below circling minimums and approaching the MAP. This is where lesser mortals who have decided all by themselves that there is no way to fly to the official MAP will go missed. The FAA takes you all the way to the runway for yet another reason. You still have one more chance to land. If you break out at the MAP below circling minimums and you have one mile of visibility and can operate clear of clouds, you can ask for a "contact approach." This gives you permission to deviate from the procedure and using normal maneuvers make your own approach to a runway.

When you fly an approach to an airport without a working tower or control zone (Class D), your lowest controlled airspace ends at 700 feet AGL. Since most approaches bring you in below that, if you have your one-mile visibility and fly clear of clouds, you can make any approach you want because you are VFR. You won't even have to ask for the contact approach. Most of you will never learn this in your training because you use a hood and shoot multiple approaches without landings until the end of the lesson. The goal of your training becomes flying approaches and not making landings.

That is why the rules on landings don't make any sense to you. Why else would you accept the idea that you have to leave the approach a minute before you have to leave the approach? You must not memorize rules you do not understand and have never used in the environment for which they were designed. You must train in the clouds to see how the rules can safely give you every chance to land. For you professional aspirants, can you imagine telling a future boss that you couldn't deliver the goods because you broke off the approach a minute early, only to find out later that the airport was well above minimums?

IFR en route altitudes are pretty straightforward: fly at the altitude assigned by ATC at or above the minimum en route altitude (MEA). For you off-the-airway folks, you need to be 1000 feet above the obstacles within four miles of your course, or 2000 feet in mountainous terrain. Tell me, do you bring along VFR charts and draw four-mile wide lines on either side of your course to insure compliance? I have a feeling that the majority of aircraft in this category fly high enough in cruise not to worry about this regulation.

A question remains, though, about those VFR charts that used to be your best friend. It is easy in IFR training to get lulled into false security from the babysitting of your instructor and ATC. You are being watched over by two people who know the training area intimately. You get used to training without your VFR charts, which are left back in your car. Once again your training distorts the reality. Since no flight is 100-percent pure IFR unless you are capable of blind takeoffs and landings, it behooves you to carry the appropriate VFR charts. Once you go off on your own, it will be really nice to know what is under those clouds, especially if you have engine trouble. What if you have a full electrical failure where VFR conditions are within your sight? A little pilotage could get you out of trouble.

Back to our altitudes. What do you do if a controller assigns you an altitude below the MEA? I was flying over the coastal mountains of California with a student when the controller assigned us 8000 feet. The MEA is 9000 feet. I questioned the controller on this point, and he explained that his minimum vectoring altitude (MVA) for the approach was 8000 feet, so everything was fine. Fine for them maybe — they weren't in a Cessna 172 over the mountains in IMC conditions, which upon descending to 8000 feet promptly lost radio contact with ATC. It seemed like a radio failure at first, so we checked and found that the comms transmitted; we just couldn't receive anybody.

Having accepted the new altitude, we decided against climbing back up to tell them about the communication loss. Pilots conflict over lost communications rules. If you strictly interpret the FARs to fly the "highest altitude," that would indicate a climb back to 9000 feet. However, the AIM clearly stresses that you operate at the "appropriate altitude for the route segment being flown." Since we were descending for the approach, that would indicate flying at the minimum authorized altitude — which is what we did. We weren't overly distraught as the latest weather report at our destination before we lost communication indicated a 3000-foot ceiling and ten miles visibility, so this became a great teaching opportunity for my student. We had all the realism combined with a way out. In the intervening many minutes of silence, we worked out a perfect lost communications contingency plan. We had it all set when the controller came blasting into our headsets.

"I have been trying to reach you for 15 minutes. (It wasn't that long.) Where have you been?"

As if it was our fault. "We have been trying to reach you, too. Could our problem have been our altitude assignment below the MEA?"

You have to have your side on the tape if for nothing else but your own personal satisfaction. The radios were okay on the airplane, but they were not great and that made all the difference. Always remember who is pilot in command and if you don't like the altitude assignment, get another one.

Radio communications are critical to IFR traffic separation after the clouds roll in. That is why we have a procedure if you lose communication. If you really take a close look at the procedure, you will notice that it is geared for only one aircraft at a time having a radio failure. The thought that in the intervening confusion another aircraft might have been cleared to my old altitude, or that another aircraft in our area may have the same communications difficulties, influenced our decision to remain predictable and stay at 8000 feet.

One thing most training programs don't cover (because you spend your time in isolation under a hood or in a ground thing) is how to help other pilots make the system work for everyone using it. One way is to report unforecast weather, like icing. You must also report anything affecting your safety, like being assigned an altitude below the MEA. You should also know how to make position reports. Now that radar covers most of the country, this skill has been reduced to yet another rote memorization exercise to be repeated before an examiner. The best way to learn this is to give a real position report to a controller, if for nothing else but the practice.

One day you might need to do it for real; one night I did. I was flying after midnight with a student down the West Coast from Seattle to San Francisco when a report came from ATC that a radar site in Oregon would be down for four hours of maintenance. All traffic within 100 miles would be out of radar contact and we would have to make position reports for traffic separation. There were about five aircraft in the area and judging by the staggering silence that followed the announcement, none had made a position report within their collective memory. By sheer luck, we had recently practiced these reports, so they were no problem for us. My student took the initiative during the dead frequency to give a flawless position report, complete with the names and estimated arrival times at subsequent checkpoints. The controller came back full of accolades for the performance. We then held a little tutorial on position reports with all the other pilots somewhere over the Oregon coastline. Never pass up a chance to help your fellow pilots.

There is a statement regarding the FARs covering lost communications which states that the FAA can't possibly make rules for every situation. Truer words have never been spoken. However, there are some glaring gaps in the FARs. If they could clear them up, many myths and misunderstandings would be cleared up as well. Somehow the rule about when to shoot the approach has been twisted to the point that students are taught that you should always commence your approach from the initial approach fix (IAF) as close as possible to your ETA. Great, now we will have pilots without radios, in radar environments where their position is intimately known to ATC, flying in early when they have unforecast tail winds, holding forever until the system is completely backed up with traffic, and there is no way to contact the offending pilot. Every controller I have ever talked to about this has pleaded with me to teach that whenever aircraft lose their comms, they should get out of the system as soon as possible, for they are only a hazard as long as they stay in.

The myth comes in applying a rule that covers only the exceptions to all situations. Read the rules. The only time you have to wait for your ETA to begin an approach is when the clearance limit is the initial approach fix, or a fix other than where the approach begins. Think about this: When you file a full cross-country IFR flight plan, you file to an airport. When you receive your clearance, unless some rare circumstance prevents it, you are going to be cleared to that airport. It is amusing that just about all IFR flight plans end at an airport, yet this contingency has been left out of the lost comms procedure by the FAA.

Now we get to interpret the FARs for ourselves. The first item of any clearance is the limit of your clearance. Think of any of your clearances and it will probably have started off "you are cleared to ___ airport." When you are cleared to an airport, you are authorized to land at that airport. That's what being cleared means. Since any IFR aircraft without radio communication is a hazard as long as it remains in the air, your first duty should you encounter VFR conditions is to get out of the system. If IMC conditions prevail all the way to your destination, your second duty is to shoot the approach as soon as you get there, land, and get out of the system.

The difference comes if you get an amended clearance anywhere along the flight that changes the clearance limit from the airport. Now you have to time your arrival based on the new clearance times. Remember that the goal of the lost comms procedure is that you can lose your radios anywhere after being cleared for takeoff and have a procedure to accomplish your goal, which is to make a safe landing. Why the FAA chooses to focus solely on the exceptional cases and leave out the rules covering the standard clearance to an airport is a mystery.

By leaving out any lost communication rule governing a clearance to an airport, they have also left out any rule governing your descent to that airport. The FARs say that when cleared to an initial approach fix (or some other fix), you have to maintain the highest of your possible altitudes and cannot descend until the appropriate arrival time. This means that you turbocharged folks who are cleared to an approach fix and assigned a flight level will have to hold that until over the approach fix. This makes no sense at all.

There is no rule for being cleared to an airport or descent to an airport, and since you cannot violate a rule that does not exist, as pilot in command you have the authority to plan a normal descent so as to arrive at your initial approach fix at a normal altitude to shoot the approach and land.

IFR flight plans in a radar environment can be an exercise in futility because the computer that grants you your clearance will spit out whatever preferred routing is in current use despite any particular route you choose to file. It is amazing how quickly you are expected to rethink your whole flight, review what they have given you on your charts, analyze whether you have the required fuel and can comply with the altitudes, remember the weather for the new route, and do this in the few seconds allowed by the ground controller impatiently tapping his fingers.

You are required by regulation, however, to familiarize yourself with all information concerning your flight. That takes planning. What is the point if ATC has the right to give you any routing they choose? This forces you into noncompliance with the FAR governing preflight. As pilot in command though, you are still responsible for that clearance once you accept it. The trick in a busy, radar environment is to call for your clearance with your engine off and your charts ready to review your new flight plan. Take the

time to check it all thoroughly before starting your engine
and calling for taxi clearance. If you have questions, get
them resolved. If you cannot comply, then get a new routing. You will learn to think quickly in this environment and
to bring along the fuel for any number of possibilities. You
will find that many times your carefully crafted flight plan
is just a mere formality.

Since we have sophisticated computers now, the FSS
should immediately give you your expected clearance at
the time that you file. Then you could take your time
reviewing the clearance before you ever get to the airplane.

For those of us who are used to radar environments,
we tend to leave the flight plan at the bottom of the flight
bag because it hardly ever means anything. The other
extreme in IFR is operating from the rural airport. I was
flying out of Nebraska one summer. It was good VFR, so we
elected to pick up our IFR clearance in the air. "Cleared as
filed" was all that we got out of the FSS. Our flight plan
was of course at the bottom of the flight bag. Neither my
student nor I had ever seen a situation where we didn't get
a complete clearance from ATC, and we had only a rough
idea what we had filed.

All you folks training at the big terminals should make
the effort to have at least one flight originate from a rural
airport. Try to find a place where there is no tower, no
clearance delivery, no FSS, no radar, none of the customary
amenities. You may have to pick up your clearance by
phone. My experience with this is that you get about 15
minutes before your clearance becomes void, which is about
five minutes less than you need.

There are airports where you have to be in the correct
spot on the field in order to reach someone by radio. At
Orange County Airport in Los Angeles after the tower closes,
you have to go to this spot on the ramp near Runway 19L in
order to reach Coast Approach. There may be other spots,
but this is the one that I found. Departing IFR from a
nontower field can be fraught with long delays because the
separation requirements are such that you can't depart

with inbound IFR traffic until they are on the ground. You will find it far better to depart VFR if possible and get the clearance when airborne.

Don't forget to close your flight plan when you land at an airport without a working tower. Don't lose the habit you developed in your private for closing flight plans. You must demand the widest possible range of experience from your instructor before you get your rating. Sadly, most students are preoccupied with getting the minimum training necessary to pass the checkride.

To a much greater degree than approaches, holding patterns as currently taught are destined to drive you nuts. This maneuver requires the most imagination on your part of any IFR procedure because you spend most of your time without a reference to any navigational aid. The exception is the NDB hold where the ADF arrow gives you a constant reference to the station, thereby making them the easiest of holdings. However, because of ADF paranoia these holds hardly ever get practiced. The VOR is extremely limited in the information you get for holding because you only know where you are when crossing the station and when within ten degrees of the inbound course. Anything after turning outbound is a guessing game.

Through a series of entry headings, wind correction angles, variations in timing, and attempts to track back to the fix, you are supposed to fly this perfect racetrack in the sky. Why? The reason is that instructors throw this unnecessary burden on you so that you will be able to impress an examiner in VFR conditions with your pretty racetrack patterns. With all your energy expended on the procedure, you completely lose sight of the objective — parking. The purpose of holding according to the AIM is to keep an aircraft in specified airspace awaiting further clearance. As long as you remain reasonably close to the fix, ATC will be very happy.

Since you are under a hood and most of the hold is performed out in space away from a navigational reference, you never know whether you flew a good racetrack or not.

You will only know how well you intercept the inbound course, so this is where you have to fix everything for the next pattern. The degree to which you have to adjust the inbound track depends on how well you guessed the timing and heading when outbound. This is where your imagination comes in. You must take into account all the variables and visualize where the aircraft might be tracking in order to make a nice pattern when away from any reference. You won't know your ground track — but your instructor will.

When you are under the hood, your instructor most likely has a skyblue view out the window and knows exactly what path you are flying. This distorts your training. Your instructor will ask you leading questions on timing and wind correction designed to lead you to a ground path he can clearly see and has mapped out in his mind. No matter how well or badly you fly the path planned for you, there is no comparable visual backup for you to confirm anything you are told about your holding. This is how instructors get so nitpicky about holds and why they drive you nuts.

You will know only if your holding pattern was nice based on what your instructor tells you, but you will never have a picture in your mind to relate what you are told, so the whole process is futile. If you want to both shut up your instructor and really learn something valuable, get ATC to clear you for a real hold in a big, fat cloud. That way no one except ATC will have any idea of your exact path over the ground. Somehow a pretty racetrack pattern will not be the priority it was in VFR conditions, which is as it should be.

During your holding training, take off the hood sometime and see for yourself what a hold actually looks like and how small it really is. In your typical IFR trainer, you will hold at 90 knots of airspeed. In a no-wind condition, you will use up a whopping 1.5 miles of airspace flying a one-minute leg. Most of your holds will be somewhere above 3000 feet to keep you above any of the instrument approach paths. From 3000 feet, 1.5 miles on the ground doesn't look like much. It doesn't look like much on a radar screen either. Therefore, the only people you can please with a pretty racetrack are instructors and examiners who can see the ground.

This is not to say you may stray far away from the airspace designated by ATC for your hold. I am saying that flying the perfect pattern is overkill, for that misses the objective and misleads the student. You won't get too far off track anyway unless you have a very strong crosswind. In that case you have to be able to think imaginatively.

I once flew a stage check where the crosswind was so strong my first two patterns became repeated parallel entries because I kept being blown back on to the inbound course from my outbound leg. I didn't have my mind cluttered with trying for the exact racetrack so I could notice that the needle of the VOR had centered while flying outbound. Remember that if you get lost, just go back to the station and enter the hold again. ATC might want to know what you are doing as well.

Too much fuss is made during training about remaining on the protected side of the hold. Why? Is ATC going to clear another aircraft to hold on your unprotected side at your altitude, where you both get to play chicken in the clouds? Will there be a mountain on your unprotected side? Besides, the recommended parallel entry from the AIM diagrams a pathway outside the protected airspace — oh no! What are you protected from anyway? Does that mean you are unprotected the rest of the flight? I have never received a hold at less than the MEA; how much more protection do you need than that? Why not call the airspace where you hold the "designated" side.

The words we use in aviation are very careless in selection. However, words convey specific meanings. If pilots are to understand the correct meanings, we have to start using the correct words.

The problem with standard training for holds is the mental workload. An instrument student is expected to: Guess when he is three minutes from the holding fix and slow down (a neat trick without distance measuring equipment — DME); figure the correct entry and appropriate outbound heading; set the navigation instruments; time the first outbound leg; track inbound and analyze the effect of the wind; come up with a wind correction angle; mentally double the correction in the opposite direction and come up with a new outbound heading; time the first inbound leg after becoming established in the hold and then either subtract half the extra time beyond a minute to fly the inbound leg from the outbound leg or doubling the shortfall in time and adding it to the outbound minute; recite the acronym "turn, time, twist, throttle, talk" (affectionately known as the "T's"); converse with ATC; fly the airplane; track the navs; and listen to the babble of leading questions by the instructor trying to create a pretty VFR racetrack pattern in the sky. Uh-huh. This is why you have problems — there is simply too much to do.

The objective is to park the airplane in designated airspace. Your training should reflect that goal rather than the current system where we are fanatical about unnecessary procedures and formulas. By changing the goal to having you try to remain reasonably close to the racetrack holding pattern by bracketing both timing and headings so that each time you aim for a better track inbound, you can fly with a far reduced workload which will give you more time to think and learn. If you try to do all that stuff you learned in training in real weather after your rating, you could forget some step and get hopelessly confused. Flying through the clouds is no time to make your life complicated. All you have to remember about holds is how to make the recommended entries and then to make your best efforts to adjust your timing and headings to intercept the inbound track and fly it for about a minute.

Learning to fly instrument approaches will probably consume the bulk of your training. This is because you learn with minute precision all the possible procedures for each individual segment of the approach without ever having the process tied together with logical objectives and goals. While overburdened and preoccupied with the exact technique for the next step of the approach, you lose your ability to control your headings, airspeeds, and altitudes. With all the mental clutter, you will become backed up and forget how to maneuver the airplane as the memory of your pitch and power settings slip away. The airplane tends to wander as your mind overloads in a desperate attempt to remember and fly with perfection the next fragment of the approach as it has been taught to you.

The next series of errors will come when resetting the radios and navigation instruments while trying to remember some clever acronym for adjusting the radios that works fine on the ground. I have had students come to me with three or more acronyms that they have been taught to repeat during an approach. This is looney. After all those steps have been completed, they were steaming past any semblance of an approach. You cannot fly an approach when all you learn are cute sayings that totally distract you from your purpose.

What is an instrument approach? It is obviously not a series of disconnected procedures all thrown at you at once, even though that is standard training. An approach is a way to get you, without hitting anything, to a place where you can visually make a landing. All the headaches you get, excess memory work, acronyms, approach segments, pitch and power settings, communications, turns, headings, altitudes, descents, and various other things are all designed to put you on an ever-increasingly precise path to the runway. The closer you get to the ground, the more precise you should be.

Therefore, the goal of the early part of the approach is to size up the conditions and learn all you can so that you will know the best way to fly the last part of the approach. You can't do that while fussing too much with needless silly things. You have been making approaches to airports since before you soloed. The difference now is that the guesswork is gone, and that some of the time you will be in the clouds and have to fly by reference to the instruments.

Think about how you make a VFR approach to an

unfamiliar airport without a tower. You fly about 2000 feet above the airport, head out on a 45-degree angle to the direction for which you wish to make your landing pattern, and start a descent. After a couple of miles and 500 feet of altitude loss, you reverse your course and fly inbound, still making a 45-degree angle to your pattern. You level off at pattern altitude with the runway clearly in sight and make your visual approach to the runway. This is virtually identical to the typical, full, nonstabilized approach. The difference is that you fly over the station instead of the airport, unless the station is on the airport, and all your courses and altitudes are published so you don't even have to think.

We could greatly advance the knowledge of our students by teaching private students to visually fly approaches to nontower fields that duplicate what they will later do on instruments. They could learn to fly procedure turns and fly specific distances and altitudes. Like learning the ADF in the private, the visual backup would solidify the procedure in their minds. You should request this of your instructors. Remember, though, that everything is backwards and you will be taught the approaches in pieces. However, you can never see the puzzle when all you have to look at are the separate pieces.

The VOR approach will be the easiest to start with because you should already have lots of practice with them from your private training. If you do not yet understand the VOR, get more training before attempting these approaches. To fly a VOR approach, all you have to be able to do is to intercept and track radials, fly level, turn and descend, use a timer, and be able to talk on the radio. These are all private skills, so there is nothing new. Get it out of your mind that there is some mysterious secret to instrument flying because there isn't.

The idea of a VOR or any full nonstabilized approach is to cross over the station at an altitude which is higher than the folks coming back inbound. You then track outbound and descend some to the published outbound altitude. Then you turn around and track in the same way you went out and descend some more. After either crossing the VOR or getting within a certain distance from the airport, you get to descend to the lowest altitude that is safe for you on the approach path to put you where you can see the airport. If you don't see the airport by a certain time, distance, or station, you climb out and fly to a place where you can hold and figure out what to do next. No big deal.

The way you make sure you don't fly too far on approach is to either have DME or to time your various legs. You folks really crack me up because most of you show up for instrument training with those fancy digital timers that you think make you look so professional. There is that image thing again.

There are few sources of amusement greater than watching a student trying to fly approaches while punching all those buttons for each segment of the approach. You cross the station outbound and punch in two minutes. Then you stop it for the procedure turn, which I will now call a "reversal turn" because that is what you are doing. After turning out on your reversal turn, you start or reset the timer again and fly for a minute. Then once you get established inbound, you try to set it for the final segment to identify the missed approach point. This assumes that you remember to do all that and you don't hit the reset button when you really meant to hit the start button, causing you to have to reset the time and start it again. Approaches are hard enough without you purposely making your life miserable. I know because I got one of those fancy digital timers when I was a student.

There is only one place where you have to time your leg precisely: from the final approach fix (FAF) to the missed approach point (MAP). There is no reason to get hung up on exact one- or two-minute legs way up high when flying outbound because you are not going to hit anything on the ground from way up there. I hated the digital timer so I went back to my old toy box and found a stopwatch from my childhood. This was standard equipment before the fancy stuff.

Back to the approaches. You know you won't hit anything at your outbound altitudes because there are planes flying inbound at lower altitudes and they have to have their IFR separation from both you and the ground. That is why the outbound legs aren't critical. You lose sight of this when you lose sight of the objective that you are doing all this approach stuff so you can land.

When you concentrate on just the individual leg, the only way to make it significant is to fly exactly a set amount of time. This is how misplaced goals screw up your thinking and mess up your approach. Relax when you fly outbound. The other goal of the approach is to achieve visual contact with the airport. You can't do that outbound even in VFR conditions.

Now that you have the right thinking, let's apply the timing. If you have a fancy timer, then set in the time from FAF to MAP on the digital timer long before you ever start your approach and leave it there. When you cross the initial approach fix outbound, hit the stopwatch, or note the second hand on the clock on the flight panel to the nearest five-second mark. After two minutes or whatever time you use, start your reversal turn. Don't touch the stopwatch; just let it run. After another minute goes by, turn back in to intercept the inbound course. The stopwatch will still measure a minute whether you start from zero, or whether you already have two minutes on it. Keep your life simple.

When you cross the final approach fix, just start your preset digital timer. You can make the approach simpler still by learning the 90/270 reversal turn, which is so much easier than the standard timed course reversal that it should be the first reversal you learn. If you don't have an image problem and never bought a digital timer, just restart the stopwatch when you arrive at the FAF. In the real world, you will most likely be vectored on to the inbound course anyway so if you learn very simple methods of timing in your training, you will remember it better when you need it long after the flight test.

Tracking a VOR is no big deal because it isn't that

sensitive. You can wander a bit and still keep the needle fairly centered. The closer you get to the station, the more sensitive the needle becomes. Sometimes it is hard to know if you are crossing the station or you are completely off course. This is why DME should be on every aircraft used for IFR flight. Like the NDBs, if you fly the reference heading, you will at least parallel the course and figure what is happening while heading in the right direction.

Recognizing the limits of the VOR, the FAA wisely keeps you fairly high on the approach. What people seem to have the most trouble with on these approaches is descending. You either forget to descend or forget to stop descending. The first one is amusing; the second one will kill you. Leave room in your brain to think about descending by keeping the approach simple and by making your pitch and power settings instinctive.

Every approach is different because the conditions are always different. You need a strategy, therefore, to make the whole approach successful. Part of that strategy includes learning as much as you can about the conditions in which you will shoot the approach. This involves listening to the ATIS and observing your outbound progress. If you get fanatical about timing and tracking your individual segments, you will burn yourself out and have nothing left for the most critical part closest to the ground. Part of your overall strategy involves conserving and managing your energy. There is no point in tracking outbound with all the accuracy you can muster when you are going to have to completely leave the radial and go off in space to turn around.

Flying the perfect reversal turn is irrelevant because that is not your goal at this point, although you would never know it from standard instruction. Since the FAA doesn't even care what kind of turn you use when the standard reversal is published, why would you care how accurately you fly the only procedure you are likely to learn? If your overall objective is to land without hitting anything, then the goal of your reversal turn is to get established on the inbound course, nothing else. Students who are taught to concentrate on the

minute details of the perfect reversal usually blow through
the inbound course, forget to descend, or both, all because
their priorities are screwed up.

If you have done your work properly, once you intercept the inbound approach course, there should be nothing
left for you to set up from here to the MAP. All that is left is
to fly the airplane. Okay, so you intercept the radial and
start down. Things are heating up on the flight deck because the ground is getting closer. The winds have probably changed from your outbound leg. The most critical
time is that last little bit when you have all the stuff to do
at the final approach fix. This is where the "T's" really come
in handy. Just remember that when crossing a VOR, it
won't give you any guidance until you get away from directly over the station, so don't go making any big course
changes. What worked to keep you on track to the VOR
should be pretty close to what keeps you on track from it.

Training with a hood has a drawback whereby you never
develop the habit of looking for the airport while flying the
approach. This undercuts your objective of getting safely to
a point where a visual landing can be made. In the real world,
you can break out any number of times in any number of
places. When I have to use a hood, I have students remove it
at different times so approaches don't get too predictable. One
day I told a student on an approach to remove the hood. He
told me, "No, we aren't at the missed approach point yet." Do
you believe this? What did he learn from his previous instructor, that peeking anytime before the MAP is cheating?

I can just see more and more pilots racing into VFR
conditions and screaming by other airplanes because they
don't want to look up until they are good and ready. Remember, you only fly approaches so you can get to visual
conditions and land. You must get in the habit of looking
for the airport anytime after you are established inbound.

Visibility is safety, so the higher off the ground you
are visual, the safer you will be. You are also responsible to
see and avoid traffic anytime you are visual, so you must
look up continually. You are looking for your landing crite-

ria as soon as possible. When you have the required visibility, the runway environment in sight, and the ability to use normal maneuvers and rates of descent to get there (which will be more normal the sooner you can get started,) unless you have to wait for a visual descent point, you can now deviate from the procedure and make your landing. Mission accomplished.

Anytime the instrument approach doesn't bring you within 30 degrees either side of the runway of intended landing, you must make a circling approach. A circling approach is whatever you want it to be as long as you remain at the correct altitude for your airspeed until adequate visual reference and normal maneuvers will allow you to descend for landing. The only time you can really take advantage of your ability to fly over any part of the airport to get to your runway is when you are the only aircraft over the airport, and that only happens when the field is IFR. Otherwise, you have to flow into what may be crowded VFR traffic patterns.

The tendency then is to drift back up to the VFR altitude so that you look like everyone else. This is a bad habit as you could drift right into the clouds on an IMC day. The cure is as always to fly in actual weather.

The power of myth struck my old flight school where they believed and taught that you can't descend for the runway until you are within 30 degrees of final approach. Somehow the criteria for when a circling approach is required got twisted into how you have to fly a circling approach. It made for some unusually steep descents at the last minute. I wonder if they still teach that?

Anyway, wait until you have to stay at your circling altitude because a solid deck of cloud is right above you, and you see how weird it feels to be maneuvering that close to the ground. This is an important lesson that teaches how critical it is to know your own airplane so that you can safely fly in adverse weather, stay below the clouds, and still make a landing.

The NDB approach is what strikes terror into the

heart of the instrument student because most of you have no idea how to use the ADF. You need to learn the concept of relative angles and to get the visual knowledge backup explained earlier. Once you have that clearly in your mind, the NDB approach is actually the easiest of them all. The instrument isn't all that accurate, so no one expects much precision from you anyway, which is why the FAA wisely sets the minimum altitudes the highest of any approach.

Here is the no-fuss, full NDB approach. When you cross the station outbound, fly the reference heading and do not attempt to track. If the arrow is within ten degrees of the tail as you proceed outbound, you will not have to track at all. If (after a reasonable amount of time for the ADF to settle) the arrow looks like it is wandering, then make a five- or ten-degree correction toward the arrow. There is no need for exact tracking because you are only on the outbound for a couple of minutes and you are going off in space to turn around.

On your way back in from your reversal if you use the traditional 45-degree intercept, you will have intercepted the final approach course when you have a 45-degree relative angle on the same side as the beacon. Once again, no in-flight math is necessary. At the point of intercept, turn to the inbound reference heading and descend. The problem with ADF arrows is that they usually wiggle, so there is no way to fly an exact track.

You have a choice of ways to fly to the beacon. If there is no wind, then go directly to the beacon. If there is only a light crosswind, you could make whatever tiny adjustments are needed to home right in on the beacon. Your goal after all is to cross directly over the station, not to impress your instructor by confusing in-flight calculations, so homing is perfectly legitimate. Depending on the strength of the crosswind, you may want to take small correction bites into the wind and regularly check your track by returning to the reference heading.

As you get close to the station, you may elect to home in as well. After you cross over the station, turn directly to the reference heading and stay there. When you are half-way to the MAP, check the arrow. If it is within ten degrees, then stay on your reference heading. If it is starting to move outside the ten degrees, then make a five-degree or so correction into the wind and hold it. The less you deviate from your reference heading, the better for your approach.

Remember that in an actual NDB approach if the wind is strong enough to blow you far off course, it is also strong enough to blow away the fog, so chances are you will be visual early in the approach. The wind is not your nemesis; you are. It is not your fault that you have been taught to correct using a ludicrous method of calculations, bearings, and mental arrow reversals and opposites. What you must do to cure this problem is find an instructor who uses relative angles, reference headings, and has no use for in-flight calculations.

If you get confused at any point after getting established inbound before the station, then home directly for the station. If you are confused after passing the station, then just fly your reference heading. If you check your reference heading anytime and find you have left the approach course, then go missed. I think that should keep it simple enough. This is the way the commercial pilots fly NDB approaches. Why we expect students who are still learning the system to master the most cumbersome and inefficient methods is beyond any logic.

For those of you old enough to remember, there was a TV show called *The Time Tunnel*. You walked down a tunnel of concentric rings ever shrinking into a smaller cone. Flying an ILS is something like that. The electronic beams that guide you in create a progressively smaller cone the closer you get to the runway. What this means is that the wallowing you do to stay on the approach outside the outer marker must be reduced to just the slightest of corrections as you approach the middle marker. This difference you can only learn with experience.

You will be told by your instructor that you are "chasing the needles." This is one of those great nebulous expressions like "maintain a stabilized approach," that really doesn't tell you anything you can use. What it means is that you are using corrections too big for the size of the ILS cone where you are flying. Some theory of ILS operation might be useful here. You should learn the operation and limitations of the ILS components, particularly the localizer and glide slope. They are extremely narrow beams that allow for great precision. If you fly only by trying to keep your localizer and glide slope needles centered, you will always be correcting for what has already happened, so as you get closer to the ground and further behind the airplane, you try harder to catch up, which leads to overcorrection and being told you are chasing the needles.

What you need is a strategy to stay ahead of the airplane. The basic ingredient is the proper pitch and power setting for your airplane. Since this is a stabilized approach, one setting should fit the whole approach. My strategy is to fly an ILS fast because the greater your airspeed, the less the moving air will divert you. For my Cessna 172 that works out to 100 knots, 1800-1900 RPM, 500-feet-per-minute descent and the ball of the attitude indicator touching the bottom of the horizon from below.

To avoid chasing the needles, use the instruments that give you the information you need faster than the needles can. To hold a localizer, hold the heading on your directional gyro that keeps the needle centered. Focus first on the heading; if it deviates, then the needle is sure to follow. By leading the localizer with the directional gyro, you can progressively bracket the correct heading. Keep in mind that this is a continuous process because the winds change all the way down. However, for the new student the problem is not the wind so much as the inability to hold a heading, which will be blamed on the wind.

The glide slope has its own strategy. To stay on your basic three-degree glide slope, there is a formula such that you can take your groundspeed, divide it in half, and add a zero, and that will give you the proper rate of descent. For my 172 that means 100 divided by 2 equals 50, add a zero gets you 500 feet per minute to hold the glide slope. My 100-knot setting for the 172 is no accident. It provides students with easy numbers while learning the ILS. To hold your glide slope, try to maintain 100 knots on the airspeed indicator and 500 feet descent on the vertical speed indicator. This will vary, of course, as the wind affects your groundspeed.

There is also a delay on the vertical speed indicator to take into account. Where the glide slope needle moves faster than the vertical speed indicator has yet to be determined. It is safe to say, however, that it is close to the middle marker.

Once again, the prominence of the basic flight instruments prove that instrument flying is neither a mystery nor anything all that new. There is not a whole lot to say about approaches because the difficulty you encounter can be traced directly to your thinking, strategy, understanding of the goals and objectives, and your ability to fly the basics. If you have problems, now you know where to look.

One approach you will not get until after you get your rating is the visual approach. This is because the FAA requires a minimum amount of instrument time that you must acquire. This leads you to always use the hood because you don't want to pay good money without building your instrument time. This is why your training is unfairly skewed towards flying by reference to instruments, which as you now know is only a portion of what you need to know. To be a good IFR pilot, you also have to be a good VFR pilot. The solution is to change the requirements so that there is no minimum requirement for instrument time. That way your instructor would be free to show you the whole system and send you for the checkride when you are ready, not at some arbitrary limit.

Back to our topic. The reason for the visual approach is to expedite the flow of IFR traffic in VFR or Marginal VFR conditions. You might be assigned a visual approach by ATC, but your instructor will turn it down because you need the instrument time. You should see a visual ap-

proach in your training just so you will know what to expect
when you get one on your own — which will be instant
bewilderment.

From the very structured world of IFR comes this
formless void where you are still on an IFR clearance, but
you are responsible for your own traffic and obstacle clearance. How you now get to the runway is your business. ATC
will simply clear you for the visual approach and you will
sit there wondering what to do because you have no idea
what the procedure is. Only if you get a charted visual
approach will you have guidelines; other than that you are
on your own.

One night I was with a student flying in a Bonanza
into San Carlos, California, a little airport hidden under
the San Francisco TCA (Class B), and right beside the
usual final approach course of SFO International. There we
were in the TCA at 4000 feet directly over San Carlos when
ATC asked if we had the airport in sight. We couldn't miss
it as it was a crystal clear night. "Cleared for the visual
approach" were the next words out of ATC. There we were
at 150 knots, in a TCA, with big jets to the east, SFO to the
north, mountains to the west, the Palo Alto ATA (Class D)
to the south and absolutely no guidance from ATC. We were
surrounded.

After a brief instant of dumbfounded looks on both our
faces, we came up with the only solution: a steep spiral
directly over the field. This is the only time I have ever had
any use for that maneuver. You feel strange on this type of
approach even though you are still IFR, especially in a
TCA. You wonder about the limitations like what airspace
are you cleared through and what you must avoid, in this
case someone else's airport traffic areas. We had the clearance of approach, but not the tower's so we stayed clear of
the ATAs. This, of course, made for a stretched definition of
normal maneuvers; however, we had been given no choice.

The point is that there are no rules for flying a visual
approach — only rules for receiving one. We were also
caught by surprise and had no time to make a plan. Many

times though ATC will tell you ahead of time to expect the visual. This is not the only time I have been dumped by ATC in an awkward position. What you should do is always expect the visual whenever the destination has VFR conditions and work out a plan in advance to land. Visual approaches are popular with commercial operators because they get you in faster and save the boss money. You will be encouraged on the job to take them when you can.

There is only one acronym for which I am particularly fond. That is *the T's: turn, time, twist, throttle, talk.* As long as you continually repeat your T's, you will always be prompted to do the right thing without leaving anything out. Now go sit on a mountain somewhere and imprint this on your brain forever. Think of it as an IFR mantra, or maybe perhaps, a modified haiku. Take any approach or holding pattern and by following the T's step by step, you will get through the procedure.

The logic of having to turn first is for holds and for when you have to turn at the final approach fix. If you have to turn that close to the ground, you don't want to wait. Long after your rating should you run into difficulty, it is the T's that will bring back all your training so you can handle your situation.

Try to do some of your training in high-performance aircraft if you are so endorsed. The IFR procedures get so ingrained that any extra step kind of gets lost or forgotten, like the gear, for example. Many pilots have trained in low-performers because it is cheaper and then hopped into a fast single only to find themselves on approach without a plan for dropping the gear, flaps, power, and where to do all the prelanding stuff. Learn a high-performance strategy that includes pitch, power, and airspeed settings for your aircraft. You can figure on dropping the gear when you intercept the glide slope or crossing the final approach fix. Make a good IFR checklist and you should be all set.

I cannot stress enough that if you want to learn how to fly in the clouds, you absolutely must fly in the clouds. There is such an infinite variety of experiences and knowledge that will be unavailable to you should you stay under the hood. They also teach you bad habits and distort your perceptions which will put you at a great disadvantage when you encounter actual weather. Considering the state of the average ground trainer, the only thing that those machines can teach you is how to fly a ground trainer. Any simulator worth using will probably cost as much as an airplane. Simulators are okay once you have an understanding of the system, but not before. They should never be used as a teaching tool; instead their use should reinforce what you already know.

You should not miss training in the clouds for they are some of the most beautiful things you will ever see. Punching out of white velvet gives you a sense of the power and speed of flight. Sliding between layers gives you the feeling of space flight. There are foggy nights where the air is so still you would swear you were on the ground with the engine running, even though the altimeter reads 5000 feet. I have seen approach lights glow in the window and appear to be moving about like a video game when in reality our airplane had the only motion. I have seen up and down drafts that pegged the vertical speed indicator while flying in solid cloud, where students learned the critical nature of maintaining attitude, maneuvering speed, and getting a block of airspace from ATC.

There are sights and experiences that will stay with you long after your training if you have the desire to seek them out. If you want to train for the checkride, stay with the hood and simulator. If you want to train for the rating, seek out the clouds.

11 • THE COMMERCIAL

✈

"Welcome to your first job. We've got a load of passengers here who have to get on their way. Since you are the new pilot, we want you to stay sharp, so why not practice a couple of lazy-8's and chandelles en route. A steep spiral to the airport would be nice, followed by some 8's on pylons. Maybe you can use the tower as a pylon if the pattern isn't too crowded."

What's wrong with this picture? Has there ever been an operations director who ever wanted you to do anything but get the passengers and cargo to the destination the fastest, cheapest way possible? You would never know it from the requirements of the commercial certificate. The commercial is obsolete. Much of the certificate is a rehash of the private, albeit to slightly higher standards. You could accomplish the same thing yourself simply with more practice. The rest of the certificate requires perfection in maneuvers that no commercial operator would ever tolerate on the job.

See if you can figure out the logic of how the FAA regulates commercial pilots. The main privilege of a commercial certificate is that you are able to carry persons and property for compensation or hire. That's what it says in Part 61 of the FARs, and that's ostensibly why you get the certificate. However, all commercial operations that carry persons or property for hire are governed by Part 135 of the FARs. You can't even use the commercial certificate earned under Part 61 for its intended purpose. The only things you can do for hire with a commercial are the exceptions to carrying passengers and cargo listed in the beginning of Part 135. They include banner-towing, ferry flights, crop-dusting, aerial photography, short sightseeing trips, fire-fighting, and the biggie of them all — flight instruction.

The real commercial certificate that allows you to carry passengers and cargo for hire is the Part 135 Airman Competency Letter. However, the only way you can be eligible for the Part 135 letter is to already have a commercial certificate. This makes the commercial the only certificate that you must acquire to earn a privilege that you can't legally use, so that you will someday be eligible to earn the certificate that does allow you to exercise the privilege granted by acquiring the commercial.

Under the current system, you need a commercial certificate to work for hire. To work as a flight instructor, you first have to have the commercial because instructors work for hire. To get the commercial, you have to be able to demonstrate the commercial maneuvers. There are only two times that you will ever need to know the commercial maneuvers after that since they have no relevance to any commercial operation. The first reason is so you can pass the flight instructor checkride. The second reason is so that you can teach students those maneuvers so they can get the commercial.

The only reason that most commercial students are getting the certificate is so they can go on to become flight instructors and work for hire. Therefore, the only justification for learning the commercial maneuvers is to be able to teach other pilots the commercial maneuvers, so they can go on to teach yet more pilots the commercial maneuvers. This is a vicious circle. Put another way, you get the commercial only so that you can teach the commercial to people who can't use it for the privileges it grants, rather only for the purpose of becoming flight instructors to work for hire and teach the commercial to other students who have to become instructors, because no one can work as a commercial pilot carrying passengers and cargo with only a commercial certificate. This is why the certificate is obsolete.

If you actually want to work as a real commercial

pilot, under Part 135 of the FARs, you will need 500 hours of flight time in order to qualify for VFR operations and 1200 hours for IFR operations. You are eligible for the commercial certificate in only 250 hours, 190 for you Part 141 academy folks. Since you won't be any use to a commercial operator until you can fly IFR, you have a huge gap of up to 950 hours to fill in the old logbook after you acquire your commercial, before you can actually use the privileges granted by the certificate. This is why you become an instructor. For most pilots this is the only reason you instruct; then you get accused of being a time-builder. You can't win because you have to build the time to be eligible for the commercial jobs.

I have some recommendations for the FAA. Since the only reason a person should ever have to instruct is because he loves instructing, the Certified Flight Instructor (CFI) Certificate should by itself allow any pilot to earn a living as an instructor, regardless of the pilot certificate he holds. At 250 hours, any pilot should be eligible to take a Part 135 competency check that would qualify him for any commercial operation, including IFR. The fact that the FAA requires 1200 hours of experience proves that they do not trust the current system of training pilots, yet they have done nothing to overhaul the system. When taught properly, any pilot should be competent for commercial operations in 250 hours. This makes the commercial certificate totally unnecessary. The pilots of Japan Airlines are 747 qualified in some 300 total flight training hours. Does that tell you anything?

To get the certificate, you have as always eligibility requirements to fulfill. You have to be 18 years old, be able to communicate to ATC through your accent, and be healthy enough to pass a second-class medical exam.

Your aeronautical knowledge requirement includes operations in high-altitude aircraft, and competence in high-performance and turbocharged aircraft. Both of these are now covered by separate endorsements, which is another reason the commercial is obsolete.

The high-altitude requirement by implication means flying turbocharged aircraft; otherwise, you won't get to any high altitudes or be able to pressurize the cabin. Since flight schools don't want students to chew up very expensive turbocharged aircraft, they probably won't be available to you until after you have your 1200 hours, so why learn about them unless you can use them? By the time you are able to fly high-altitude aircraft commercially, your employer will train you very well in their use because he doesn't want his aircraft chewed up either, so once again why learn about them now?

My solution to this is to beef up the knowledge requirements for the high-altitude and high-performance endorsements. That way the commercial written could be scrapped because all that would be left would be the same old questions from the private written. Who needs a written test on a certificate that is covered by other endorsements or certificates and can't be used for its intended purpose? Since my recommendations are not yet in effect, I shall now offer you my turbo/pressurization mini-course:

"Turbocharging blows air into the engine. Pressurizing blows air into the cabin."

The meteorology requirement is stricter and more involved than the private certificate. This makes sense on the surface, but the logic falls apart for four reasons. The commercial operations available to you strictly VFR are so limited that your private knowledge will easily suffice. You have already flown much greater distances to get your private than your operating range carrying passengers for hire with a commercial. For most high-altitude operations, you will still need a separate endorsement. Lastly, if you desire any kind of aviation career, you will have to have an instrument rating, which will teach you far more about weather than will the commercial.

What the commercial does to help you the most, is to require a wide range of flight experience before giving you the certificate. You need some night, some high-performance, some instrument, lots of cross-country, and lots of pilot in command time. The key to success in any type of flying is your varied experience, for that is where the knowledge comes that will quickly yield the solution to any problem. This is especially important when you later operate for hire with passengers and carry students. Try to get as much solo time as you can during this period. It is likely to be the last block of solo time you will have for quite a while.

This is a time of great confidence-building as you are forced to face lots of new situations on your own. Since the commercial maneuvers are useless, any dual time spent learning them is equally useless. Therefore, it is the varied areas of experience required by the certificate that will do you the most good. Well, I had to find one redeeming quality.

My solo practice for the commercial was some of the most valuable time I ever spent in an airplane. I learned how to get a Cessna 152 off the ground in a 45-degree crosswind of 25 knots and crab across the Mojave Desert. I learned how easy it is to misjudge the distances of aircraft at night after executing evasive maneuvers to avoid an aircraft that revealed itself to be more than 30 miles away. I learned how quickly I could become lost at night when the beacon of my home field went out on my first solo night flight, and all I could see were miles of unfamiliar city lights. I learned on that same flight that something as simple as lifting a wing to look below can solve the problem when an obvious landmark comes into view. I learned how to borrow some of my instrument training to intercept radials from VORs and navigate through mountain passes that all looked the same.

I learned the value of resourceful thinking while flying one of those valleys in the mountains and lost my exact position. There were two possible airports I could have been over, and the one below me had a name that was too small to read from my altitude. The runway numbers, however, were in plain sight, and the two airports had different runways. By referring to the Airport Facility Directory, I was able to identify my position without having to interrupt the flight. I learned from my long cross-country flight that even though you can't see the destination, if the check-points all arrive at the planned time, and the airport has a VOR on the field, you can descend at a preplanned time and fly right into the traffic pattern and have it work out perfectly despite marginal visibility. I learned that there is always more to learn, and that if you can keep your head and draw upon your expanding knowledge and experience, anything can be accomplished.

Most of your dual time will be taken up with the new maneuvers, with a review of some private maneuvers that you must once again demonstrate to an examiner making up the rest. Some of the names have been changed from your private to make you think there is some lofty new level to commercial training. For example, short-field work becomes maximum-performance procedures, but it's still the same old stuff.

Here is the new stuff. "lazy-8's": This is a maneuver that can only to teach you about lazy-8's despite being passed off as an advanced coordination maneuver. If you can't coordinate, excuse me, balance an airplane after 250 hours, then lazy-8's won't help you. They are, however, lots of fun, as are most of the commercial maneuvers, so you might as well make the most of them as your money pours out.

The lazy-8 looks more like a lazy-U when pictured in the handbooks. It will only resemble an 8 if you imagine a large paintbrush coming out of the nose painting figure 8's in the sky. Since there is some aircraft balance involved, this would be a great postsolo maneuver for private students. They would be really fun for students to learn, especially if there is no need to ever be tested on them. We do that already for maneuvers like "dutch rolls."

For commercial candidates, the lazy-8 just isn't relevant because of the tiny bank angles and small range of airspeeds. Most examiners require you to enter these at maneuvering speed. The only way to both enter these maneuvers that slowly in high-performance airplanes and not exceed the maneuvering speed during the descent portion is to start with a power setting so low (around 18 inches usually) that you couldn't possibly get a feel for the maximum performance called for in the *Flight Training Handbook*. If you are lucky, you might gain 300 feet of altitude during the climb — some performance. Maneuvering speed is reserved for violently turbulent air, maximum load factors, and full or abrupt use of the controls; none of which you will come even close to encountering practicing lazy-8's, so why the arbitrary restriction? It's not like you are going to practice these in a thunderstorm, so all maneuvering speed does is defeat the purpose of what the maneuver is trying to demonstrate. The result is really just a "baby-8."

If balance is an issue with the FAA, then commercial candidates should be required to perform wingovers at 60 degrees of bank, at climb power, with an entry speed at the top of the green arc. By the time you have slowed to just above the stall speed, you will have most likely gained about 1500 feet of altitude. Now there is maximum performance without exceeding any legal or aircraft limitation. It is amusing how we baby commercial students with reduced power and low bank maneuvers yet put presolo students through full power departure stalls. The one thing you will learn from the lazy-8 considering the incredibly slow roll rate, is patience.

Just a note here on your choice of training aircraft. If you train in a low-performance aircraft because you think you are saving money, think again, because you are only going to have to learn those maneuvers all over again in the high-performance aircraft that is required for your CFI checkride. That makes for two complete sets of different procedures because of the two types of aircraft and the fact that for the CFI you fly from the right seat. Flight schools love this because they get you to pay twice for your training.

The "chandelle" is a maneuver that according to the *Flight Training Handbook* "demands that the maximum flight performance from the aircraft be obtained." Your objective is to gain the most altitude possible at a given power setting and angle of bank. What a contradiction. Like the lazy-8, it is undercut by low entry speeds. Even though you add climb power, you can't possibly achieve the maximum performance or maximum climb. Any bank angle you induce increases the load factor which also gives away climb. Any speed less than your maximum lift to drag (L/D max) increases your induced drag which again cuts into the climb. You can't get maximum performance from a maneuver that by design gives away all the performance. To do a chandelle properly, you have to end up on the verge of a stall with maximum drag and minimum airspeed which leaves you no performance at all. Getting your performance back will require changing your angle of attack, which means lowering the nose, and there goes your altitude.

Just out of curiosity, is there any practical commercial use for flying a high-performance, turbocharged aircraft at maximum power, ending a chandelle with its nose pointing way up in the air, where the engine is overheating because of the minimum airspeed? The usual justification for learning this maneuver is to get you out of the proverbial box canyon. Why would you ever fly into a box canyon? Does this mean that because we don't teach this maneuver to private pilots they will never fly into canyons? Commercial pilots are supposedly good enough to work for hire, so wouldn't you have more sense not to fly into a canyon than a private pilot?

Even if you were stupid enough to get into a tight spot, are you going to have enough room to take off and accelerate to maneuvering speed before you initiate your restricted performance climb to chandelle out? How close to the ground would you throw your airplane into a 30-degree bank? How much of the country has box canyons anyway? How many canyons have airports at the bottom? Finally, how many pilots are going anywhere near a mountain airport in a box canyon without a good course in mountain flying? If you want a maximum-performance climb, consult the appropriate performance chart in the manual. If you want an introduction to acrobatics, then you should learn chandelles.

What about the "steep spiral?" The idea here is to lose altitude quickly in a gliding turn while remaining over a point. The *Flight Training Handbook* calls this a great way to build up proficiency in power-off gliding turns. What a useful skill. I wonder if passengers appreciate this maneuver as much as the FAA? One rationale given is that you can lose altitude rapidly in the case of an emergency using this maneuver. The question is, which emergency?

If you lose your engine, you want to come down at your best glide speed to give you the maximum ability to glide to a landing spot and the maximum time to troubleshoot and plan your landing. If losing altitude is required after an engine fire, a medical emergency, or rapid cabin depressurization, you are going to use the emergency descent procedure which will bring you down much faster than any steep spiral. You also learn this maneuver to be able to correct for wind while falling out of the sky. However, this may not be the most prominent thing on your mind during an actual emergency.

Since there is no purpose for this maneuver that can't be bettered by a more useful technique, you might as well just sit back and enjoy unsettling the people on the ground looking up at you. My personal theory is that when pilots learned to fly in Cubs and Champs which didn't have retractable gear and flaps to hang out, the only way to lose altitude quickly without building up excessive airspeed was to load up the wings in a steep spiral. All we need to do now is update the *Practical Test Standards* to represent the second half of the twentieth century rather than the first.

There are a couple of tricks you can use in your favor when being tested on steep spirals. You can keep the bank just steep enough so that the examiner never has a clear view of the point you are tracking. The other thing to do is to pick about three or four points that look the same so that if you get blown off one, just tell the examiner you were really tracking the other one. It also works to have at least three fields, preferably in a row, to roll out on for your simulated forced landing. Aim for the middle one and you are covered for overshoots and undershoots.

If the examiner doesn't like where you are doing the steepest bank, just explain that because of the Coriolis

Force you never know exactly where the wind is coming from on the way down and you were just trying to compensate for what you extrapolated to be the actual component effect on your flightpath. For the last circle, you can treat it like an 8 on a pylon because sometimes lining up the wing gives the illusion that you are flying a perfect ground track when in reality you are not. On checkrides, perceptions are far more important than reality. Unfortunately, this includes the examiner's perception of how this or any maneuver should be flown, the PTS notwithstanding.

I had to perform a steep spiral for my CFI checkride where I was told to maintain a 60-degree bank throughout the entire maneuver. So much for wind correction. Anyway, the directional gyro tumbled during the second turn. I pushed the button to stop it turning, which reset the directional gyro 180 degrees to our actual heading. The other practice of this inspector was to have me roll out at the same heading as we entered, not to any simulated landing field called for in the *Standards*. The inspector screamed at me for what he thought was rolling out 180 degrees from our entry heading. Fortunately, I acquired the habit long ago of always having a visual backup (in this case I used a mountain), so I knew the aircraft was back exactly where it had started. It became very quiet in the airplane after I reset the directional gyro to the compass.

At this point in your career, you should be noticing that instructors, examiners, and inspectors don't fly very much; they only watch. This leads to some interesting situations when mistakes that they refuse to admit put you in a delicate position from which you must extricate while still proving you are right, so you can pass the flight test and retain your dignity. As a private pilot, you always assume you are wrong, even when you think you are right. This sets up an internal conflict which causes confusion and resentment. With enough experience, you begin to know when you are right, which means you have to learn a skill that isn't covered by the standards — diplomacy.

My favorite diversion from the real world is the

"pylon-8." This is a great maneuver if you have dreams one day of racing P-51's at the Reno Air Races. The purpose here is to divide your attention along a flightpath you can't see, other traffic you hope isn't there, and hazards on the ground you hope don't get too close, because you are totally preoccupied with lining up your wing on a pylon.

The confusing concept here is the idea of a "pivotal altitude." Most of you will translate that in your mind to mean the most important or critical altitude for the maneuver. What it really means is that is where the airplane appears to pivot about the pylon. Considering that a pivot is a fixed point about which another object rotates, this is another example of language that doesn't fit. Helicopters can pivot; however, airplanes have to turn. Just call it your "line of sight altitude" from now on and you will understand.

The big problem with 8's on pylons is picking the pylons. You can't judge distances from 3000 feet as you fly into a practice area, so picking appropriate pylons from that altitude will be sheer luck. Try looking for just one good pylon, fly around it, roll out for your three to five seconds, look under the other wing for another good pylon at the proper distance away, and then fly around it.

I often wonder where these maneuvers originated and what were their real purposes? For example, flying around pylons might have had a practical use in early navigation, or perhaps in carrying and delivering mail where a bag was dropped around a pylon marking a change of route. I just don't know. If we really want to get back to basics like that, why not require commercial candidates to perform at least one night landing using a bonfire for the approach and tiki torches for runway lights. There are still places in the world that do this.

As much fun as it is to line your wing up with a water tower and watch a perturbed farmer shake his fist at you, I don't believe it teaches anything meaningful. The knowledge to be gained from any of these maneuvers is not transferable to any commercial operation. With incredibly rare exception, I have never seen a student relate any skill to anything but that individual skill.

What is equally interesting is what the commercial does not test, but you see regularly in normal operations. For example, ATC frequently assigns the short approach to aircraft for the expeditious flow of traffic, yet it isn't on any flight test. Your commercial flying will take you through ARSAs and TCAs (Classes C and B); yet there is no requirement to demonstrate that you can fly through them properly. Many of the commercial maneuvers would be great for club fly-ins and competitions instead of being required for the certificate. What we should also consider is teaching procedures that should be learned just for their own sake, that are lots of fun, with maybe some peripheral benefit to be derived, where we do not require the time, practice, or precision necessary for testing them on a checkride.

When you were learning cross-country skills during your private, the emphasis was on safety, building your experience slowly, and deciding when it was safe to go. Commercial operations are different. Your training should be geared to reflect the fact that a commercial pilot fills a critical niche in the overall operation. Private pilots operate on their own. The commercial pilot is part of a team that flies the planes, fixes the planes, markets the service, and administrates the company. Your superior pilot skills are critical to move the people and goods as quickly and efficiently as possible to make a profit for the operation.

This is lost on most instructors who would have you believe that performing a pretty chandelle will somehow land you a good job. The commercial pilot has to think about what will save the boss more money: the lengthy wait for an IFR departure with the assurity of getting through or a quicker special VFR departure out of the airport area with the possibility of having to return because of the weather? If you are on a VFR flight, can you find a more direct route that bypasses the need for wandering from station to station using VORs? These are the questions that you should be digging into during your training rather than the current system which amounts to little more than getting your private all over again.

The commercial could be a useful certificate if we restructured it to reflect the needs of commercial operators and the airlines and prepared pilots to meet those needs. The first thing to change is the eligibility requirements. The total flight time required for the commercial should be amended so that all applicants must have 250 hours, for experience is the best teacher. In order to be eligible, you must also possess an instrument rating. In the flight time between your instrument rating at 125 total hours and the 250 needed for the commercial, you would be required to spend 50 hours in solo IFR cross-country flight, 35 hours of which must be accomplished in actual or at night. It staggers the mind that there is no requirement to either acquire or practice using the instrument rating before becoming a commercial pilot. After the cross-country time, the remaining hours would be dedicated to a totally new training program to include high-altitude operations, hands-on experience with turbocharging and pressurization equipment, and training in crew coordination, cockpit resource management, and line operations flight training.

There are many procedures that could be added to the Practical Test Standards that would make for better commercial pilots. The first is the short approach mentioned earlier. Besides being useful to ATC, it gets you on the ground quicker which saves the boss money. Short approaches require keen judgment as the aircraft is in a bank at very low altitudes with little time to maneuver into a stabilized, modified final and land. The next thing to add is the long straight-in landing. Straight-in approaches require judgment of speed and distance and the ability to plan a descent from quite far out.

You should also get the feel of high-performance aircraft at gross weight. All of your training once you stopped dual lessons in two seaters has been in aircraft that were loaded far below capacity. Aircraft feel very different with only half the seats and tanks filled and no baggage except for your flight bags. The safety and speed margins are cut way back when you practice maximum-performance take-offs and landings with aircraft at gross weight. You should also be required to recover from accelerated stalls and spin entries in properly certified aircraft.

When you do acquire your commercial certificate, you should be entitled to operate for hire, just like it says in Part 61, by any Part 135 commercial operator for any operation VFR or IFR. That should do away with the separate hour requirements of 500 and 1200 hours and the need for a separate Part 135 competency check. It would also allow pilots to move straight into commercial operations without the need to instruct. This would force the flight schools to stop treating instructors like indentured servants and create the need to hire only dedicated professional instructors and pay them a livable salary.

We now need a new checkride to reflect our new training regimen. The flight test would consist of two cross-country legs of 150-200 miles each way. One leg to be flown under IFR, the other VFR. They will be planned and flown as an actual commercial operation. You will use a high-performance single or multi-engined aircraft, with a full passenger load, with baggage, and enough fuel for the first leg. The fuel and baggage will be loaded so that the aircraft can depart at or about gross weight. Your examiner and volunteers will be briefed properly and treated exactly as passengers on a regular charter flight. The IFR leg of the trip will include one actual hold assigned by ATC, one stabilized, and one nonstabilized approach at a tower-controlled field. One of those approaches will be flown to minimums, followed by the published missed approach procedure.

For pilots requesting single pilot IFR privileges for hire, the use of an approved 3-axis autopilot will be demonstrated en route. After the first leg, the oral will be given. The volunteers can go to lunch. The VFR leg will include relevant maximum performance maneuvers including short- and soft-field takeoffs and landings from actual short and soft nontower fields along the way. The simulated emergency procedures during the VFR leg will include: for a single engine aircraft, an engine failure to a landing at an airport within gliding distance from the en route altitude; for the multi-engine aircraft, a single engine landing from cruise altitude; for both aircraft, a full electrical failure, a gear failure, and a no-flap landing. The simulated IFR emergencies will consist of lost communications, an encounter with unforecast icing, and a vacuum failure.

Upon completion of the test, you would be able to work for compensation or hire for any operation as a commercial pilot. Should you only desire to work for hire in VFR operations, a restricted commercial certificate would be issued limiting you to 50 miles in day VFR conditions for any operation.

By effectively combining the instrument and commercial into one certificate and throwing out the irrelevant maneuvers, we can resurrect an obsolete certificate and make it significant for the needs of aviation today. We can test pilots in realistic commercial situations that could and do occur in both VFR and IFR operations and turn out well-rounded, capable, commercial pilots.

12 • BECOMING AN INSTRUCTOR

———————————————————————————➤

So, you want to become an instructor. Do you have any idea what you are about to go through to make it to your goal? If you knew in advance, many of you wouldn't even get started. What gets most of you through is that you do not know what to expect going in, and by the time you find out, you have too much time and money invested to quit. You are about to realize that up until now, you have been able to virtually bluff your way through oral exams.

Now your knowledge will be challenged as if you were defending your doctoral thesis before a board. Your level of knowledge going into this challenge is woefully inadequate, such that you will feel you are learning to fly all over again. Your brain will churn with the burden of infinitesimally trivial factoids desperately crammed into each lobe of your cranium in order for you to appear brilliant before the FAA inspector holding your whole career in his bureaucratic hands. Your standard of knowledge is this: When you know what there is no reason to know, you will know enough to become a flight instructor. What a pisser.

It is said that those who can't do, teach. If that were true, there would be aircraft strewn all across our landscape. People do not become flight instructors because they cannot fly. On the contrary, there are many exceptional pilots engaged in flight instruction. The irony that excellent pilots should be forced to become excellent teachers in order to watch other people fly for two years, such that they will build far more teaching expertise while giving away all their flight skills in order to become supposedly better qualified commercial pilots, is completely lost on the FAA. People become flight instructors because it is the *only* pathway, not a better pathway, when they set out to become commercial pilots. Why else would people tolerate spending a fortune learning to fly, only to give up flying. Flight instruction is therefore reduced to a ritual time, a rite of passage, or an apprenticeship, if you will. It is rarely, if ever, voluntary.

The contradictions build because this certificate has by far the highest standard of flight training because of the awesome responsibility the FAA puts on instructors. Why would the FAA force the greatest responsibility on one of the least desired and most required professional certificates? Unlike those who build their time in more narrow fields such as pipeline patrol or glider towing, the instructor is exposed to an incredible variety of situations requiring immediate analysis, solutions, and actions. This is because students have no knowledge usually and certainly take no responsibility for the creation of their disasters. Therefore, the instructor has to be the best trained and fastest reacting pilot of any of the certificates and ratings. Over the course of two years teaching all those possible certificates and ratings, the average instructor will have seen enough action to be prepared for almost anything that can come up. So the time an instructor is most likely to get actual flying experience is when the aircraft is in imminent danger.

The reason pilots have to build the flight time through instructing to operate as commercial pilots — is money. If given the choice of instructing, or the money to fly a light twin around the country for 1000 hours to build the time necessary to get that low paying commercial job, pilots will always fly rather than instruct. There is no choice.

There is no connection between the ability and desire to fly, and the ability and desire to teach. Many flight instructors absolutely hate teaching, are great pilots, do a wonderful job anyway, and can't wait to leave the profession. There are lousy pilots who love teaching, can't imagine doing anything else, and turn out reasonable students in spite of themselves. The desire and competence to teach vary infinitely among the individual pilots. Since almost everyone is forced to teach, you get almost every kind of pilot teaching. If you must instruct to further your career, then you must do what it takes to become an instructor.

Contradictions abound in flight instruction. The FAA treats instructing in their literature on the same level as a calling to the priesthood, where the CFI is devoted to some lifelong mission to instruct purely for the love of teaching. Perhaps that is why they care not that instructors take the equivalent of a vow of poverty as they enter the profession. The reality created by the FAA is that the CFI certificate is your legal entitlement to set up your own pyramid scheme to attract enough students to pay for the hours you need for your real career. Some of your students will go on to become instructors to do the same thing. All the sugar-coated glamor the FAA can put in their instructor materials can't cover the cold, hard reality that they themselves created and maintain to this day — that flight instruction is nothing more than a feudal arrangement of servitude.

Take a look at the math from an instructor's perspective. The average pilot can be instructing with about 300 hours, but won't be qualified for a job these days until he has about 1500 hours. If you take the 300 for training and another 200 for actual flying you do in training for further ratings, student demonstrations, flying out of tight spots, and the occasional flight you actually take just for fun, you will spend 1000 hours or more doing nothing but watching people do things incorrectly that you already know how to do. How can this possibly prepare you to be a better commercial pilot? There is a conspiracy here to get cheap labor out of new pilots. Maybe that will be the topic of the next book?

Although you now have several hundred hours of watching people give you flight instruction plus several years of education in some type of school system, you will now be treated as if you have never seen a teacher of any kind in your entire life. So that you may know how the whole process of learning works from the point of view of a stranger to the concept, the FAA has dissected the learning process into a set of theories called the "fundamentals of instruction." To help you become an effective instructor, the education psychology presented in the *Aviation Instructor's Handbook* is appropriate for analyzing the motivations, frustrations, failures, and successes of your students — when you get them.

Having likely spent the better part of your life so far in school and all of your flight time as a student, you know only too well how this psychology works from the wrong side. To counter this, the FAA has made a valiant effort to present the learning process from the instructor's point of view. The problem is that seeing instruction from the student end is so ingrained in you, that you cannot make use of the teaching theories you have to learn to become an instructor. The FAA has not yet created the bridge from the psychology you have internalized in school and flight training for avoiding learning to presenting the theories of learning in a form that allows you to use them before you become an instructor. Consequently, the theories of learning and teaching become nothing more than an exercise for the checkride, to be quickly forgotten as soon as the certificate is in your hands.

Anytime you work with the mental and emotional processes people use to learn, an in-depth knowledge of the workings of the human mind is a tremendous teaching advantage. The *Aviation Instructor's Handbook* is the FAA's attempt to explain those workings to the instructor candidate. The levels of learning described in the book are rote, understanding, application, and correlation, with correlation being the highest and most desirable level. Correlation is the ability to transfer the knowledge gained from learning one skill to solving a problem from an indirectly related situation.

The most common example is how the rectangular course teaches students how to fly a traffic pattern. We already know that flying rectangles outside your references only teaches students how to fly outside a pattern and parallel to a runway, so their prime example is shot. This is a perfect example, however, of a negative transfer of knowledge, the diametric opposite of correlation.

We also know that students cannot engage in any positive transfer of knowledge; for if they did, instructors would be out of a job as students could soon teach themselves. They don't because of the fundamental flaw that students neither have nor take any shred of responsibility for their training. Therefore, no level of learning above the most basic will ever be attempted by most students.

Since the whole theory of instruction is built around the ability to raise students to a level of understanding that overwhelming evidence and experience have proven impossible, the entire theory of the levels of learning collapses from the top down. Working back down from correlation, the theory holds that student pilots will apply what they have learned on their own. This isn't true either because of the need for instructor approval, the need to avoid embarrassment and criticism at all costs, and the need to hide the fact from everyone that they are student pilots. Therefore, virtually no application of a procedure will take place without prompting by the instructor.

As for understanding what has been taught, the next level down, most students are content to understand a maneuver only to the extent necessary to pass a flight test. We know this because when asked the procedure for a short-field takeoff, most students will demonstrate an understanding of the procedure. What they won't understand is that they have to use that technique if they ever land on a short field and want to takeoff again. This is why they can't advance up to the next level and apply the short-field technique when faced with an actual short field.

This leaves us with the lowest level of learning — rote. Rote is the least desirable level of learning according to the theory, as it is simply the parroting of information, or the mechanical repetition of mechanical skills, without any objective or purpose. The only possible conclusion is that our entire flight training system is built around rote learning, despite any claim by the FAA to the contrary. This is proven every time students say they did something "because my instructor told me to." At this point, you have an internal conflict that will never get resolved because you are being forced to accept flight training theories that go against every emotion and experience you have ever had throughout your own training.

Rote pervades everything we teach. In your gut you know it. Even the FAA's comprehensive attempt to delve into the inner workings of the mind of a student pilot is sabotaged by rote thinking and analysis. You must learn the theories of learning by rote, repeat them back to your stage check pilot by rote, fill in rote answers to questions on the CFI written exam, parrot back the information to the FAA inspector by rote, and finally you will teach these theories to other CFI candidates by rote. Because of the conflict between learning everything by rote and the disdain for rote training given to you from the FAA, anything you learn by rote is something to be discarded after it has served its purpose.

This is why it is so easy to forget the theories of learning so promptly after the checkride. You will even be encouraged by CFIs during your training to forget the entire *Aviation Instructor's Handbook* when you become a CFI as incentive to get you to learn it for the checkride. Everything is backwards. You will forget these theories at the very time when you should be studying them the most, which is *after* you begin teaching.

Just as the instrument student has no visual back-up for the ADF, the instructor has no mental back-up for the theories of learning until they are applied to real students. You can't see this in your instructor training because you never get to work with real students; only CFIs can do that and they don't use this material anymore. Everything is still backwards.

Although seldom directly relevant to any particular student or situation, great general insight can be gained by referring to this handbook long after you have started teaching. This book is the most useful when it is not even used at all. When you go over the fundamentals of instruction, try to relate them to the motivations, successes, and faults of your own training, as you are the only direct experience with a student pilot you will get. There is so much to be gained from these theories of learning that it is a waste just to learn the stuff in rote fashion to pass the checkride. Since we know our system of training only teaches the first level, the trick now will be to change our students so that higher levels of learning are demanded by them from their instructors.

As for the process of learning I, of course, have my own interpretation. Learning, according to the book, is a "change in behavior as a result of an experience." Here is how it works in the real world. You tell the students what to do and they change their behavior. If they do not change to what you have told them, then you give them more experience.

"Learning is purposeful." The students will do what you say because they want their pilot certificates, that is their only purpose. Students will do anything you say because they figure that you know what it takes to pass the flight test; therefore, no further justification need be made in their minds. That is how procedures are retained by rote with absolutely no effort to raise the level of knowledge to understanding, application, or correlation. Those levels just aren't relevant in the mind of the student. This is why flying a traffic pattern and practicing rectangular courses have absolutely nothing to do with each other as far as the student is concerned.

The FAA says that learning cannot be achieved by spoonfeeding; yet this is the only possible method of learning under the constraints of our current system. The vast majority of student pilots limit their own potential by learning exclusively by spoonfeeding. Spoonfeeding is encouraged as a desirable quality in commercial pilots. In my experience, the large flight academies and commercial operators have all graduated from the spoonfeeding school of instruction, and therefore promote new pilots with identical qualities. This systematically weeds out the small percentage of pilots who refuse to submit to rote learning, think for themselves, ask questions, and try to analyze the real purposes behind the training, if there are any.

Rote learning is perpetuated by advancing pilots who learned by rote to teach through rote to their students. "Learning is an active process." Wrong. Nothing is more passive than our current method of teaching because rote learning is not just accepted, it is required. Now you know why you have a mental conflict during your training.

Much of aviation we stole from somewhere else. We stole wings from the birds, navigation from the mariners aboard ships, governors for our propellers from factory machines, and passengers from the railroads. So we might as well steal the Laws of Learning from Professor Thorndike of Columbia University. These laws were written early in this century and have been only slightly changed, added to, and amended since then; so we are still dealing pretty much with turn-of-the-century thought. What is really needed is a complete overhaul to make these laws relevant to our current flight training system. They haven't been, so let's see how they stand up today.

The Law of Readiness states that "Students learn when they are ready to learn." The truth is that if they show up for a lesson with money in their pockets, they'll be ready. Since they are paying for a service where nothing is required from them, they will only be ready to be spoonfed. You tell the students what to do and they will be ready to do it. If they do not do what you tell them, it is going to cost them much more money, so depending on their wealth, sooner or later they will be ready. Writing checks without getting closer to the checkride is an experience that changes their behavior.

The Law of Exercise states that "Things most often repeated are best remembered." The problem is that students become very exercised when they have to pay to do things that they already know. This ties in with the law that "Learning is strengthened when accompanied by a pleasant experience." One of the hardest jobs in instruction is to keep things fun for the student. Students have fun when they know what they are doing. This is rare in a lesson because you have to continually work on what they do not know, since nobody really learns much from something they already know how to do, however much fun it is.

Your job as an instructor, therefore, is to walk the razor edge of challenge, balanced between boredom and frustration. Unfortunately, students have an aversion to in-flight models or demonstrations by instructors. Whether because they are paying for the flight and want every penny's worth, or whether they don't know what to expect and therefore don't quite trust the instructor, the net result is that the things they see most often repeated are their own mistakes. You, therefore, have to make things pleasant by inspiring their sense of accomplishment in meeting challenges, not let them repeat their own mistakes, and not let them get bored or frustrated.

The Law of Primacy states that "Unlearning is harder than learning." Rote learning causes the belief that the first rote way you learned to fly is the way all your students should fly as well. When procedures are learned without question, they are thought to be the only correct method. When you fly with other instructors' students, you will see great variations in their rote methods compared to your own. Since you believe that you have the only correct rote methods, you believe all your students should fly exactly as you do, and you will try to convert your fledgling students. This is where you will discover that unlearning is harder than learning.

Your real goal in becoming an instructor is not teaching flying, but learning through teaching how to fly yourself. You must learn how to evaluate these other methods and, if superior, make them your own. That is the way to build the best repertoire of teaching material. You must also let students stick to a method they know, if it is safe, until they have the skills to adapt to possibly better methods which you may have. As an instructor, you should be flexible enough to teach many methods. However, your students, wherever they come from, should be afforded consistent training.

The Law of Intensity is my personal favorite. "A vivid, dramatic or exciting experience teaches more than a routine or boring experience." I live for this rule. You must not interpret this as an excuse to ever scare a student. No one can ever sanction putting an aircraft in any potentially dangerous situation just to make the lesson more lively.

Having made my qualification, if direct experience is the best teacher, how can the FAA justify leaving the spin requirement in the knowledge only category? Why can a simulator substitute for real clouds? How can a private pilot be given a certificate without the need to ever fly from a short or soft field? Coddling students eventually breeds curiosity to find out why any given warning is handed down from an instructor. What we don't want (which all too often happens) is a situation where students unintentionally create their own vivid experiences. The big problem with all these laws is that the FAA makes no effort to live by them, preferring instead to perpetuate a brainless system of rote learning.

There are a few laws of which you should be aware before entering the profession of instructing that are not covered by the FAA. Most, if not all, of your students will prove to be stellar examples of the "Law of Laziness." This I define as the sincere desire to acquire a certificate while exerting the least possible effort in the endeavor. Most pilots are addicted to rote learning because it is easy. As long as rote serves the purpose of teaching flying, the student will be happy. In taking no responsibility for the material presented in their training, the leap is made to take no responsibility for guiding their training. Students make no decisions affecting their training for fear of then having to do something about it which would require effort; nor for the same reason will they exercise any initiative that might accelerate the acquisition of a certificate. This is why they can't demand changes in any flight program and why you have total blame for any failure on the part of the student.

The only perceived options open to a student with a training problem are to change instructors and/or change flight schools. Since all flight schools and instructors are tested and held to the same standards and both operate in the same system, the differences to be found, except for some very independent operations, are minimal. If neither of these options brings satisfaction, the student must either mold to a flawed system or give up aviation; giving up requires less effort.

The "Law of Serendipity" applies to those rare students who automatically brief you on the weather, bring questions from unassigned readings, preflight before the lesson is scheduled to begin, plot diversions to critical weather without prompting, and do everything possible to be the best pilots they can be. These are the people who will make your job worthwhile. They are a rarity in a society constructed to breed out personal responsibility and replace it with passivity and allegiance. In a system that rewards rote flying, these aberrations will be your best pilots while at the same time having the least chance for commercial success because of their independence. The only rugged individualists flying today are on the movie screens or working for themselves.

As an instructor, you have to live with the "Law of Jesus." This is because students induce situations from which they, and you, must be saved. Like blind followers of a religious cult, students pilots offer their unconditional obedience in return for guaranteed salvation and safe deliverance to the ground. Any action a student takes contrary to safety, whether you are aboard the aircraft or not, is your responsibility, for the FAA has determined that you are a deity and they are lambs.

The dark side of training takes the form of the "Law of Anger." Students learn techniques more to please their instructor rather than to become good pilots because for most techniques they haven't the slightest idea of any purpose beyond pleasing an examiner the same way. When the approval of the instructor is the only possible justification and reward, the student ceases to fly for himself and seeks only to please the instructor. When the student cannot please the instructor the first time, they become embarrassed, frustrated, and eventually angry.

Many times it is the anger directed at the instructor that causes the student to exert the effort required to make the breakthrough and learn the procedure. Talk about a back door to knowledge, this phenomenon proves very successful a disturbing number of times. I have never known an instructor to intentionally provoke a student, so what happens is an accidentally positive outcome from a decidedly negative pathway.

The bulk of your preparation for the CFI test will be spent on the ground. You have already accumulated several hundred hours so you already have flight skills, even if they are from the wrong seat. One way around this is to alternate seats throughout your lengthy commercial training so that you are comfortable on either side. One of the best things you can do is to find an instructor who has recently passed other applicants through the CFI test at your Flight Standards District Office (FSDO). The unwritten standard is that CFIs have to know everything about everything. Since you can't do that, you have to take every advantage you can, which means using an instructor who knows what any particular FSDO or even inspector is stressing at the time.

The mental training for the CFI is exceptional because of the depth and variety of subject matter that must be mastered and instantly available for recall. Most commercial pilots have the luxury of developing an expertise in their operating sphere, whereas the instructor must be familiar with all subject areas, from all the ratings and certificates, all the time. The constant research and study required to get the certificate and long hours of work afterwards to maintain that knowledge make instructors the best informed group of pilots around. Your quest becomes the pursuit of the trivial factoid, where you develop the habit of seeking the answer no one knows to the question no one has bothered to ask. Get a group of instructors together and they will start quizzing each other.

Knowledge is essential to acquiring the certificate and

power after it is obtained. Somewhere along the way, you will be struck with the realization that CFI training is geared to train you to retain and dispense information. Although you sit in airplanes, your flying days for the next few years are over. When you can't fly because your students are doing all the flying, you will fall back on the only thing that makes you more important than your students — your superior knowledge. This is how you begin the practice of knowing what there is no reason to know.

The FAA wants you to be able to teach from a lesson plan. This one-page, detailed outline for a single session with a student takes more time in preparation than it does to teach the lesson. The FAA wants to see a comprehensive lesson plan for every scheduled lesson you teach, presented in such detail that any instructor could pick up your lesson plan and teach that lesson. This should make you wonder why the FAA has taken such pains to describe the uniqueness of each student, the learning situation, and the quality of the individual instructor/student relationship.

Now they say that one lesson plan fits all. If this were true, then any instructor should be able to endorse a qualified student rather than the specific one who gave the particular training. If this is not the case, then why do we have lesson plans open to all instructors? There is no flexibility to a lesson plan nor the ability to adapt to current conditions. What if you are done with your topic and have more time to fill according to your schedule on the plan? Your student may wonder if you forgot something. Should you then recalculate and borrow some time from somewhere else? What if you stay up all night making lesson plans, and for the next day you have made an intricate plan for one student involving steep turns and slow flight, where by lesson time the weather is perfectly suited to ground reference maneuvers?

Since the FAA says every lesson must have its plan, you either have to stick to the plan or make up a new one. On your low pay, get real! According to the book, generic lesson plans won't do. Each instructional period must have a lesson plan, so you can't just yank the ground reference plan from the old file.

There is such universal disdain for the whole concept of lesson plans that chances are the first time you ever heard of them was well into your CFI training. CFIs have neither the time, energy, nor incentive to spend even one minute on lesson plans; only CFI candidates do because it is necessary to get their certificates. That is why you have never seen one in your own training.

Why does the FAA require you to be able to do something you will never use in your job? The only use I have ever found from the stack of lesson plans I wrote out by hand was as a quick reference the first couple of times I taught a particular topic. In the long run, you will find that the *Flight Training Handbook* has better information without the need to write out a thing. The reason for writing out all those lesson plans is so that you can show your FAA inspector that you know how to do it and can teach from one, even though you both know that you never will. Sounds like needless hypocrisy with two co-dependants, doesn't it?

My flight school actually had all the possible generic lesson plans on file for the CFI applicants to photocopy. For some unexplained reason, I felt the need to write out all my own plans. This actually worked out because on my CFI checkride, my two inspectors made me write out lesson plans for them exclusively from memory. The lesson plan is nothing but a joke in the real world.

What you might find useful is a syllabus to organize and outline a whole course of flying in a logical sequence. You academy instructors will have to work from your syllabus as a condition of employment under Part 141. As a freelance or club instructor, you might be free to develop your own program, or not. I teach straight from the Part 61 requirements and the *Practical Test Standards* and that works out just fine. What you will do after a while is to develop a style of teaching that accomplishes your goals during a lesson by roughly following a fluid mental outline. When you acquire that skill, you can instantly adapt to both the individual student and the current conditions. The only way to develop the sense of working from a mental outline is to practice.

This is why the FAA should immediately abandon the farce of lesson plans and have CFI candidates practice teaching lessons to real students, while being supervised by the CFI giving the training. The CFI candidate will receive immediate feedback from both the CFI and the student as to what worked and what did not. That way applicants could bypass that time after the checkride where they have to develop their ability to teach from mental lesson plans.

In ancient Greece the value of the oratory was highly praised. The spoken word in Shakespeare's time was a measure of a person's worth and status. The ability to implant wisdom and knowledge through speech has all but lost its value in these turgid times as we are awash in a flood of sound bites and visual sensations. Before you can walk like Socrates in a garden imparting the secrets of flight to eager students through the power of your speech, you first have to know what the hell you are talking about. You can bluff your way through the private. You can bluff only slightly less easily through the commercial. You will not bluff your way through the CFI. Besides, don't you think it is about time you learned how an airplane works and how it flies?

Your training will now combine the ability to speak with a knowledge of the material. Unless you have a good deal of public speaking under your belt, you are now going to learn how to speak clearly, present ideas, and test whether the message got through to the receiver. In school you write all your tests; rarely is an exam given orally. Public speaking is one of our greatest fears. Corporate executives who have to present complex information hire overpaid consultants to create verbal dynamos out of meek managers.

Once again your background has ill-prepared you for an essential skill in aviation, in this case the ability to speak to a student or a classroom of students and have them understand what you are talking about. Your lesson plans are useless on the job, but before the checkride they allow you to develop your skill of speech and your ability to lucidly flow through a topic. A simple outline, however, would do the same thing. You will find that what separates you from the instructors you find so knowledgeable and intimidating is that they have developed their communication skills, nothing more. Anyone with practice can do the same thing.

Behind the starched uniform and epaulettes is a person, just like you, who had to learn to talk authoritatively, just like you. The mystical quality that shrouds CFIs like gods from Mount Olympus burdens the CFI candidate with overwhelming doubt that he will ever be good enough to join the exclusive club. Over time, though, you find your voice; every day you practice, you will sound a little more like an instructor. You will discover that sounding like you know what you are talking about is as important as knowing what you are talking about. Be careful because student pilots will believe almost anything, but FAA inspectors will not. When you can absorb and relate information with authority, you will know that you can make it.

Take a good long look at the *Practical Test Standards*. The FAA says that you only have to fly to commercial standards. You will notice that for every possible maneuver you will not only have to fly it, but simultaneously teach it, and know all the common errors associated with it, which is tough considering you haven't been allowed to teach anyone, unless you have made all the common errors yourself. For any other test you only have to know one way, the correct way; for the CFI you have to know all the ways (both correct and incorrect) that your inspector knows. This is the toughest test there is. Every other test has specific flight tolerances, but not the CFI; for this you have to be perfect. It is a completely unreasonable standard for a completely unreasonable test.

The lack of any specific flight tolerances in the *Standards* allows for incredible subjectivity on the part of the inspector, which may account for the very high failure rate of CFI applicants. You do not so much pass the CFI test as beat the taskmaster (or in my case masters) who put you through hell. Somehow we make it through, but we never forget the experience as it indelibly burns into our memories over the course of the test.

However, the *Standards* are only the beginning. Any minutely detailed microbit of information locked in the deep recesses of the inspector's mind is fair game for your test. You must know all the details of every maneuver from the private and the commercial to be able to fly them perfectly every time. You must be able to read the mind of the inspector to fly each maneuver as he wants it flown, or be able to instantly adapt to any newly imposed conditions or limitations. The *Standards* are rigorous and exquisite in tasking you. You will fly your best for the checkride that you can possibly do. On the job you will watch with irony as you try to raise and inspire your students to the standard of flight you had when you became a CFI, while your own flight skills begin the downward slide towards that of your students.

Review what constitutes a satisfactory performance, and you will see that it is not enough to have a vast technical knowledge and be able to communicate effectively, nor is it enough to fly with perfect precision. You must be able to continually execute both skills simultaneously. Yuk. It is going to take a lot of work to be able to both teach and fly at the same time. However, it is just another skill for you to master. All your instructors have flown and taught like it was nothing. After many hours of dual given, it will become nothing for you as well.

You have to have the ability, though, before the checkride without the benefit of practice. This is not a skill that comes naturally. It is the ultimate division of attention. Your flight skills have had a few hundred hours to develop, but your teaching skills have only had a few hours practice. You need to make your teaching voice come as naturally when you are flying as when you are teaching on the ground. As always, the way is practice. Go home and teach a wall or a chair how to fly. Perhaps a mirror will work if you are not too self-conscious. After the furniture knows how to fly, find willing victims who know nothing about flying. See if they can make any sense out of your blabber. When you find your voice on the ground, your voice in the air is soon to follow.

What is really going to freak you out is learning to fly from the right seat. It just feels weird. All the switches are in the wrong place. The feet work the same, but the hands are reversed. You lose your instinct for flying and have to think about all your control movements and power adjustments just like back in your private days, except that now you are frustrated because you think you should be doing better. The problem is that all your sight pictures are new and all the instruments are at an angle. What you must do is readjust your eye/brain orientation.

This problem will manifest itself most acutely on landings because you won't know where the ground is anymore. Your speed and height judgment will be completely off. What this means is that you may never have known where the ground is; however, you compensated by finding a sight picture that gave you a reasonable landing most of the time. If you were never taught properly where to look when you first started landing the airplane, you will now have to go through that same process of trial and error sight picture by guess that you used the first time. Once you know how to look over the nose directly in front of you and keep the nose on the end of the runway, you will be able to fly equally well from either seat.

Pilots with hundreds of hours just like you still swing the airplane around to line up the spinner with the centerline. Of course now that you are in the right seat, you swing the airplane around to the right. This is how we can tell who are the student pilots and who are the instructor candidates. The advantage for the flight school to having a lot of instructor candidates is that the aircraft tires wear more evenly.

It is amazing how a habit you acquired from your first

few hours (when you were taught to taxi with the nosewheel lined up on the centerline) can come back to haunt you when you want to become an instructor. Now you see why it is so critical to establish good habits early. You also see that a simple error may have a long history of complex factors and that treating the symptom will not always cure the problem. When you go through training and learn the rote common errors to please the inspector, remember that there are no common solutions and that any longstanding errors may have their origin as far back as a student's first flight.

Another thing that will feel funny is teaching someone who knows far more than you at this point how to fly a particular technique. Once again the avoidance of experience in the real environment will distort your training. Your instructor and the FAA inspector are the hardest people to deal with because you know they know, yet you have to treat them as if they do not know. They have to listen to you teach as if they do not know as well. It all gets very confusing. Students are much easier to teach because you know they do not know. They don't even know if you do not know, but the inspector will know even while pretending not to know. This is why you have to know as much as they know, if you can.

The CFI is part scientist, part engineer, part psychologist, part babysitter, mostly a sales/marketing representative, and once in a great while when you get to touch the controls — part pilot. Because CFIs for the most part have stopped flying, you are set apart from all other pilots into a select group by virtue of your knowledge in all these other areas. In a back-handed way, teaching flying will make you a better pilot because when your brain is trained by thousands of hours of instructing, your hands and feet can be quickly taught to fly the procedure printed out by your brain. Through a love and disciplined commitment to knowledge and excellence, the CFI builds a storehouse of wisdom that separates you further from other pilots the larger your wealth of wisdom becomes. To get to this point requires only an inquisitive mind and lots of hard work.

But for those pilots who have not made the journey, the CFI is just a bit different, a little special, on a slightly higher plane, if you will. CFIs cultivate this image through continuing to increase their storehouse through study, higher qualifications, and teaching in more prestigious aircraft. The gratification bestowed by students puffs your ego. You begin to feel it is completely deserved considering what you have been through to get here. It is like doctors who feel justified in gouging their patients and the government with exorbitant fees. The difference is that you have only your ego because instructing doesn't pay.

The CFI checkride is the FAA's version of fraternity hazing and initiation. Once through this rite of passage, you will have joined the club of all the CFIs who have gone before you. There is an instant recognition and bond among CFIs and a communal snobbery over any perceived lesser pilots. There is a mutual respect for any CFI; that is, until you encounter a clash of personality, ego, style, knowledge, or someone cuts into your meager profits. You join the club by surviving the checkride. You remain in the club so long as you continue to teach.

This club has a definite hierarchy. Stage check instructors are more special than regular instructors. Chief instructors are the most special of all. Your status in the club is tenuous as any serious mistake will bring the wrath of the FAA, and worse — your peers. Since you must instruct in order to achieve your goal of getting paid to actually fly airplanes, you must survive this time with your record clean. Instructing has the best and the worst that flying has to offer, and it will forever mold you as such.

Because knowledge is everything to the CFI, let's take a run through the *Standards* and see what you are up against. You would think that every commercial pilot should have the same level of aeronautical knowledge, but that isn't the case. As you gain this knowledge, you will begin to feel like you never really knew how to fly, that you really did bluff your way through those early checkrides, and that your survival in the air up until now has been largely luck

In order for you to pass on the gift of flight, you must know exactly what flight entails.

By analyzing every possible bit of flight knowledge, you will be able to take students through the various ratings and certificates because you know when they are correct, you know when they are in error, and you know how to correct their errors. You can't do that until you know all the possible deviations. Students have to learn only your right way; you have to know all the ways. That is the theory anyway; the reality is something else. During this process you will learn to fly all over again.

The first thing in the *Standards* is the learning process. How do people learn? What gets in their way? What is really behind a hostile attitude or unreasonable fear? Anyone who makes it through this training should get a counseling license along with his CFI certificate. You should also qualify as a motivational trainer, although keeping students motivated has as much to do with paying your rent as being the consummate professional. Since the fundamentals of instruction are treated as just something to get passed, it may be more than a year into your work before you realize just how much time you wasted with students because you didn't delve into learning the human factors beyond the rote level required for the checkride. Also, you redundantly hammered the correct technique into students during lessons futilely trying to provoke revelations, when in reality it is a human factor blocking any progress.

Sometimes more can be accomplished by just chatting over a cup of coffee with a student, away from any peers, than can ever be beaten out of a student in an airplane. Where our system breaks down is not in the availability of information; it is in the mindless repetition of abstract theories separate from their application to real students with real problems. Since the theory is what you will be expected to know, that is what you will memorize. For your success as an instructor, you should find out how the learning process, the teaching process, and the human factors apply to real teaching by challenging your own instructor.

If you can put a real world meaning to these theories, you will not only have a handle to understand them better, you will be a far better instructor after your checkride.

Start this process by analyzing your own emotional growth as a pilot. What were your fears, frustrations, ego boosts, guilt trips, shining moments, and crashing bores? Student experiences are painfully similar. If you do not analyze how the system treated you, you will perpetuate all its worst faults on all your students. Like the leader who refused to study history, so you will miss the real value of the fundamentals of instruction.

What you cannot know during your training, but what you will come to fear the further you progress with instruction, is the awesome responsibility placed on our shoulders. You are literally responsible for life and death. You are responsible for your students as long as they fly on your endorsements, for everyone they fly over, and everyone who flies with them after they are rated, for an indefinite period of time. This in such litigious times would normally require instructor malpractice insurance. CFIs, however, are protected by poverty. You can sue a CFI for everything he has, and the instructor will still come out ahead.

The FAA takes great pains to point out instructor characteristics and responsibilities and devotes a whole task in the *Standards* to test your cognizance of the subject. You have to be a true professional. In my book that means anyone who gets paid for a service. The FAA, however, views the instructor as some missionary cheerfully suffering for the cause of aviation. I guess that's how they tolerate such abysmal working conditions in the flight schools without objection.

Selling yourself into bondage to become the FAA's idea of a professional looks like this: you must be well-prepared and well-trained; you must provide the highest level of service; you must continually engage in further study and research; you have to perceive all relevant human factors; you must be logical and be able to think fast; you have to have good judgment; you must be ethical, sincere, honest, and most of all knowledgeable. This is not a job description; it is a canonization for sainthood.

You would think such lofty qualifications would command a commensurate salary. The truth is that you must be all of those things and more while scraping by on far less than the federal minimum wage. You will find that you can only bill about half the time you actually spend with your students unless you charge ground time from the moment they walk in the door, which you cannot. Still want the job? Read on.

You must be a psychologist and flush out any personality defect that may cause a hazard to passengers, property, and people on the ground. Due to economic pressure, you may have only one flight in which to accomplish this. You have to council the feeble through their anxieties to enable them to realize their goal of flight — if possible.

You are also responsible for keeping the dangerous pilot out of the air. Some calls will be obvious; but what about the pilot who does most things right most of the time, and you have a strange lump in your gut with no rational evidence nor anything concrete on which to base your feelings. Who will back up your decision not to let that pilot fly? Will you be able to hold your ground despite vociferous objections to management? Will the boss back you up? If you let the pilot go and nothing happens, you will look foolish. Something bad has to happen for you to be proven right. However "I told you so" just won't suffice if something bad should happen.

You cannot win with the responsibility you are taking on. It gets worse. You will have to solo people. You let them fly for the very first time all by themselves, and you are absolutely responsible for this action. Students will pressure you into soloing because they are not responsible. When you let students fly cross-country, you are even more responsible given the greater distances and potential things to run into. All your students, endorsements, checkouts, biennial flight reviews, in fact anything you sanction or put your name to is your total responsibility. There are times when you will be responsible for the actions of people over whom you have absolutely no control. Most instructors survive this period in their aviation career primarily by luck, because there is so much that can go wrong. What a system.

The CFI must have a knowledge of human physiology way beyond the basic aeromedical knowledge required of other pilots. It is one thing for a pilot to deal with himself or a passenger; it is quite another for an instructor to educate students to recognize and deal with themselves and passengers with aeromedical problems way into the future. The most important factor is not even covered in training — denial. Knowing the facts isn't enough; teaching pilots to ground themselves or their passengers when they have a problem is what counts. Even instructors who should know better fly with ailments like colds. Whether ego or economic motivation, it can still cause problems.

As a CFI candidate, you have to learn to deal with your own limitations. For any lesson, once you are scheduled to fly, there is a tremendous pull to complete the action despite obvious physiological evidence why you should not. Denial is the same reason pilots fly into weather they can't handle. Pilots rationalize away problems when the emotions say to fly by changing the question from should you go to how should you go. You have to learn how the physical and mental processes interact even though it won't be covered in the *Standards*.

As an instructor, you are very likely to witness problems like motion sickness. You will feel responsible when your students suffer, because that is what instructors do. Unless your students are regulars, they will also hold you responsible, even though they may have just downed two greasy cheeseburgers and a chocolate shake right before their first flight. If you consider all the human factors, detractors from learning, and aeromedical problems together, you have a huge parcel of blockages unrelated to the technical difficulties of mastering any skill. You must always look for the root cause of problems beyond the obvious. The FAA should do a study to see what percentage of learning problems are unrelated to technical learning, yet affect the flight training process, and then let CFIs in on the secret.

When you become an instrument instructor or whenever you fly through the clouds with students, you must be acutely aware of spatial disorientation and illusions. I was flying with one student who suddenly came down with vertigo right at the decision height on an ILS that he had flown pretty smoothly up until that point. We knew this would be a missed approach going in because the reported visibility was well below minimums. It is my standard practice to show every instrument student at least one real missed approach. However, without warning the aircraft went from a stabilized approach to a 30-degree right bank. This probably happened as a result of my student looking for the runway environment and not getting back to the instruments fast enough.

Approaching 200 feet AGL, I had time to ask but once in my most commanding voice to level the wings. Observing no immediate response, it became my airplane. After applying full power and re-establishing ourselves on the localizer from wherever we were, I executed the missed approach procedure. Not having a clue as to what had just happened, this student was angry with me.

"Why did you take the airplane? Everything felt fine!"

Felt? This was a person who knew from ground school all about trusting the instruments; yet when faced with a real situation, the knowledge and discipline went out the window, and it was back to the basic instincts. The point here is that as an instructor you have to know when the student is being taken over by an illusion without looking at the student.

Flying that low in IMC conditions, you will be riveted on both the instruments and the potential runway environment. You hope the student reads the instruments and sees them the same way you do. What this means is that an instructor has to be able to read the student's state of mind through the aircraft instruments. I wonder if the FAA has a course on this?

There is only one topic that brings a bigger scowl to the face of a student than weather — aerodynamics. This is amusing because what would flying be without these two? You really haven't covered aerodynamics well in any other rating because there is always so much else to do. You know that airplanes fly; why should you have to know exactly why? Just kidding. You can't become a CFI until you can convince an inspector that you can teach a student exactly why an airplane flies by using the same words that the inspector would use. Remember back in your first few hours how we discussed that knowing why an airplane flies isn't important, but knowing the words the examiner expects to hear are critical, and that allows generations of pilots to pass on the same explanations without any need for understanding?

Well, in your early tests it was easy to just tell an examiner what he expected to hear because you didn't know any better. Now that you are engaged in heavy research, you may start to question your teachings. Many pilots have an argument with some aspect of the official theory because it just never made sense to them, yet they have to repeat whatever is necessary to pass the flight test. The question, though, never gets answered. As an instructor, that isn't good enough.

The question that always nagged at me was the idea that the wing creates a venturi effect which causes negative pressure above the wing and that somehow holds up the airplane. The best example of a venturi are those conical, metal vacuum tubes on the sides of old airplanes. You will observe that the airflow is constricted by a conical metal casing, thus creating the venturi. What is not so obvious is how a flat wing with no constriction anywhere else can create that same venturi effect.

No explanation is given in the FAA handbooks for any force that might create this venturi. It is just assumed that you will believe it and accept it — which I have not. The FAA does believe it and expects you to accept it without question and to teach it as stated if you want to become a

CFI. Therefore you could be forced like me to commit personal heresy and teach a theory that you know is full of holes simply because the people who do believe the theory give out the certificates. This is a blatant contradiction to the FAA characteristics of an instructor which requires you to constantly engage in research to improve your knowledge and teaching.

The question then becomes: What if your research leads to new questions, revelations of the inadequacies in current theory, or even completely new theories? What do you do when the theory doesn't match up to the observed reality? You should know that under our current system, only research that leads to already accepted conclusions will be tolerated. This is why teaching has been at a standstill for decades. The *Flight Training Handbook* was last revised in 1980. Have we learned nothing new since then? No one can challenge the accepted theories of teaching aerodynamics — that is until now. I am giving you permission to question any theory until it makes sense to you.

My research on why an airplane flies always takes me to "Stick and Rudder." It makes sense to me that a wing is an inclined plane that flies at a higher angle than where it is going. The air is then forced down by the bottom of the wing and accelerated down by the top. This action through equal and opposite reaction creates the buoyancy that holds up the airplane. You cannot say that however and hope to pass the CFI checkride. The force of air on the lower surface of a wing can be easily observed by holding your hand flat out your car window to the relative air like a wing and tilting it up slightly. All the force of air to raise your hand comes from the bottom.

This still leaves the question of how the air is accelerated down from the top, why low pressure exists, and why the air upon being deflected upward by the wing doesn't just keep on going up? The official assumption is that a venturi effect does exist above the wing and the illustration given is a garden hose. (As if those two have anything in common.) My research then led me to my front yard and our garden hose. When you constrict a hose by pushing in with your fingers, you will find that the flow of water is reduced. If you press further, a hissing sound emits and the flow really slows down. Press still further and the flow stops completely. Hmmm. According to Bernoulli's theorem as currently taught, the flow should remain unchanged when you create your venturi.

There is a hole in our theory of lift. Something is missing that could explain the difference between the theory and the observation. If anyone at the FAA ever tested their material with a real hose, they would know this. No one, from student pilots all the way up through FAA inspectors, either questions or ever tries to make sense of the theory. They just learn it, repeat it, and expect to hear it back, all by rote.

My observations with the hose raised many questions. The water in the compressed venturi area of the hose actually exerted more pressure than the rest of the hose because when released, the hose returned to its original shape. If lower pressure is created in the venturi, why doesn't the hose collapse further, or at least hold its constriction? Since the hose restores its shape, how can we say a lower pressure area exists at all? If not, how can the FAA tolerate this example as a way to explain the lift created by a wing?

How can you as an instructor teach a theory based on assumptions from examples that cannot be observed? The system has created a theory that fits the assumptions. I just don't buy those assumptions because they cannot be proven. Why is there no method for review, challenge, and correction of the theories we have to teach?

Let's apply the theory to the wing. It states that the negative pressure from the venturi pulls the wing up. Considering the difference in weight between an airplane and the surrounding air, why doesn't the low pressure just pull the air above the wing back down instead? Why doesn't the air just get pushed up and over the wing the same distance as the wing chord, as there is nothing around the wing to constrict the airflow? When you go for your checkride, you can't dwell on any of this for you will surely fail. I just want you to know that if you have doubts about anything you have been taught, especially this theory, there is ample cause for your feelings. In our current system, however, there is nothing you can do about it.

The other idea that never made any sense to me was that some mythical vector force called the "horizontal component of lift" springs from the wing and turns the airplane. You can't say that on a checkride either. The theory as it is taught says that as you bank, the lift on the wing is broken into vertical and horizontal components, the horizontal component turns the airplane, and the loss of vertical lift from straight and level requires back pressure to maintain altitude.

Let's play with the theory. If lift is defined as a force perpendicular to the relative air, then the only possible lift the airplane can feel is the total lift perpendicular to your bank angle. Gravity always acts with constant direction and force regardless of the attitude of the airplane, so it makes sense that the steeper the bank, the greater the total lift that must be created with back pressure on the elevators to maintain altitude. You can observe this directly because the opposing force to total lift is the load factor resulting from the creation of the extra lift which, assuming the airplane is balanced, always acts to press your body straight down into your seat perpendicular and proportional to the angle of bank.

If the airplane only feels total lift, what is all this component stuff doing in our theory? Could it be that over time, a way of explaining the turn with vector arrows has evolved into a perversion of the theory where those arrows are now believed to be actual forces? If lift is always perpendicular to the relative air, it will always act perpendicular to the wing regardless of the angle of bank. Then how is it that the horizontal component can stay horizontal when the only lift the airplane and you can perceive moves directly with the varying bank angle?

Aha you say, the force opposing the horizontal component keeping it horizontal is the centrifugal force. Okay, check out your own observations. The control surfaces move the airplane about axes that are constant in direction; therefore, the forces created must be constant in direction as well. All side forces on the airplane are created around the yaw axis which is controlled by the rudder. Any change in the bank angle takes the yaw axis along with it as long as the tail is attached to the airplane.

If you are in a turn and unbalance the airplane such that the ball moves to the outside in a skid, or to the inside in a slip, you will feel your body either forced to the outside of the turn parallel to the bank angle, or forced inside the turn parallel to the bank angle, and perpendicular to the total lift and load factor. That is a real force and there is nothing horizontal about it. The only time you can have any horizontal force is when the total lift is vertical and directly opposite the force of gravity and the rudder is vertical, which can only happen when you are straight and level, which is the one time the theory says there is no horizontal component of lift.

It can be plainly observed that the centrifugal force and its direct linear opposite "centripetal force," the real inward component force that allows you to slip, are both parallel to the bank angle. If the force that throws your body when the aircraft is unbalanced always acts in direct line with your bank angle rather than in any horizontal direction, how can centrifugal force be represented by a horizontal line? If centrifugal force throws you to the outside of the turn in line with the bank angle rather than to the horizontal, doesn't the opposite linear inward force do exactly the same thing? We know centripetal force exists because we can both slip an airplane and feel the force pulling us into the turn.

Centripetal force is conveniently left out of the theory of turns because it proves that there is no horizontal component of lift. What has happened to make the theory work is the abandonment of observable forces in favor of imaginary arrows in order to make a flawed theory believable. This is disgraceful. There is no horizontal component force at all in a turn, only the centripetal force acting inward parallel to the bank. If Newton's Laws of Motion are applied to every other aspect of aviation, how can we tolerate the omission of centripetal force from the theory of turns, and its replacement with some unexplainable, undefinable, unobservable horizontal component? If there is a horizontal component of lift, shouldn't its opposite be a negative horizontal component of lift rather than centrifugal force? Since there is no horizontal component, it can't turn an airplane, and we need a new explanation.

Think about this: If there really were a horizontal force, what would prevent the airplane from simply sliding sideways in a horizontal direction in a bank while maintaining the heading? Every force I have documented here to dismiss the current theory is clearly observable in an airplane, yet we are asked to teach about mythical forces that no one can either feel or prove exists. The theory has holes; are you still satisfied to teach it as is?

What makes sense to me is a definition derived from "Stick and Rudder." An airplane when banked goes into a sideslip and loses lift. It turns because the airplane is lifted by the elevators and weathervaned by the tail which guides the airplane in a curved flight path. Lastly the rudder corrects for the adverse yaw. The turn is really just a lifting and weathervaning action.

You may have noticed the conspicuous absence of the tail from any aerodynamics discussion save for yaw stability. Did you ever wonder why? Most pilots are content to take every theory at face value, blissfully indifferent to glaring inconsistencies and gaping holes, and are caused no distress at spouting these theories of dubious foundation to other pilots who hold the same misconceptions, or students who are trained by rote to accept them without question as well. If this is unsettling and you find yourself asking questions, then you have just crossed the bridge to the other side of flight training. Welcome aboard.

My search for the truth about why and how an airplane flies led me to two of the most wonderful, white-haired, aviation geniuses that any instructor could ever have the pleasure to meet. They reside at the NASA Ames Research Center at Moffett Naval Air Station in California. For an entire afternoon, I sat transfixed at the wisdom two of this country's top research scientists graciously shared with me. I learned that Bernoulli and a garden hose have very little to do with each other. Bernoulli's theorem is an expression of the conservation of energy where the fluid is assumed to have no viscosity, compressibility, or resistance.

This is hardly the situation in a hose. A hose has friction (among other things) that reduces the flow of water whenever the diameter of the hose is reduced. The hissing noise is caused by cavitation, which is the sudden formation of bubbles resulting from the mechanical force of restricting the hose. The low pressure exists, but the air bubbles created because the water is actually boiling greatly slows down the flow.

Where the fluid dynamics are similar is in a discussion of real world laminar and turbulent flow. Air is a fluid we are told, but what we are not told is that it has properties of viscosity and electrostatic attraction between the molecules which gives it fluid properties. This is one reason why the air can flow over the wing and maintain its cohesive properties without any overhead constriction.

Another factor left out of current teaching is the effect of atmospheric pressure. On a nonmoving wing at sea level, there is one atmosphere of pressure acting on all parts of the wing. When a wing is in flight, there is still the atmospheric pressure before and behind the wing. However, below the wing because of collisions with the air, the pressure is slightly above the ambient atmospheric, and immediately above the wing because of the camber and increased speed of airflow, there is slightly less than atmospheric pressure. The collisions push the wing up and the lower pressure pulls the wing up. There is still the ambient atmospheric air pressure from above the boundary layer of lower pressure air which contributes to the venturi effect. Air accelerates over the wing and expands after the wing to contribute to the downwash.

There is so much more to tell, but the scientists can

explain it better than I can. Unfortunately, I never had the time to ask them why an airplane turns. The conclusion for our flight training system is that the FAA is not so much providing completely incorrect information, rather they are providing information completely incorrectly. NASA puts out a great book that should clear up any of your questions. It is called *Introduction to the Aerodynamics of Flight*, Publication #N76-11043, available from the National Technical Information Service, Springfield, Virginia, 22161.

It is time for you to acquire bureaucratic skills. The FAA runs on paper, so shall you. You have to memorize and be able to apply any regulation in Parts 61 and 91, even though the inspector gets to keep the book open. You will have to know all the privileges and limitations of all certificates and ratings you can teach, and your own responsibilities as a CFI. You should be able to write out any endorsement. All documents pertaining to the legal operation of an aircraft may be perused with fine shades of definition. The maintenance logs will be examined so carefully you would think an A&P license will accompany your CFI. You must know what there is no reason to know. You have to know what anyone else would normally be able to look up.

Take an airworthiness directive (AD) number for example. You have to know what each number in the series stands for, only to promptly forget it all after the checkride. Students won't care about AD numbers unless you tell them it is important, which you will not. Make sure that you know which is the current weight and balance form for your aircraft. You must review all paperwork ahead of the test so that you can appear knowledgeable in front of the inspector. Many CFI candidates are sunk before they even get started because their paperwork is not in perfect shape.

The reason for such painstaking analysis is not to insure your recordkeeping skills; it is to cover the inspector's butt so he won't test you in an illegal airplane. Since they are the FAA, they are supposed to know better. That is why you must prove your aircraft is unimpeachably airworthy. Inspectors are equally paranoid about violating airspace.

Anytime we approached the boundary of our local TCA (Class B) on my checkride, they both whined like little kids, insisting I prove over and over again that we were not in the TCA. Do yourself a favor before your checkride and index with labels all significant pages, entries, statements and endorsements, in all your required paperwork.

The actual checkride is more a test of will and mental combat than it is a serious evaluation of your ability to teach. There really isn't any new block of information to be learned once you have a commercial-multi-instrument rating, so the CFI becomes the pinnacle of teaching what you already know. The only way to make the CFI different from the other certificates is to vastly increase the depth of knowledge for what you already know. This is how the standard of knowledge becomes perfection and how any slip greatly detracts from your performance. How can you be trusted with helpless students if you don't know everything?

A successful checkride largely depends on who gives it to you and how your knowledge stacks up against their pet questions. In my experience FAA inspectors are a mixed bag. Some are former commercial pilots with a real world perspective, a heart, and a desire to see you succeed. They will test you to your maximum capability while giving you very fair treatment throughout your test. If you fail with one of these folks, it will be no one's fault but your own.

Some inspectors, however, are extraterrestrials who have no knowledge of how human beings think, act, or learn. You will never know what to expect from such an inspector and will feel tricked and manipulated throughout the exam. If the questions you are asked have anything to do with general aviation, it will be pure luck. Since examiners form questions around prefabricated answers, you have to be a mind-reader to know what this inspector wants.

The worst group are the inspectors who feel it is their civic duty to keep you from becoming an instructor at all costs. These folks deal with the hopelessly trivial and irrelevant in order to catch you in such intricate detail that you have to fall sooner or later. I know of one inspector who used to routinely fail every CFI applicant from my old school simply because he didn't like our program and didn't believe that any candidate could be adequately prepared to pass the first time.

You should also avoid the types who give the word bureaucrat new meaning. Robots who have no judgment, discretion or flexibility might get through. However, people who do not see the world in stark black and white colors will not make it. Avoid those inspectors who deal only in absolutes. This is your test, find the inspector you want. This may be impossible if your FSDO will only schedule examiners at random. In this case you may have to cancel flight tests until you get the one you want.

You will need all the advantages you can get for this test. Get a really good night's sleep before your test — if you can. Think of it as war, where defeat is not an option. Your inspector has all the advantages, makes all the rules, and holds all the power. There is no such thing as justice or fair play. You will roll the dice, give it your best shot, and if you beat the odds, you win. You have to believe that you are a CFI before the test or you will never get through. In the morning when you walk out your front door, take a big breath, and say out loud that today you are coming back with a CFI certificate.

To give you an idea of what you might expect on a checkride, I dug out the old notes written the day after my test. Here are the opening paragraphs of my own CFI test:

Dear CFI Candidate:

I hope somehow these pearly words of wisdom wrought by torment and exhaustion through the day-long exam will benefit you on your quest for the elusive CFI certificate. Hindsight provides a couple of hints to make your checkride less frantic. Get all your materials, essential papers, forms, test scores, endorsements, current charts, and any stuff you need at least a week in advance. Spend some time with the folks in maintenance. On more than one occasion, I poured over the maintenance logs with the chief mechanic and received a detailed tour of the components of all our aircraft. Reach beyond your training and prepare for the unexpected. Questions were posed to me that I had never heard, but through a logical thought process of what I did know, a satisfactory answer could be deduced. Practice your teaching at home. Teach your lessons out loud to sharpen your technique.

I had the rare and distinguished privilege to have two FAA inspectors rake me over the coals. Like inquisition guards, they each took turns going out for coffee and beating every topic to death. There was always one tormenter in my presence. Cool heads filled with organized knowledge will prevail. The first inspector was new to the area but not to aviation. Straightforward and fair, I was struck that this was someone who should be leading weekend scouting excursions. Beneath a friendly and easygoing manner lies a treasure trove of aeronautical knowledge and a mind quietly waiting to spring like a steel trap at any hesitation or fumble. Take your time to think and prepare your responses. Although not out to catch you with trick questions, never attempt to bluff anything.

My second inspector is the old hand out here. A dark, sinister demeanor and sarcastic wit lie barely covered by a facade of professionalism. You are shadowed by a feeling of impending doom that continually intimidates. Be strong and persevere. Although I see this person sipping martinis on a yacht when not grilling CFIs, more has been forgotten about airplanes than I will know for quite some time, and in a back-handed way cares enough to find out exactly what you know and how well you will teach. The primary concern of both of them is safety. You will be corrected to minute detail and will learn from the experience.

The test began with an explanation of the ground rules. "In the commercial test, the object is to say as little as possible, to answer only the question, and not say anything to get yourself in trouble. This is the CFI test. We will not sit here and question you to draw out what you know. We will give you topics to teach, and you will completely explain each topic. If we are not satisfied with your teaching of the topic, we will make a determination then whether to continue." Ominous and foreboding, no?

I was under the impression that the entire bookcase of aviation factoids I had lugged with me throughout my training would actually be of some use during the oral. Hardly; I could not refer to anything. All explanations, regulations and procedures came strictly from memory. "Imagine I am a person on the street corner and verbally explain it to me" was the request. As if someday when I least expect it, while walking down a dark alley on a wet and stormy night, a knife will be pressed to my throat on a street corner followed by an urgent demand to explain "P-Factor."

All of my carefully prepared lesson plans were useless as all my lessons had to be taught without them. Their only function was a quick glance by the inspectors to see that I could write one. The bulk of my oral was teaching those lessons from memory.

We began with the fundamentals of instruction. Do you know all those squishy, psychological factors that make instruction possible? You better. You also better be able to apply all the laws, techniques, methods, and skills of learning and teaching to real life examples. The *Practical Test Standards* were kind of an afterthought since the inspectors picked topics from the Table of Contents of the *Flight Training Handbook*. I never really knew what to expect.

You can't prepare for everything, so all you can do is your best. Then we dissected the sectional chart. You will get a lot of scenarios to solve. For example: How would you determine if the weather was above VFR minimums in a control zone of a nontower field with a working FSS and your aircraft was not equipped with a radio? It gets better. You are at a nontower field at 7:00 p.m. and you witness an aircraft accident. What is the first thing you are going to do? Save the occupants was the first thing that came to my mind. They wanted me to say notify the NTSB. And how could I do that at night with the local FSDO closed? Ask your instructor if you don't know. The questions were simple, the answers were not.

"Teach me everything about weight and balance?" After 45 minutes of teaching weight and balance, my answer would be critiqued. We covered in detail aerodynamics, stability, spins, systems, constant speed propellers, turbochargers, forces on the aircraft, load factors, Vspeeds and their meaning, flight fundamentals, basically everything in the *Flight Training Handbook*. I was done with the oral in just under 4.5 hours . . .

13 • MOVING UP TO MULTI-ENGINED AIRCRAFT

Oh boy, are you in for some fun now. Just think, you now have two engines, just like a *real* airplane. Twins are full of wonder. Twins are full of fun. Twins boost the ego. Twins have two of everything. Twins require real skill. Flying twins is a quantum leap above anything you have seen until now. There is such a big difference when you move up to a twin-engined aircraft that I could never figure out why my old flight school put students with their private certificates still wet, straight from a Cessna 152 right into a Beechcraft Duchess? Once you figure out how flight school owners think, the answer is all too simple — money.

The dark side of multi training is that this is by far the most expensive flight time per hour. Twice the number of engines, twice the fuel and maintenance, and a couple of dollars extra for the instructor really ratchets up the cost to you. Flight schools, if they can afford the equipment, can make a killing on multi training. I don't know all the games out there, but here is how my old school played with us. If you take private pilots with minimum flight experience, all of which is in very low-performance aircraft, and then suddenly expose them to two engines, constant speed propellers, retractable landing gear, and much higher airspeeds, they will be completely lost for many of their first flight hours. Those early flights will be strictly limited to high-performance training, with any multi-engine aspect pushed back later in the training. The flight school gets all that extra money from a student, who has no business in a twin when he has not even received any high-performance experience in a single.

My old school was a Part 141 academy. All the students were career pilots in training. The next rating they were told to get after the private was a private multi. Nothing is more useless to someone who has to get commercially rated anyway than a private multi. No student could ever use the rating at my school because the twins were unavailable for renting.

The next phase in the game was to put as much other flight time in their way before training for the commercial multi, namely the instrument and commercial single training. By the time the students were again scheduled for multi training, they had pretty much forgotten everything and had to pay the exorbitant multi rate to learn much the same stuff all over again. What a racket — what a gold mine!

Since multi training is by far the most expensive type of training, try to get it when it will do you the most good, where you can learn it in the shortest amount of hours and at the cheapest cost. I never touched a multi-engined aircraft until after my CFII (which means instrument instructor.) My school, of course, advised against it, and my peers thought I was crazy for being different, but I walked away with a commercial-multi-instrument rating in 15 hours (including the checkride) without ever spending one dime on a private multi I could never use.

This strategy was not the result of any innate brilliance; I just could not afford to do it any other way than the cheapest. Never get a private-multi rating unless you intend to spend your time in aviation flying privately. My fellow students felt cheated after hearing of the low-cost investment possible for the commercial multi, and some kicked themselves for not doing it that way as well. Never forget who is the customer, who is paying for the training, and who gets to make the rules. Since the test for the private multi and commercial multi are virtually the same, the question becomes how many checkrides do you want to pay for? If you already have an instrument rating, then your multi-instrument rating is a mere two approaches away; one using both engines, and one on a single engine.

Besides delaying your multi training as long as possible, the best thing you can do to prepare for the upcoming transition from singles is to get as much high-performance time as you can. When adjustable props, retractable gear, and fast aircraft are second nature, your basic training twin is an easy step up. The more factors your intended twin has in common with high-performance singles, the easier will be the transition. The normal procedures will be nearly identical except for the duplication and synchronization of the engines. The only block of new stuff will be the emergency single engine operations.

Compare the performance of the Cessna 210 and that of a Duchess or Seminole, some of the most popular training twins. You will find that the weights, airspeeds, and handling qualities are quite similar, which is interesting considering the dramatic differences in the aircraft. Any high-performance single time, especially in the heavier models, will be great preparation for twin flying.

Single-engine airplanes are all reasonably similar; a Cardinal is a Sierra is an Arrow . . . However, multi-engined aircraft can have vast differences in the specialization of each model requiring far more familiarization before you hop in and go. You can't say that an Apache is a Seneca is a 310, even though they are all light twins. Recognizing this, the FAA requires 15 PIC multi hours and 5 hours make and model before you can teach in any particular airplane.

This experience can be expensive. Multi-engine time is a commodity like gold. It can be horse-traded, bummed, acquired as a safety pilot, or even (God help me) given away as an instructor. Try when you are rated to get as much multi time in as many different types of twin as you can. That way when you are in a position to teach, you will have more aircraft open to you.

If your goal is to be a commercial pilot (and why else would anyone learn to fly a multi-engined aircraft?), you might try to find the twin closest to the ones used in commercial operations. Pricing for twins can vary widely and deals are available. In my area one school I know charges more for a Beech Duchess than a local flying club gets for a Cessna 310. Which aircraft would you rather fly? The problem with most of the training twins you see at flight schools is that they are only used in training. They have no commercial utility. You will be ahead of the game if you have experience in aircraft that either are, or closely resemble, commercial twins in weight, speed, and complexity.

The twin is a completely different animal from the single and requires new thinking and a new strategy to fly. Because of the FAA's continual fascination with retesting you on checkrides for your presolo maneuvers, namely stalls, steep turns, and slow flight, most flight training programs are forced to pass over giving you the change in thinking required for you to maximize the utility possible from twins. You should be introduced to the stall, steep, and slow characteristics of each aircraft, but there is no earthly reason for them ever to be demonstrated on any checkride after the private, other than for the CFI certificate and add-ons because of the teaching aspects involved.

One way to start thinking differently about twins is to take a look at those two little lines surreptitiously inserted in your aircraft specifications called "wing loading" and "power loading." Not defined in any detail or properly addressed in any handbook I can find, a comparison and analysis of those numbers from what we can glean from our regular training tells the real story of twin flying. Wing loading is measured in pounds per square foot or the amount of weight that each square foot of wing has to lift. Most pilots know that twins have a high wing loading because some other pilot has told them. Pilots who may have no clue as to what that means are perfectly willing to mention it in conversation with other pilots because they know it is appropriate. The other pilots may not know what it means either, so they don't question it, but mention it themselves in subsequent conversations for the same reason. The myth of sounding like you are knowledgeable is most desirable.

As for power loading, it is the measure of pounds per

horsepower or the amount of weight each horsepower has to move. We know twins have greater performance than singles, and this can be clearly shown by their power loadings. Twins are usually bigger, heavier, and faster than singles, yet their power loading is lower. You have proportionally more power moving less weight.

Twins also have a higher wing loading than singles. This means each square foot of wing has to lift more weight.

Compare the wing and power loadings and other characteristics of the Cessna 172 and 310. The typical 172 has a 36-foot wingspan lifting a 2300-pound airplane using a 160-horsepower engine. Your average 310 has 520 horsepower lifting a 5300-pound airplane, but get this — with only a 37-foot wingspan. This includes two tip tanks and two engine nacelles and compartments, which take a lot out of the wing. Short of getting out a tape measure, you can easily see that the 310 has far less lifting area hefting a far heavier aircraft. That is wing loading. The 310 has a wing loading of 29.6 pounds per square foot compared to 13.2 for the 172. Each square foot of the 310 wing is lifting over twice the weight and doing over twice the work of the 172 wing.

From the lift formula, we know that lift is a combination of wing planform area, air density, a whole bunch of factors best stated as angle of attack, and velocity. If we reduce one factor for the equation to balance, something else will have to increase. The 310 has a small wing area; that is the factor being reduced. We can't increase the air density to balance the equation; that is up to nature. We could increase the angle of attack at which we fly, but that would increase the induced drag, so that won't help. The only possible factor left is an increase in velocity. Therefore, the only way to efficiently get the required lift from a smaller wing is to pass more air across it.

You may have heard the expression that altitude is the friend of the single and airspeed the friend of the twin — now you can see why. The smaller wing of the 310 has less frontal area and skin for friction resulting in less parasite drag. This helps the aircraft to go faster, which puts more air around the wing, which enables the wing to carry a higher load.

The next question is how else can we get the extra airspeed? The 310 has a power loading of 10.2 pounds per horsepower compared to 14.4 for the 172. The 172 engine has to move almost half again as many pounds of aircraft per horsepower, so it needs to get more lift for less work at a lower airspeed from the wing. The pieces should now be falling into place. The superior performance of the twin comes from more power moving less weight, faster, with less drag, with a wing that works harder to generate more lift. However, the harder your wing has to work, the faster you will have to go to generate the necessary lift. This cuts into your short-field and low-airspeed performance — that is the tradeoff.

Does this change your thinking? Are you now able to take wing loading and power loading for any aircraft and get an idea of its flight characteristics? Up until now, performance has been a pretty nebulous concept somehow related to airspeed and rate of climb. Now through a comparison and analysis of wing and power loading, you can more accurately judge the real performance of any airplane. You might say that the greater the wing loading and the lower the power loading, the greater the potential performance. The new Learjet 31A has a wing loading of 62.4 pounds per square foot and a power loading of 2.36 pounds per pounds of thrust — not directly comparable to horsepower, but you get the idea.

Performance, however, depends on many factors including your utility and mission requirements, so don't make wing and power loading your only criteria. We know twins usually have greater rates of climb than singles. Now you know that results from the greater acceleration and airspeed made possible from a lower power loading and the reduced drag from less wing area. This allows the higher wing loading to generate the lift required by a heavier airplane. Wing and power loading are inseparable as concepts.

Cruise flight is also faster in the twin because of the

reduced drag, although not generally all that much due to advances in high-performance singles and the extra frontal drag of wing-mounted engines. The best glide speed of an aircraft will be higher with the higher wing loading because of the need for the wing to work harder, although the gliding distances are not all that different. A Cessna 172 at gross weight with a windmilling propeller can glide 1.5 nautical miles per 1000 feet of altitude at 65 knots. The interesting thing is that a Cessna 310 with a double engine failure (highly unlikely), both propellers feathered, and the gear up can actually glide 1.8 miles per 1000 feet, but at an airspeed of 96 knots. The reduction in drag probably accounts for the greater distance possible from the 310, but you can also see that by increasing the velocity of air over a highly loaded wing, you can get a comparable glide distance to that of the lightly loaded, bigger wing.

You can also see that slowing down a higher wing-loaded aircraft takes away proportionately more lift than the lower wing-loaded aircraft, which makes it possible to achieve greater rates of descent in a twin more easily. You will find that power management is much more the province of the twin since the disadvantages of the harder working wing must be made up with an increase of available power. All of this contributes to a completely new thinking and appreciation for the inherent difference of multi-engined aircraft. By understanding these concepts, an entire strategy can be formed that clearly teaches you the real theories and practical operation of multi-engined aircraft. If wing and power loading are not in your multi program, insist that they be added in.

There is all the difference in the world between wing and nose-mounted engines. The slipstream of the single does little for you but give the airplane left-turning tendencies. In the twin those engines blow lots of air around the wing, creating much additional lift. How much additional lift I have yet to find documented. A question arises that since the slipstream is straight into the wing from the engines rather than from the relative air at the angle of attack, the major benefit probably results from the airflow over the wing. At high angles of attack, the slipstream may actually hinder the impact air below the wing by deflecting it away. This, however, is pure speculation on my part.

Anyway, in a single the way to control lift is with angle of attack and airspeed. With the engines out on the wings, your lift is also regulated by changes in power, thus making your training in power management in the twin even more critical. Everything is a tradeoff, so what you gain from proper power control is extra airflow to get more work from the highly loaded wing, more airspeed from the low power loading, and more flexible lift control through the power for better takeoff, climb, and descent performance. All of this contributes to much greater flexibility in controlling the aircraft. This will be lost on most multi students as you are forced to dwell on presolo maneuvers and rote procedures and never learn the strategy to utilize the extra capability.

Just as the engines give you more lift through the greater slipstream, they also give you much more drag and higher rates of descent when that slipstream is removed and the propellers are windmilling. Those big prop disks can produce some incredible drag, so idling both engines can help produce amazing descents should you have the need. Managing your power, therefore, is managing both your lift and drag. This is where the increased flexibility comes from. Proper power management must be added into your overall twin strategy. One of our big problems in instruction is introducing you to all of these concepts separately, if at all, with no effort to tie them all together. Therefore, you will have to do this on your own.

One of the most interesting comparisons of our 172 and 310 is when looking at the maximum-performance, short-field takeoff and landing distances. They are remarkably close for two such different aircraft. However, by taking into account all of our previously discussed concepts, there is no mystery at all.

On a standard day at sea level using aircraft at gross weight, your basic 50-foot obstacle will be cleared in 1440 feet by the 172 and 1795 feet for the 310 according to their respective manuals. This means that an aircraft which is almost 2.5 times heavier, which has to go some 20 knots faster to take off, can be over the same obstacle with only an additional 355 feet of runway. This is possible because of the extra performance of the power and wing loading, reduced drag, and increased slipstream over the wings. Your basic 172 can clear a 50-foot obstacle on landing under the same conditions in 1250 feet while the 310 can do it in 1697 feet, which is an additional 447 feet.

On the landing roll, those big props windmilling plus some aerodynamic drag and good braking allow a 310 to still come in about 30 knots faster and not take up that much more runway. You can amend your definition of performance to include the ability to accelerate and decelerate a heavier aircraft over a greater speed range without using significantly greater runway lengths. Add this to your total picture and you should be well on your way to changing how you think about twins.

Whereas concepts like wing and power loading are usually ignored, ideas like the "critical engine" are completely overblown. Imagine classifying one particular engine as critical. Does that make the other one superfluous? Since most of you now use training twins complete with counter-rotating props, the whole idea is irrelevant. This leads to distorted training because when you finally do encounter a conventional twin, you will approach single engine emergencies with great apprehension as you imagine the failure of that critical engine pulling you right out of the sky. Fascinating the things we put in pilots' minds, no? Chances are that none of you would notice that there even was a difference in performance on a conventional twin, unless this concept had been brought to your attention and you were looking for it.

The point with the critical engine is to show that the P-factor of the right engine is greater than the left because it moves the thrust line further from the center of gravity, which increases the yaw, which requires more rudder, which reduces your performance slightly — big deal. Since the loss of either engine so degrades performance, the curiosity of how operating on the right engine makes a bad situation just slightly worse is a point of interest better explored on the ground when there is nothing better to do.

I have yet to ever hear of a pilot flying a multi-engine aircraft lose the right engine, make an emergency landing, and report to everyone how ecstatic he was that it wasn't the critical engine that had failed. The simple truth is that once you lose an engine, the critical one is the engine that is *still operating*! You had better learn to think that way and treat your remaining engine as such, least you become a glider. The only time I have even noticed a difference between the two engines was on a hot and humid summer afternoon where the left engine of our 310 gave us an ever-so-slightly better rate of climb over the right. This was most likely due to the relative humidity and air density, concepts not usually considered in most critical engine discussions.

By the overuse of the word critical in applying it to so many situations (like crying wolf), it loses all impact on pilots. We should reserve its use for conditions that really are critical. My personal theory is that the concept of a critical engine came into wide use with the advent of the Piper Apache, which has the hydraulic pump that works the gear and flaps mounted on the left engine. It makes lots of sense to think of the critical engine as the one that takes out the gear and flaps as well.

Like the old fraternity initiation practice where you have to repeat the Greek alphabet three times while holding a lighted match, so the twin candidate must memorize and repeat the ten conditions of Vmc (minimum control speed). They both hold about equal significance. Every multi-engined aircraft has a red line on the airspeed indicator clearly displaying the highest value of Vmc. Keep flying safely above that speed, and the legal aircraft will be able to maintain directional control.

Of all the Vspeeds, this one seems to hold the greatest

fascination for how it is derived. There is no equivalent interest for memorizing the derivation method or conditions for the stall speeds, maneuvering speeds, maximum cruising speeds, best rate or angle of climb speeds, or best glide speeds. You have to know those speeds and what is their significance to your airplane, but you don't have to recite the conditions under which they were set. Why then is Vmc singled out for regurgitation of its ten or so conditions before an examiner? You should be able to read and learn about its conditions and then leave it alone. Once again rote learning and repetition block out the potential for so much extra wisdom possible from exploring concepts like wing and power loading.

Vmc (minimum control speed) should be renamed "minimum directional speed" so it won't be confused with minimum controllable airspeed. When you leave out all the jargon, it is nothing more than the lowest speed where you can keep the aircraft straight on one engine. All knowledge is helpful, so it doesn't hurt to see how it is derived and understand the factors involved.

Vmc is fluid just as Va (maneuvering speed) is fluid. Your actual Vmc for any given situation contains so many possible variables, including such intangibles as pilot technique, that the exact calculation of Vmc (even when you can repeat its conditions) is impossible. However, the observation of when the aircraft reaches Vmc is easy — you can't keep the nose straight.

Va can't be observed so easily nor calculated in flight unless you know the exact weight of the aircraft, which is constantly changing as long as the engines are running and burning fuel. Therefore to be safe, you calculate your operating Va based on the lowest value for your flight. This occurs at the lightest weight. You can be safely below the lowest value of Va by using the landing weight maneuvering speed for the whole flight.

Vmc works the same way, except to be safe you want to be above the highest airspeed value. By knowing the airspeed of the worst case scenario, indicated by the red line on the airspeed indicator, it matters not what your changing Vmc value is because you will remain above it.

It's like trying to gauge when an aircraft will stall based on the published stall speed. Your actual stalling speed is dependent on factors like weight, center of gravity, load factor, acceleration, and of course, angle of attack. There is no point with stalls in assigning one speed value when buffets and horns are your best indication of an approaching stall, but by assigning the value under particular conditions, you should be able to avoid them.

The same holds true for Vmc, where your best indication is a gradual loss of directional control. Since avoiding Vmc is your goal, the best chance is to avoid the worst known case. The nice thing about airplanes is that for stalls and directional control they take the infinite variety of variables into account and still give you simple, unmistakable indications of trouble should you wish to find your actual airspeed value.

The logic of teaching Vmc somehow gets twisted because for each condition you are warned how badly it affects your performance and at the same time how critical it is for establishing the highest Vmc, which makes the individual condition somehow seem desirable. These conditions are only desirable in order to present the worst scenario to make you safe by avoiding it. When your instructor talks about how wonderful it is to know the most about the worst conditions, keep this in mind.

Having said that, let's look at some of the conditions. Maximum power on the operating engine and nothing from the other is the greatest factor in directional control since it is the imbalance of thrust that yaws the airplane. Fortunately the only time you should need full power in a twin is takeoff. Even a go-around can be done with climb power thanks to the low power loading. The time we are most at risk is, therefore, on takeoff. Although the aircraft can fly below Vmc, you never want to do that close to the ground. This is why you have higher takeoff speeds to keep you safely above Vmc.

Because it is the imbalance of thrust that causes the yaw, anything to reduce that difference will lower your actual Vmc. Any power being developed on the problem engine or loss of power on the good one due to density altitude or engine condition will lower the differential of thrust and therefore the yaw. It isn't always the all-or-nothing proposition described in your handbooks as there are any number of possible engine thrust combinations. When you perform Vmc demonstrations up high, you will find that directional control can be maintained sometimes quite a bit below the red line. Think of that as a safety cushion for those times when you inadvertently slow down to near the red line, like when you learn landings and go-arounds.

The most rearward center of gravity is another biggie for Vmc, but you are not likely to see the effect of this in your lightly loaded, forward C.G. lessons. For more realistic training, load up a twin with volunteers and baggage near gross weight with a midrange C.G. and then see how the aircraft reacts. This is the way the aircraft will be loaded during your later commercial operations.

Most twins have two possible flap settings for takeoff: one for normal conditions and one for maximum performance. Vmc is calculated with the flaps up if that is normal for your airplane because that will get the highest Vmc value. This is interesting because extended flaps lower your Vmc. Therefore, the maximum-performance takeoff (which is usually more precarious in the single than a normal one) becomes actually safer in the twin as far as Vmc is concerned because of the extended flaps. So why aren't all twins required to use flaps on takeoff? Just a thought. Maybe because any benefit from the flaps is canceled by the lower airspeeds required for maximum performance. If you have the runway, though, to accelerate well past Vmc, the safety of extended flaps doesn't really matter.

Extended landing gear provides the stabilizing drag which lowers your Vmc. It also lowers your speed and greatly lowers your rate of climb. Since airspeed is your friend, the logic of twin flying dictates that you raise the gear at the first positive indication of a climb. For me, that is as soon as I see the ground move and can feel that the aircraft is out of ground effect. Just like the flaps, there is no Vmc benefit to leaving the gear down once you have passed the red line. Since you don't rotate in most twins until you are above Vmc, leaving the gear down anytime after you leave the ground effect is wasted performance. There are a lot of twin pilots who never learn any new strategy and still want to leave their gear down until passing the departure end of the runway.

Since parasite drag increases twice as fast as your airspeed, and since twins both accelerate, climb, and fly faster than most singles, leaving the gear down hurts the twin far more than the single by the greater proportional loss of performance. The extra speed and climb of twins makes the possibility of gliding down on one engine and stopping highly remote but for the longest of runways.

Remember that high-performance illustration of how much runway is actually needed for any to be usable after takeoff? Imagine how much runway is required to be usable in a twin at 200 feet while trucking along at 100 or so knots. Should you lose an engine at that point with the gear down, you will probably be coming down somewhere way past the runway. If you had raised the gear at a positive climb, you may have the option to climb out. What you really lose by delaying retraction is the safe altitude and airspeed that you should have attained. Now we are getting into takeoff planning where your individual conditions will determine when you should raise the gear. Just don't get locked into one fixed rule because that rule will only apply to one set of conditions.

Another condition is the windmilling propeller. This creates as much drag as leaving the gear down, and when combined with full power on the other side probably sets the two strongest factors in a high value of Vmc. The good news is that engines don't generally go from full power to nothing instantaneously. Many times there are warning signals from the engine instruments, the sound of the en-

gine, or a partial power loss. Any power behind the engine
will eliminate the drag from windmilling caused by flying
through the ambient air. Feathering, therefore, is not always your immediate reaction to engine trouble, even
though your training by considering only the total failure
would have you believe it.

Now combine the possibility of a windmilling propeller
with the decision to leave the gear down until the end of the
runway. If you have both the gear down and a failed engine
with a windmilling propeller, you have pretty much given
away all your options except a controlled glide on one
engine. What you should always consider is the combination of factors and your possible options. With the gear up
at a positive rate of climb, you may only have to deal with
the troubled or failed engine while still having the option to
feather and take it around for a landing.

Two of the interesting conditions for Vmc are the ones
the FAA assumes. It is assumed in the determination of
Vmc that the aircraft will be in a five-degree bank into an
engine while flying the aircraft at zero yaw. Well, doesn't
everyone bank five degrees into the engine that is most
likely to fail while holding the nose straight with the rudder? This is the one contradiction in what is otherwise a
comprehensive explanation of a most dangerous situation.
You cannot set the most adverse value for Vmc by assuming that a beneficial condition will exist immediately upon
engine failure. It will not exist at all during the most dangerous time for loss of directional control, the initial climb out
after takeoff while the pilot is deciding what to do.

What we have then is a red line set at *almost* the
highest value of Vmc. It would be interesting to redetermine
Vmc for some light twins based on level wings and a centered ball and see if there is a significant difference. Here
you have memorized all these conditions only to find that
the FAA snuck in something that improves the situation.
Think about that the next time you fly a twin.

For some unexplained reason, the *Practical Test Standards* insist that you show an examiner once again how

well you can stall or almost stall a twin, perform steep turns, and fly slowly. Although it is useful to see these maneuvers to get to know the airplane, there is still no reason to learn the procedures well enough to demonstrate to an examiner unless you are going for an instructor rating. Once you are familiar with the characteristics of the airplane, endless repetition to make them all look pretty is ludicrous. You can do that on your own after you are rated. No commercial operator cares how well you can get a twin into a stalled condition, only how well you can recognize and avoid them. Twins are so expensive to operate — why would anyone want to fly them as slowly as possible?

As for steep turns, they are fun at first to see how the momentum works in heavier aircraft. Beyond that, however, they aren't going to help your understanding of multi-engined aircraft. When so much flight time is devoted to presolo maneuvers, precious time is lost from things that new multi pilots really should see. For the commercial multi checkride, see my chapter on the commercial certificate. The multi requirements could go well beyond their obvious duplication of single engine certificates and ratings by including such things as night cross-countries, night touch and go's with simulated single engine approaches, lots of instrument practice, VFR and IFR arrivals to strange airports, short approaches, emergency descents, gross weight loads, long flights requiring fuel management, high-altitude flight, multi-turbo operations, or any other experience likely during private or commercial operations.

We teach multi students how to handle all the weird stuff and emergencies very well, or as some would say, to excess. The *Standards*, however, allow hardly any opportunity to teach pilots how to fly the aircraft normally. From this training many contradictions arise. Think about abnormal situations like critically slow airspeeds. There you are, roaring along with both engines somewhere about climb power, while flying as slowly as possible, most likely below Vmc, which is something you have been warned never to do. How's that for a mixed message? Granted your actual

Vmc will be lower should you lose one during very slow flight because of power settings below full power, but if you are told that flying below Vmc is abhorrent in most situations, these exceptions only serve to confuse the new multi pilot. Maybe the takeoff is really just an exception as well? Don't even consider that thought.

Power-on stalls are a special case and deserve some exploration. In your basic single, stalls are no big deal. When practicing them at a safe altitude, the most likely adverse condition is a secondary stall or spin entry, both of which you can be taught to avoid. In a multi, should you have an engine failure while attempting a full-power stall, you could fall into an unrecoverable situation. For those pilots and instructors who engage in this risky activity — are you nuts? You cannot impress upon new multi pilots the dangers of Vmc with the duty to memorize all its conditions and then go fly below it with both the engines blazing.

What can you learn from pitch angles so high you would never try them anywhere near the ground? It comes back to how we are not taught to develop a new strategy for multi-engined flight. The multi has a better thrust-to-weight ratio than the single; what we call power loading. You have induced lift from the engines, creating a slipstream over the wings. Your power-on stall speed is therefore quite a bit below the power-off stall speed and well below the Vmc speed. What do these maneuvers teach other than a dangerous habit?

The idea of the departure stall is to demonstrate the danger of trying to clear an obstacle when the aircraft performance and local conditions won't let you. Now, your basic obstacle for test purposes is 50 feet. That is not an obstacle; it is a bump. Anyway, at sea level a Cessna 310 climbs at 1500 feet per minute. That rate of climb would put you over a 50-foot obstacle in two seconds. You would never get that insanely high pitch attitude and low airspeed even close to an imminent departure stall until approaching the pattern altitude and heading for the next county. Maybe you would if you yanked as hard as you could on the controls, but that isn't realistic either.

At sea level the power-on stall, a condition with clear visual and aural warnings, isn't your big concern. Loss of control during an engine failure that happens without warning is your concern because it can put you in immediate jeopardy. While climbing over an obstacle at your best angle of climb speed, you may be near or at your Vmc. Where the departure stall will be a concern is when reduced power is all that is available from the engines, thus lowering the Vmc, but where you still have to clear an obstacle. This condition is to be found at high-density altitude fields. Even a 50-foot obstacle at the end of an 8000-foot runway can be a hazard under the right conditions.

If you can set the power about 20 inches and pitch up so painfully slowly that you induce a high angle of attack without inducing a high pitch attitude, you will get indications of the imminent stall at attitudes much lower than you would otherwise expect. That will teach you something valuable.

This same scenario is duplicated when you try to clear an obstacle that remains in your way as you get closer, requiring more climb and progressively higher pitch attitudes. Power-on stall speeds get confused in some pilots' minds where the bottom of the green arc is considered the power-on stall speed rather than the power off speed in a clean configuration. Power-on speeds are not marked on airspeed indicators.

In thinking about stalls, think about how twins fly based on their wing and power loading. The higher lift requirements of the twin can be met because of the slipstream forcing air over the wing at a greater velocity than your airspeed. Therefore, airspeed does not have the same impact on the power-on stall speed as it does in a single. This makes for proportionately lower power-on stall speeds in twins.

Should you reduce your power prematurely, you will observe how a highly loaded twin wing can produce immediate and rapid descents when you take away the induced airflow. You can quickly raise your stall speed to the power-off value simply by reducing power. This can result in bone-jarring landings when an unstalled power-on condition becomes a stalled power-off condition as the slipstream induced lift leaves the wing. What you should really be studying with twins is the interaction of power management and its effect on stall speeds in all flight conditions rather than demonstrating your presolo stalls again.

Twins have more Vspeeds for you to understand and know because of the ability to operate on a single engine. One of the most important of the new airspeeds is the "safe single engine speed," Vsse. Due to the outrageous number of training accidents from operations at Vmc speeds that were set optimistically low, the FAA sought to provide a cushion of airspeed where training in single engine operations and intentional engine cuts would be far safer because of a greater measure of airplane control. We can further improve on this idea by putting a red arc on the airspeed indicator that extends from Vmc at the low end to Vsse at the top.

This could serve as a kind of takeoff decision arc. You never want to be below Vmc with full-power, yet after you accelerate past the top of the arc, you will be in a good position to control the aircraft and attain reasonable performance. Think about that when a maximum-performance takeoff is required. Your basic Cessna 310-Q has a Vmc at 75 knots, an obstacle clearance speed of 78, Vx at 81, and a Vsse of 90. Having never found a good definition for the obstacle clearance speed, I can only speculate that you would pop-over your 50-foot obstacle at that speed and then immediately lower the nose to accelerate to a safe climb speed.

Anyway, a red arc from 75-90 knots would alert pilots that any maximum-performance takeoff is in a potentially dangerous area. Anytime a climb speed is used that is less than Vsse, there is a potential control problem should you lose an engine. Which is not to say that you should not use those speeds for they may be required for your takeoff. As long as both engines operate normally (which they usually do), there is no problem. The risk you accept is that should you have an engine failure you will have a far more difficult control situation than had you operated above Vsse.

The closer the speed you use is to Vmc, the more you are betting your life against your engines. I often wonder when I read about the various Vspeeds which were set by test pilots in perfect aircraft hoping to set great figures for sales performance and how many are set by lawyers avoiding manufacturer exposure. It seems like the obstacle clearance speed and safe single engine speed are perfect examples of each category. There is a balance here where you may be able to clear any obstacles and still maintain your safety by flying at Vsse, until you are in a position to accelerate to the best single engine rate of climb (Vyse), which is marked by the blue line on the airspeed indicator, and then to Vy. The answer as always is to think it through.

What about that blue line? Conventional wisdom holds that you climb out no lower than the blue line and approach no lower than the blue line until the runway is made. Since airspeeds like Vyse are fluid and vary with aircraft weight, does it make sense to mandate the blue line speed in all situations when a slightly lower speed in a lightly loaded aircraft could give much better performance? The blue line is the best place to start when operating on a single engine because it is set for an airplane at maximum gross weight, your worst legal case. Try experimenting, though, with your lightly loaded trainer by flying at slightly slower airspeeds and monitoring your performance on the altimeter and vertical speed indicator. You will find many times that a climb can be maintained below blue line and the aircraft is under good control, where flying at blue line may result in level flight or even a descent.

We could incorporate the calculation of a modified Vyse into the takeoff planning based on the aircraft takeoff weight. Once again the arc idea could be included on the airspeed indicator by providing blue arc limits that would for consistency extend through the same three aircraft weights used for determining various maneuvering and best glide speeds. While we are at it, why not end the green arc at Va. The blue arc would have the highest airspeed value at gross weight, a white slash for the middle weight, and end at the lowest weight. The green arc would top out at Va, have a white slash for both the middle and light weight airspeed values and end at Vs. The best glide arc would be orange and would extend through the same three weights with a slash for the middle weight just like the other arcs. This arc is a great way to set the landing approach speed for your weight. This way you could easily compare airspeeds for a particular weight for Vyse, Va, and best glide as well as have a red decision arc from Vmc to Vsse.

If you think the airspeed indicator is becoming clogged with arcs, just take out the miles per hour arc and the true airspeed wheel. You should use the flight computer anyway for calculating true airspeed. Besides, who needs statute miles in an airplane anymore? My favorite contradictory airspeed is Vxse, the best angle of climb on a single engine — as if anyone would even contemplate climbing over an obstacle with only one turning.

What makes flying twins fun and will give you much satisfaction is learning to master all the systems. Twins are much more complicated than your basic single, especially when you reach beyond the training twins. You may not be aware of this, but as you fly progressively more complex aircraft, you develop a certain ability to teach yourself about new aircraft and spend more time doing ground work to master the systems so that less time is required in the air. Take this quality and make the most of it. As planes become more expensive, you have great incentive to learn as much as you can on the ground before you ever fire up an engine. There are ample manuals and handbooks available to feed your knowledge.

You are supposed to be a professional now and should not be coddled anymore. This means that your training time is now dependent on your personal motivation. Pour over every page of the aircraft manual. Sit in the cockpit until you know where everything is and have rehearsed all your procedures until they become instinctive. When you can do it on the ground, you can do it relatively quickly in the air.

The fuel systems can be much more complex than what you are used to, especially in an aircraft like the 310. Your twin may have as many as four to six tanks. One of the major causes of engine failure is fuel starvation resulting from fuel mismanagement, so it is startling to find that you can get through an entire multi course never having switched a tank.

The systems are getting so complicated that your instructor may not have all the answers anymore. Since all pilots seem to revert back to presolo students whenever they get with an instructor, the tendency is to be disappointed in the instructor when they cannot immediately spoonfeed you the answer. You should now know where to look for the answer yourself. When you have really detailed questions beyond the scope of your training, find an instructor mechanic and make an appointment for some ground school. This is how I learned the level of technical knowledge that got me through an Airline Transport Pilot oral without undue difficulty.

Once the engines were mounted on the wings, the noses became very short. With short aircraft noses, all those great sight pictures derived from screws and markings on the cowling disappeared. You will need your instruments more in the beginning until you develop new pictures based on different sources of information. The big problem, as usual, will be landing. You will need to find the variation on the nose at the end of the runway idea that works for you in your new airplane.

What really becomes critical is looking ahead of you and putting yourself on the centerline of the runway, not the airplane. When you look out the window from the left seat, the left engine appears further ahead than the right, causing you to try to swing the right engine forward to make them look even. In the same way you once tried to center a single using the spinner, you will now subconsciously do the same thing, this time in an effort to balance your distance from the engines. The result is the same also; you will swing the aircraft around to the left before it touches and squeak the tires. You will swear everything is fine right up until touchdown and then not understand why the aircraft is crooked.

The windshield on twins can be more curved than the singles you are used to, especially if you are a Cessna pilot. This also leads to distortion on landing as to what straight looks like. You will have to find out for yourself what sight picture allows you to land longitudinally straight, for no one else can look through your eyes to see what you see.

One capability that is often underdeveloped in the twin pilot is the use of differential power. During both takeoffs and landings in strong crosswinds, the twin has an advantage over the single by its ability to overcome the weathervaning tendency by supplementing the rudder with additional power on the upwind engine. You will probably not be taught this because chances are your instructor was not taught this. Differential power for anything but taxiing has not made it into the instructor/student evolutionary process. Also, there is no time for this when preoccupied with making those good old steep turns nice and pretty for the examiner.

I only know about differential power use from my own experimentation. You will most likely have to ask your own instructor to experiment with differential power — then you will both know how. I have found it most useful on takeoff. It really helps to carry extra power on the upwind engine until they both have full power and the rudder gains its effectiveness.

For landings, however, I still tend to rely on the old faithful sideslip method while carrying just a little extra speed. You can forget when you move into a twin that it is just another airplane and the flight controls still obey the same physical laws of nature, so, yes, you can rush down the runway on one wheel in a twin. Differential power will keep you longitudinally straight, but it won't keep you from being blown across the runway. This isn't a problem on takeoff where the friction on the wheels helps to keep you centered. You can't use the crab method on landing with differential power to keep from being blown because they will cancel each other out.

However, when using the sideslip method, differential power on approach aids the rudder in keeping you straight allowing the twin to handle higher crosswinds than the single, which is limited only to rudder and airspeed for crosswind correction. You could get into trouble trying to keep straight as you idle for the flare if you have been using both power and rudder in some outrageous crosswind because as the power comes off, you will need extra rudder. This is a skill that should be developed slowly after you gain much experience way beyond the checkride. When you are good, you will be able to land in howling winds on one wheel while still carrying power on the upwind engine. You should see this added capability in your training. Ask for it if it is not included.

The greatest waste of your time and money in all of multi training is learning how to show your examiner a really great drag demonstration. Once again, the only people who need to be able to show this are instructors. Regular multi students should be able to simply see the effects on drag and performance using various combinations of a windmilling prop, a feathered prop, and the gear and flaps. Do you really need to spend some $200 an hour perfecting a demonstration so you are able to convince an examiner that you know a twin will in fact descend faster on a single engine with the gear and flaps hanging out? Is this a surprise to anyone?

What is of interest is the difference in performance between a windmilling propeller and a feathered one for this is a new concept. It is important for you to see this because most pilots acknowledge the effect of a windmilling propeller without quite believing it until they actually see it for themselves. There is nothing you could gain from learning how to demonstrate drag that couldn't be better observed while learning how to deal with an engine out en route. This is also a good time to note any difference in performance between the left and right engine to see how critical your left one actually is.

Learning the drag demo overshadows learning about the effects of drag because most of your energy goes into memorizing what to drop and when rather than observing the effects of what you have dropped. If you have the willpower to look, you might note your performance on the vertical speed indicator. Chances are, though, you will be religiously holding the blue line airspeed so you can't check the effects of lower aircraft weight on your performance by varying your airspeed in different configurations. Most multi students don't have the wherewithal to challenge their instructor and ask to try different airspeeds below blue line, thus rendering the drag demo largely useless, unless you like to talk about what should have happened in the air when you get back on the ground. See if you can calculate the best Vyse for your takeoff weight and then see how your vertical speed performance compares to your calculations and how the controllability of the airplane is affected by the lower speeds at its lower weight.

Learning to fly twins is some of the most fun you can have in an airplane. Try to get as many varied experiences as you can beyond the narrow and inappropriate limits of your training. Fly in all kinds of weather and get lots of night experience. The extra speed of twins will take some getting used to, but if you fly slightly wider patterns and leave more room and time to do things, there won't be any difference from high-performance singles.

When you do return to singles, they will seem painfully slow. New twin pilots have delusions of grandeur and feel that those slow single airplanes are now just bug-smashers in their way. This royal image is reinforced by ATC who tend to give multi pilots better service because they think you know what you are doing if you fly a twin, except of course for those training twins with the counter-rotating propellers. Remember that no one is any more important in the sky. You will just have to learn to be flexible enough to fly the faster twins in all airport environments. You will be humbled again when you step back into singles.

Read as much about twin flying as you can after your training to be fresh when you get to fly them again. Twin experience, like instrument skill, is fleeting. You have to make every hour count because of the cost, so after you are rated, always try to fly in twins at every opportunity to maximize your exposure and retain your knowledge and skill.

14 • YOUR FIRST INSTRUCTOR JOB

————————————————————————————————✈

There was a great old television show from the 1970's called *Kung Fu*. David Carradine played a monk from a Shaolin Temple. Cloistered within its confines as a boy, he led a life of study and hard work. His head was shaved and he wore a uniform. He was no better or worse, nor in any way distinguished from any other student for they all looked the same. He had the same strict and disciplined program to follow as every other student, and all his decisions were made for him. From this he eventually graduated, charged with the mission to spread the Tao throughout the corrupt, decadent, and debauched "Old West." When graduation came, he was put through a brutal final test after which he was shoved out the back door of the temple which forever closed behind him. It was time for him to try and survive in a completely strange new world for which he had all the basic skills, but with none of the necessary experience or judgment. Many adventures were to follow. As a brand new and untested CFI, especially you Part 141 graduates, this is pretty much where you are right now.

The fun times are over. The leisurely pace of student life is gone. Time to leave the nest, hit the road, prove your worth, and start paying back those student loans. Time (if you can do it) to convince an employer to actually trust you with real live students. Should you be one of the fortunate who gets to teach at the institution from which you trained, you will already be indoctrinated with whatever bizarre system drives them. Consequently, you will fit right in. You cannot count on getting a job at your school, try as you might to buy your way in, for it is purely an economic decision. The cold, hard reality is that upon graduation they either need you or not. All the carefully cultivated brown-nosing in the world can't create an instructor slot.

Flight instruction is nothing but a business, and this could be your first real business lesson. Flight schools dangle the possibility of a job in front of you throughout your training so that you will spend the most money possible at that school. It works because you believe yourself to be a superior pilot such that when you graduate they won't be able to resist you. The school knows this psychology, which is how they continually get away with this game. You can even watch several of your buddies hit the highway without a job and still secretly delude yourself that the school will automatically want you to teach when the time comes; maybe yes, and then again, maybe no.

I think that teaching at the same school where you trained severely limits your depth and experience. You begin to believe that only graduates of your school know the right way to fly. This is especially true if you work at a big academy staffed by other young hot-shot clones, teaching foreign students who disappear after graduating. You never know how successful your program really is in the long run because you never hear from the folks once they go home. In such an isolated microcosm of ivory tower aviation academia, you completely buy into the particular myths being passed down through your school because there is nothing in your experience to cause you to question anything. My ivory tower never hired me, causing me to go out and find another school, where I had to learn the new set of myths that they wanted me to teach. Only by breaking out of the insulation of one school will you see the power and permeation of myth that infects all schools.

The advantage of teaching at a big Part 141 academy is that you will be handed students. You will have to teach from a strict syllabus with no room for originality or style. Planning for your lessons will be nonexistent as your job is to spoonfeed the same lessons that were spoonfed to you.

You won't have to spend any of your time selling yourself as students will just show up on your schedule. This is a great advantage. You will probably get a cute, little uniform and become nothing more than a dispenser of procedures. You will not have to make any decisions as everything you do will have already been preapproved by the FAA.

On the other hand, your flight experience will be limited to whatever few types of aircraft are in the fleet. Large schools like to standardize everything, including the types of aircraft. You will fly the same, tired lessons to the same approved airports. If your school really wants to keep the insurance down, count on flying from big runways, with no soft fields or mountain airports permitted. You will fly the program with no possibility of individuality, creativity, or tailoring your instruction to the student. It can't work of course because students are individuals.

Many of you will end up in flight schools or flying clubs. The disadvantage here is that you constantly have to sell yourself. You are subject to the whims of students who gladly let you starve in the winter. Most of your students will not have paid the huge lump sums of money to cover a complete program, like the academy students, so their individual motivation and dedication are variable and inconsistent. You will, however, meet pilots of all ages, occupations, backgrounds, and experience levels. You will fly with recreational amateurs and senior airline captains.

Try some persuasion with the chief instructor and you can teach off gravel and dirt strips, mountain fields — in fact, anywhere you can convince a student to go. You will fly a wide variety of aircraft as many private owners require your services. You can get an incredible variety of flight experience while having total control to design your own program.

Flying schools come in all the shapes, sizes, and variations mentioned in earlier chapters. You just have to see for yourself what's out there. The economy sucks these days. Aviation is in its biggest slump ever. Many airlines have folded, tossing jet jocks out on the street and into the market to scramble for the commuter and cargo jobs you are preparing to get one day. As you have traveled this journey, you have consistently watched lesser mortals fall out of the process. Making it this far has taken pure resilience and an indomitable will to reach your goal. You cannot let anything stand in your way if you are to ever become a commercial pilot.

By becoming an instructor, you think you have accomplished something. However, to any prospective employer you are just another baby instructor. Someone, somewhere has to be willing to take a chance on you for you to proceed any further. At this point you need a job — any job. Which brings us back to the flying schools. It is nice to be able to sit back and think about the kind of school where you want to work. You cannot dwell on this thought for a minute, for such a luxury will never be available to you until you have given at least 500 hours of dual instruction. As long as the current situation in aviation remains, it may never be an option.

Gather your resume and recommendations for it is time to go visiting, just like you did when you were selecting a flight school. Unlike the good old days when you had money in your pocket and flight schools bent over backwards to convince you to take lessons at their schools, when you return to some of those same places looking for a *job*, the treatment you receive will be just the opposite. The glut of pilots and instructors on the market only aggravates the situation. "We really don't need any instructors right now, so please just leave your resume and we'll be in touch." Sure they will.

Here is the problem. Although you feel like you have just captured the "golden fleece," to prospective employers you are the biggest risk they face. Your initial impression after rejection is that you have been prematurely and unjustly judged. You have done all the right things, passed all the required certificates and ratings, the FAA says you can teach, so what is the problem?

As checkrides have nothing to do with flying, so acquiring the CFI certificate has nothing to do with actual instructing. Instructing according to the FAA is a calling; to the flight school, it is nothing but a business. Think about how you got the CFI. You did everything but teach students. You have absolutely no experience in your field. Why should anyone take the risk of letting you practice on a good paying customer when not even you can predict the outcome of the lesson? Sure, you can simulate teaching people who know what they are doing for that is how you passed the CFI test.

You also know the purely theoretical laws of learning, all the flight standards, endorsements, plus endless facts and figures, but you still haven't the slightest idea how to teach. The CFI like the commercial, requires that you have no actual experience in exercising the privileges of your certificate. The student pilot at least gets 20 hours of solo time before the private. The instrument candidates can experience flight in the clouds, if they so choose.

To solve the new instructor dilemma, I propose that the CFI requirements include ten hours of practice teaching to volunteer students who should at a minimum have their private and be rated and endorsed for the training aircraft. Split up the practice time into five hours of flight and five hours of ground instruction. Each lesson should be reviewed by the supervising instructor, CFI candidate, and the volunteer student all together. That way you would have real experience before the CFI test as well as some real experience to present to an employer.

Since the current rules are not going to change any time soon, the best thing you can do is to cultivate prospective employers while you are still working on your ratings. These people are more likely to give you a chance when you become a brand new instructor if they have gotten to know you over time when you were a student.

You are in for a big mental adjustment as you look for a job. While still attending your flight school, you were rising to the top level of your peers. The acquisition of the CFI certificate is seen as the pinnacle of achievement. To your employers, however, you are nothing but a fledgling instructor competing with far more senior instructors for jobs. It's like when you go from high school to college. When you do this, you switch heaps, from the top of one to the bottom of another. You have to adjust from being the big person on the high school campus to the lowly frosh in college. You cannot begin to rebuild your stature among your peers until you gain recognition as a working instructor. To do that you first need the job.

Worse than this adjustment, your orientation to aviation is all wrong. Everything you have experienced up until this point has been academic. From here on out it is commercial. You will find that is all the difference in the world.

Your first 100 hours of instruction are a critical time for you. Not for any safety problem for you or your students because your flying will be just fine. New instructors generally are overcautious. You won't let students make bad mistakes because you will be right there.

The problem with new instructors, who have up until now done all the flying themselves, is that now they have to get used to letting students make inconsequential mistakes which are not likely to carry any chance of incident, but are very desirable for the students because they can learn from them in safety. You can clearly see these mistakes coming because of your experience. Most new instructors are too quick to take the airplane because they feel strange not doing any flying anymore. Your checkride training was designed to impress an examiner by both flying and teaching the FAA way. This, of course, does nothing to prepare you for real students who don't understand you, don't always listen to you, and still want to do all the flying anyway.

You also have to learn to tell students the right way even though showing them yourself would be far quicker. That is not what instructing is about. What is critical for you in this time is to make the adjustment to real instructing as quickly as you can to start building a good reputation. As nothing up until now has prepared you for this, it is natural to make mistakes at first. That is expected and even tolerated for a short time by students and bosses who know you are engaged in on-the-job training. You must improve almost immediately as patience from all parties is very short; otherwise you will be blackballed before you even have a chance to improve.

What you never want to get in this business is a bad reputation. Your reputation is everything in the closed community of aviation. I guarantee that your students will talk, and they will not talk to you the same way. They will talk to each other and to the chief instructor.

Being too quick on the controls is one of the most common problems for new instructors. It was also one of my problems. The boss hauled me aside one day to enlighten me to a condition for which I was completely unaware. It takes a conscious effort to change a pattern like that. Soon I was hauled aside again and told that my students were much happier now. Who knew? They never tell the instructor. The point, though, is that anyone can learn and change their patterns.

Here are some other problems common to new instructors: having too little patience; raising the voice quickly and inappropriately; saying one thing while doing another; not admitting mistakes; not being honest with the student; setting impossible standards; forgetting what it was like to be a student; not looking into human factors; getting on a power trip; and worst of all — thinking you know what you are doing. Your employer is aware of all this. It may be your first job, but you are surely not their first baby instructor.

One way to cut down on your learning curve is to seek out the chief or a senior instructor and make him a mentor. This person will be aware of your situation in detail because your students will have already talked to him. Get an honest appraisal of your teaching as soon as the word comes in. If you make an effort to be the best teacher you can, most people will respect your wish and make every effort to help you in the process.

Attitude is everything. Should you demonstrate your desire to improve, you will be valued. Should you prove a disaster with no desire to cooperate because you think you know it all, you may find yourself paying off your loans selling used cars.

Aviation is a small club. Your reputation will precede you wherever you go. I was sitting around this old picnic table with an instructor friend of mine at my club one day. We were talking about some of the horror stories which occasionally come up in casual conversation. He paused for a minute and said, "You know, I have never heard a horror story about you." To me, this is the perfect comment on a couple of years at instructing, considering how much can go wrong. I returned the compliment for it was the truth. This instructor was one of the few I really respected. The worst thing that can happen to you in this business is to be talked about. Oscar Wilde is wrong.

New instructors have no idea how much time they can spend waiting compared to how little time they will spend teaching. New instructors don't get a flood of business. Flight schools rely on senior instructors who have proven their profit-generating capacity, so they get their schedules filled first. That is unless management is playing favorites to disrupt the troops. Usually, though, new instructors, like juvenile predators around a kill, wait for the scraps. The established instructors love teaching the glamor stuff: commercial, multi-engine, instrument, things like that. Consequently, the student pilots will generally go to the recently hired instructors.

Everything is backwards, so it should come as no surprise that flight schools frequently match people who have never flown with people who have never instructed. I began teaching at a different field from where I learned to fly, where much training of controllers was going on, at a flight school under new ownership, where I was the only full-time instructor. There I was a new instructor, teaching new students, at a new field, with new controllers, at a school under new management, with new rules and procedures to follow. Somehow we all survived that first summer. New students can be pretty flaky and over half of them will drop

out, even from the old-time instructors, so there is no loss
to the school usually in pairing new student pilots with new
instructors who need teaching experience anyway. The system is clearly geared to advance the school's profit rather
than cater to any student need.

When you start your first job, you will just sit and wait
at first, and wait, and wait, and wait some more. In my first
two weeks, I flew about ten hours. The rest of the time I
waited and tried to sell my talents to prospective students.
Slowly, drearily slowly, drudgingly slowly, as the spring
lumbered into summer, more folks came dribbling into the
school and I started to fly more. As soon as I was barely
able to call myself busy, the next instructor came on line.
He waited, and waited, and waited some more. Eventually
he had a crop of folks to teach as well. This brought on the
third instructor who waited, and waited . . . Get the idea?
This is how we built the flying school. Waiting is common
throughout your instructing. There will always be slow
times, like when all your students are doing solo work, or
the weather is bad for long stretches. Your biggest wait will
often be when getting started at your first job.

Unless you work at a big academy that does your
selling for you, chances are you will do a great deal of demo
rides. Now we all know the demo ride has nothing to do
with introducing a student to the world of flight training
and everything to do with selling the product, so here is a
chance for you to prove your worth. How well can you sell?
Your value to most flight schools directly depends on your
ability to generate business for the boss. You have to put as
many people in airplanes as you possibly can.

There is an inherent conflict of interest here because
the FAA determines that you are the officer of the skies
keeping the riffraff out of everyone else's way. If at all
possible then, you have to find prospective students, demo
them, sell them, and keep them coming back for more,
unless they have absolutely no business in the sky. If you
don't find students for demo rides, if you can't attract and
keep students, you can kiss that commercial career good-

bye. Whether or not you enjoy selling, or even teaching for that matter, is irrelevant. You have to do it. How else are you going to get the hours necessary to leave instructing? This is the only facet of aviation that most people get into with the sole purpose of getting out as soon as possible.

That contradiction means that your first priority is to yourself through the acquisition of flight hours, not some messianic pilgrimage to save the customer from the realities of the business. And they call us "time-builders" — ha! The hypocrisy that instructors are here for the cause of aviation is only for the lifers. Your justification for becoming the best flight instructor you can is so that you can gain the best reputation, so that the greatest number of people will know you are a good instructor, so they will seek you out, so that you may get better at selling yourself, so that you can acquire the hours to get out of the business.

The art of the demo ride then becomes critical to your future career. Here are some secrets. You have to sell without selling. Put them in the airplane and let the wonder of flight, whatever the individual perceives it to be, work its magic. Have you ever sold used cars? Just a thought. The product you sell is the pilot certificate with all the wonderful inherent joys therein, where the only pathway to that glory is through you.

To be successful, you cannot just sell flight; you have to sell the belief that you are the only person your demo rider can imagine as an instructor. Can you excite the trepidatious? Can you motivate the "wanna-bee," as in I wanna-bee a pilot? Can you take someone who just walks in the door for a brochure, to ask perhaps a few idle questions, put him in an airplane right then, talk him through a lesson, and sign him up for the whole program? That is what flight instructing is really all about. This may sound cynical to the puritan who wants to teach flight in the ideal FAA world, and it is a contradiction to the clear warnings issued to prospective students from the beginning chapters.

You have crossed a bridge never to return. The system

requires you to do whatever it takes to get your flight time, for that is the cold, hard reality. My job now is to tell you how to go about it. One way is to use all the tricks I warned you about earlier.

There are presently too many pilots on the market. Many former commercial pilots with instructor certificates they kept for emergency income are now exercising that privilege and squeezing out new instructors struggling for those jobs. Most of the other jobs in this world are at best a pain in the ass, that's why we got into aviation and put up with its abysmal conditions. Every employer knows that, so they can pick and choose carefully among the torrent of unemployed yet highly qualified pilots. Why should they waste any time with an untried commodity like you — unless you can sell.

If you can bring your own real live, guaranteed students to a potential employer, your job prospects will rise dramatically. To them it is free money. Employers know that unless you plan to end your career right here in instructing, you will be moving on in a year or two, or as soon as you get a better opportunity somewhere else. They will try to squeeze every bit of profit out of you they can while you remain under their employ. They also know that you need them more than they need you. Any instructor can be replaced in the time it takes to make a phone call. You are forced to satisfy the needs of your employer, your students, and the FAA, all at the same time. None of this will occur to you during your CFI training because you were preparing for something analogous to a tenured college professor rather than the snake-oil salesperson demanded by your employer. You still have to instruct well, but to succeed you have to remember that the bottom line is always the *bottom line*.

New instructors are startled to find out just how much sales work is required of them. You can't teach if you don't have any students. You can't get hours if you can't teach. The sales job falls on the instructor because your employer knows you are just slightly more desperate than they are to have students. The boss does the advertising and marketing, gives you a desk (or in some cases rents you a desk), keeps a large portion of the dual rate for themselves, and then tells you what a great opportunity they have bestowed upon you. Oh boy!

The future of general aviation depends on flight instructors putting people in airplanes, yet nowhere in the CFI standards is there any section on ethical sales and marketing. How are you supposed to know what is expected of you and what is legal? Although the FAA should not in any way interfere with legitimate and fair commerce, it might be an idea for them to periodically pose as prospective students and check out various flight schools. They will find schools where instructors are forced to openly lie and mislead students in order to keep flying and keep their jobs. Some guidelines on sales would not be a bad idea for it would protect the instructor as well as the student.

I have no sympathy for school owners and managers, having worked for enough of them to see the same disturbing patterns. Here are some common lies and deceptions that many instructors are forced to use in their sales pitch. The most frequently asked question is "How much will the certificate cost?" As you might expect, the answer to this question will be the most frequently misleading. The game of quoting a price based on minimum hours that no on can meet continues to draw in student pilots because they don't know how the game works. You didn't get your private in the minimum time; how can you tell someone else that they might? Some schools get around this question by simply telling instructors not to give any estimates or quotes at all, just keep avoiding the question. That way the students can't come back later when they exceed their own unrealistically low estimate of how long it should take.

You may be asked to sell a package at a particular price which on the surface appears to the students as a guarantee, but is really just a lump sum for an account that gets drawn down with every flight. Should the students exceed the projected hours, you may be told to show them the fine print and give them the news that the boss wants more money to continue the training.

Instructors are a vital part of any flight school, and you have a vested interest in its success because your personal goals are directly tied in. You may be asked to do many things that you do not like, but will be in your best interest.

I was very fortunate to find a club made up of instructors from regular schools who were disgusted with the standard practices. We set up a school that allowed the instructors to teach to the highest standards, to teach our own way, and where the students were treated as we wish we had been treated. Students received honest and realistic training costs and times, in writing, with never the implication of a guarantee. Our students didn't feel they were misled over the programs. There is no excuse for resorting to deception to attract students by presenting the uninformed with inaccurate information.

Some schools work to insure that your students won't make anywhere near close to the minimum times. Remember that game one school played where they stacked up lessons every two hours, the lessons backed up, the aircraft were late, and after traveling to and from the practice area, there were about 15 minutes actual teaching per lesson? What do you tell your students in this situation when they ask why their training is taking so long? You have to serve the employer to keep your job, so you can't tell the students that their money is being funneled away in return for a minimum of actual training.

Do you tell your customers that you are a Part 141 academy when in fact the application process is still ongoing? I was asked to do that. I know a school where they charged the students a dual rate six times higher than the amount given the instructor. This exorbitant rate was justified by telling the customer in the sales material that only the best and most experienced instructors were on the staff. In reality the only instructors who could afford to work for such low pay were the brand new recent graduates of the flight school. They already knew the system and didn't have to go through the process and expense of looking for a job that may not be out there. The flight school knew them and didn't have to take the chance on unknown instructors.

What do you do if you are a beginning instructor in that situation, your student asks you how much dual time you have given, and then shows you the sales material that states how experienced you are supposed to be? You have to live with your employer, your students, your FAA, and your conscience — usually in that order. You may have to get a job elsewhere should you ever feel compromised. The choice is yours.

How do you sell yourself successfully and not violate your principles? It isn't all that tough, although sometimes you may have to do a balancing act. There is nothing wrong with enthusiastic selling as long as you are honest, so let's talk sales. Your first contact with potential customers is often the only chance you will get to hook them on having you teach them to fly. Nobody calls or visits a flight school without a lot of thought and a desire to at least try flight. Your job is to complete the process and close the sale. You must give people the permission to fulfill their desire to learn to fly. You must believe that they will demand you for that purpose. You have to take that first contact as far as you possibly can. If you can get them off the ground, you should have a student.

Instructors fail to attract business because they have no formal sales training. You would think that any school owner with a brain would offer a professional course in sales to all the instructors rather than churning through people to see who survives by making the boss the most money. The big academies have no need for any of this because they overcharge for flight training and have the reserves to hire professional salespeople to relieve the instructor of any sales responsibility. These are the folks hidden away in the cushy offices with the word "counselor" written on their wooden desk plaques. Get to know these people; you might get more students from them. It is not efficient for the big schools to have instructors selling a service when their talents lie more in motivating students to stay in the system.

If you do have to sell the service, assume the customer has already made the decision to fly and you are just going through the motions. Never give up. When someone calls for information on the phone, don't give away all your hard-earned secrets for nothing. This is your living; get a name, get a number, invite them in for a visit, schedule a lesson or a demo if that is all they will go for, talk about flight with authority to build credibility, and lead them to the airplane. When someone drops by for a visit, your job is easier because he is already midway through the process. Don't let him walk out the door with a brochure in his hands. Put him in an airplane right there and then.

If you do not do this, at the next flight school he visits, should I be the one he meets, with a little more sales effort than you made, I will put him in my airplane. You will have done my background work, and I will be flying with your student. People want to be sold. They love it. If you can't sell, you won't fly, you won't eat, you will have to take another job, and worst of all you won't get to fly big jets. That is how the system works.

Okay, so there you are a new instructor with a prospect staring you right in the face — now what? Take him on a tour of the school. Look at those cute pictures hanging on the wall. They aren't there for decoration. They are there to give your prospect something to look at so he doesn't feel threatened by direct eye contact, so you can make him feel comfortable. Thus you will have something other than lessons to talk about so you can fill him with flying wonderment, which is to break down his resistance and help you put him in an airplane.

Find out your prospect's particular needs. If he wants to fly for fun, then tell him all the things you can do with a private certificate. If he desires a career, tell him of the ratings, certificates, time and money involved, and the fun of getting paid to fly an airplane. While you are doing all this, work your way out to an airplane. You know how demo rides work from the instructor's point of view; it is your duty to be the great tour guide through the experience. Grab the keys on your way out the door and go fly.

When you return, be sure to take the time to answer all his questions truthfully. This is buyer resistance time. If you want a really nice touch, stick around to answer some questions after he has paid the bill. This shows that you actually have an interest in his flying and don't just want his money.

You must always tell the truth because somewhere down the line any lie will catch up with you. Even if the boss implies that you have to tell a certain story, by remaining distant from the student, he will not be the one who catches the flack. If you are the one telling the lie, the responsibility falls on you. Should you shave the realistic time to get a certificate in order to hook a student, as soon as he acquires that amount of flight time, guess what he will ask you, "Why am I not ready for the checkride, you promised?" Always, always, always be straight with your students. You can fill your customers with all the wonders of aviation, put them in the airplane, let them see how much fun it is, and let them take full responsibility for continuing their training — all this while knowing that you have told them the truth about what flight training entails. That is how you succeed.

How do you handle a situation where you are directly told by the boss to lie? If I get a question that requires as a condition of my employment an answer contrary to what I can honestly give, I send the prospective student to the boss. Let him tell the story. The trick is to do this without the boss knowing you ducked the question. The boss will find a reason to dump you if you do not toe the line even if it is dishonest because then you are not a team player and can be replaced.

You will also lose students if you lie to them when they find you out. From this your reputation will suffer, your business will drop, and the boss will dump you anyway. The only way you can win is either to quit (not often a very viable option) or take yourself out of the process by passing on the question to the boss.

Okay, so now that same student comes back from the boss with an answer he doubts is truthful. You will have credibility because you are not the one who told the lie. Things now get very delicate. You have to answer the student in a way that is truthful, that keeps your job, and that keeps the student flying with you. People are funny. The contract is really made with the instructor, not the boss or the school. It is not unusual to have students fly with you because of your honesty when they know damn well the boss has lied to them. What a game that is to play, yet I have seen it many times.

Let's say that you teach at a school that publishes a brochure which quotes certificate prices based on the FAA minimum. Your potential student may have done his homework, talked to other students, or somehow learned that no one makes it in the minimum. You may be asked a question for which your prospect already knows the answer, just to test your honesty. Many times the question asked has nothing to do with the real question being asked.

What you can tell someone who wants a private is the national average number of hours to get it and that he will do better or worse depending on his ability, his commitment, and how hard he is willing to work. Tell students that you will do your best to get them as close to the minimums as possible and then outline what they will have to do because only they can make it happen. Explain that the price quoted is based on the minimum time, and should they be ready at the minimum, that is what it will cost. The minimum is therefore a goal, not a guarantee. Should the student fall short of your plan (as they all do) as long as you do your part, they have only themselves to blame. The school is happy because they acquired students and you didn't make them out to be liars. The students are happy because they were fairly and honestly dealt with by you. By not promising or implying any contract, they have no case to press on you when their training extends however far beyond the minimum. Although not included in the CFI standards you can now add negotiator, facilitator, and mediator to your list of qualifications.

Just because you are a new instructor doesn't mean there is any less responsibility lumped on your shoulders. You have the full weight of the FAA just waiting for you to goof and none of the experience to make those critical decisions. You are kind of like a new bartender around closing time hoping all your customers get home in good shape.

New instructors are particularly vulnerable to a character I call "the brute." Your basic brute intimidates you into signing them off for whatever they need whether ready or not — most times they are not. You are not so much an instructor to these people as you are an obstacle. These pilots shouldn't be in the air anyway, but they are so you will have to deal with them. Since the instructor is the last line of defense to keep the skies safe, you are the one who has to "just say no."

Here is the conflict. The school doesn't want to rent an aircraft to anyone who will cause any damage. Anything short of that, they want the business. This can be an incredibly fine line. You will generally find that the richer the pilot, the higher the performance they desire; the more money they have for the school, the more the school lets them slide, and the more they are allowed to become brutes. You have to try to satisfy the customer and the boss.

I have had many pilots show up on my schedule who have walked in the door, told me exactly what they will do for their checkout, usually a couple of bumps around the pattern, how much time that will take, and how much money they will spend. Implied in all this is that any deviation from their plan will be immediately reported to the boss, with whom they are, of course, intimate friends. Why do people insist on playing power games? Besides, these are the folks who often need the most remedial instruction. The new instructor, though, tends to regard these bums with awe. You think they must know what they are doing to be so sure of themselves. It's all a bluff. The old-time instructor will see these blowhards coming miles away and deal with their egos accordingly.

This is also a good test to see if you are working in the right place. When you see how the boss backs up your decisions, you will know whether he cares about you at all. Some bosses will respect your decision over the customer every time. You can ask these folks to evaluate your decision process and learn from the experience. Should you be blamed for sending a perfectly good meal ticket out the door regardless of the pilot's skill, you are in the wrong place. You may put yourself into a situation one day that may jeopardize your career and standing as a pilot. If something happens and the FAA wants to talk to you, they have absolutely no interest in hearing about any pressure put upon you by your boss or your customer (even though it can be real pressure) because it is your name that goes in the logbook. The FAA still treats instructing like a privilege. Welcome to the real world.

There is no reason for a new instructor to be subjected to a brute without some measure of seasoning. Your best recourse (if in doubt about a pilot) is to get a second opinion. Have the chief pilot fly with your customer; he knows how to handle such people. Let him take all the responsibility for denying a customer the airplanes; after all he isn't going to fire himself. You will make better decisions with experience. However, there will always be times when a second opinion is warranted because you just aren't sure about a person but don't have any concrete evidence to back you up. Remember how as a new pilot you thought you should know everything? The same thing happens again because as a new instructor you think you should have perfect judgment. It doesn't work that way.

If a pilot can get any angle to show that you personally kept him from flying an airplane, he will. What you have to do with a bozo customer is to turn it around so that he, not you, is keeping himself from flying because he either isn't safe or needs more instruction. Don't say "I am going to check you out and decide if you can fly our airplanes," because then it is you who becomes the obstacle. Instead say, "Here is our checkout in writing. When you meet these standards, you can rent the boss' airplanes." You have to take yourself out of the blame loop. People can't fight anonymous standards.

Your judgment, however, is purely subjective, which can still put your word against theirs. You may check out some egomaniac who flies really well but his attitude sucks. Document the checkout in writing stating how well they flew for you. Judgment and people skills will also have to become part of your repertoire.

It was a cool Saturday morning out at my new school. I had been a working instructor for all of a week. In walks this thirty-ish yuppie with the appropriate lime green and pink shorts and shirt. He sat down, stared me in the face, and began this prerehearsed tirade.

"I know all about you instructors who are just time-builders. You just want my money and won't teach me to fly. Do you have any career instructors who know what they are doing? How many hours do you have?"

A lot more than you schmuck, I thought to myself. What an idiot. He thought he knew it all. I wanted to say that since he was such an expert already, why even bother looking for an instructor? Why not just hop in an airplane and teach himself to fly? I wanted to say that his arrogance would be his worst enemy and cause huge delays in training. Even if he found an instructor who could tolerate him, he wouldn't listen to anything the instructor had to say.

I really wanted to say that such an ass has no business in the air. Years later I have done just that. This, however, was still my first week. I must have been dealing with a lawyer. Well-briefed on all the buzzwords for some feeble attempt at credibility, he nonetheless had no knowledge of flight or flight training. I wanted to say all those things, but this was my response:

"I am a brand new instructor. I have just over 300

hours in the air. I have no intentions of making instructing my career and every intention of flying big jets one day. I am fully qualified to teach you to fly and will do the best job I can if you want to try. The airplanes are just outside, and I am available to fly *right now*."

I was accused once again of all those nasty charges made earlier before he stomped out the door. A silence fell on our humble school as the black cloud faded away. Someone like that would go tell your boss that you mistreated them even if you didn't say anything because that is just the way some people are. There I was taking all this abuse for nothing. I promised myself not to let anyone abuse me again unless he has paid me for the privilege.

Any potential problem customer is good cause for you to get witnesses. Invite the chief or a senior instructor in on the conversation. A supportive working environment can do wonders for your confidence after an episode like this. I have daydreamed of having that jerk come back into my office; maybe even to go flying with him. Through some caustic sarcasm, well developed over the years of instructing, you can reduce such animals to pitiful rubble in short order.

The secret, though, is to do it in a manner that precludes any repercussions. If you aren't going to get the students, because they have no business being students, then you might as well enjoy yourself with some creative wit. Remember, you are hired to teach flying; you do not make enough money to suffer through any abuse — that is what your management is for. New instructors are told and unfortunately believe that there are no bad students; therefore, anyone who gets away is your failure. Hardly. You will soon learn that some types are best avoided, rather than you (with Herculean efforts) getting them in the air, only to see them back out anyway after a few flights.

Sooner or later — usually sooner — you will run into a personality conflict. Many times you can't work with a student simply because of a clash in temperament. You will learn with experience to recognize this in people ahead of time, but at first you still think you can teach everyone. It really is amazing just how much the personalities have to match in order for effective instruction to take place. You will lose a student at some point in your early employment who will start with another instructor. You will think of this as your failure just like when you lost the unsafe pilot. Later you will learn that it is a relief to be rid of students who are uncomfortable with you for your lessons without them will be infinitely more enjoyable.

If you are in a club situation, try to work with the other instructors so that you all get the students who are best suited to you. Sometimes a good chief instructor will do this for you. Should you be teaching in a shark environment where all instructors jealously guard their catches no matter how they match up, nobody wins and you may want to seek employment elsewhere. You must analyze yourself to see which students do the best under your style of teaching.

You must also discover what kind of instructor you want to be to your students. I am the type of instructor who needs hard-driving, motivated students who are not afraid to take initiative, ask questions, and take responsibility for their training. I overwhelm the timid and bug the hell out of the lazy. I have no use for any student who requires spoonfeeding. I seek out students whose personalities I admire. When we make a great match, the lessons are a joy. Those students often come in below the average training times by a significant amount, so everyone is happy. This is not to say that you cannot teach people who are not great matches with you, nor is it to say that you have to be buddy-buddy friends with your students — just that the most effective teaching can only take place between the most closely matched sender and receiver of that instruction.

You will also run into instructors who challenge what you are doing. This can be tough when you are just getting started in the business. I was new to a school and had to work with an instructor who was about as different from me as two people could be. He had far more patience than I did. He also fed all knowledge directly to the student. The students attracted to this instructor were sheltered types who did exactly as they were told and no more. Those pilots were trained slowly and carefully and came away with good skills. However, their experience was so limited that they were overcautious and incapable of the resourcefulness necessary to handle anything out of the ordinary. They were safe pilots because they would stay on the ground if there was any doubt about flying.

We had a running philosophical argument as long as we worked together because it was my contention that by limiting the students exposure only to nice days, those students had no knowledge of the real dangers of clouds, limited visibility, adverse weather, or the possible danger from those hazards aggravated at night. The counter-argument was that my students (having flown through the clouds with me) would be tempted to do it before they were properly rated or they would take unnecessary chances in the air. It has never happened, nor do I expect it to either. I contend that my students and pilots know exactly the limits of themselves, their flight skills, and the airplanes they fly. They have on many occasions come out safely from unforecast adverse conditions because of their experiences in training. This is why I believe I am right.

The other instructor is convinced that he is right. Two different instructors; two different philosophies. We each attracted completely different students who went with the one they thought was right. Who do you think is right? We each turned out good pilots (mine were better).

This difference in style gave our school variety and options for the individual students. Our school was therefore able to attract more students. I have sold students on the school realizing we were totally unsuited to work together and because of our differences could place them with other instructors. I have been given students the same way.

This is no reflection on any instructor's ability to teach. What it does demonstrate is that if you are a new instructor and have the courage of your convictions, it doesn't mean that you are wrong just because you are different. Nobody has the monopoly on good judgment. You might just have something going for you that can benefit your students, your school, or maybe the whole profession of instructing. The more instructors are forced to be clones, the less any diversity can flourish to inspire any improvement of the system.

Teaching students on the ground generally isn't a problem as you slowly gain experience. However, giving ground instruction with your fellow instructors listening is something else again. How do you fit into the pecking order? What do your peers think of your teaching? What do you think your peers think of your teaching? Your students may pay the checks, but your self-esteem comes from your fellow instructors. Unless you are very sure of your material, or you are high on the pecking order, you will be nervous when another instructor can hear you teach. This is one way to prove you know your stuff. It will also make you a better instructor because when you can defend your information to a peer, you know you have good stuff to teach.

What is also happening is that the eavesdropping instructor is checking out his own information against yours to reinforce his confidence. Some of you may have the experience of teaching behind inadequate partitions arranged to resemble a maze to study rat behavior. The constant din of endless briefings drones on, punctured occasionally by the comic face of the chief instructor peering like Kilroy over the wall. Every instructor listens and competes with every other instructor for status on the ground. You may have joined the staff, but you haven't been accepted into the instructor fraternity until your peers are convinced that you know your stuff well enough not to make them look bad.

The general tenor of a school is set by the chief instructor, unless the place is so huge that everyone gets lost in the shuffle. Anyway, say you are new to a school and the chief instructor has some off-the-wall ideas and policies that you are to accept and follow. It is one thing to have a philosophical difference with a fellow instructor for you can each do your own thing. However, when it comes down from the top as a requirement, you have a problem.

This happened to me once when I was working under a chief instructor whom I considered foolish at best. As a new instructor filled with self-righteousness and arrogance, I disagreed with his ideas publicly. Big mistake. I had a student removed from my schedule on a fabricated pretense just so the chief could throw a little tantrum and get even. The student had no problem with my instruction. Shortly after this, I started teaching at a club nearby and that student came right back, so this was at worst an inconvenience.

I talked about the chief to the other instructors, and the chief did the same to me — whenever I was off the ground. Everyone talked to and about everyone else. Pilots love gossip. Some would say the chief just made mistakes, had some interesting ideas of flight instruction, or came with a few peculiarities. I thought the chief was an idiot. Everyone makes and is entitled to make mistakes, and no one thinks of himself as an idiot. Therefore, whether you are an idiot or not depends on the point of reference. Judge this one for yourself.

The chief wanted every student to be able to draw from memory the fuel system of a Cessna 150 before they could solo. No other system, just the fuel system. Imagine the poor students on downwind on their first solos wondering if they drew the thing correctly. We ignored our chief as much as possible.

The other arbitrary rule that came down was that every landing had to be accomplished with full flaps — no exceptions. We were informed that some FAA study had concluded that all aircraft land best with full flaps. Uh-huh. I never could find that study, not that I looked that hard. I have found other pilots who confirmed this idea that all landings should be done with full flaps, so there was reason to give it some thought. There are questions to be asked. Why do the flaps have many intermediate settings? Why not make flaps have only three positions: up, takeoff, and down? The reason, of course, is that different situations demand different settings. Any pilot who is trained to think will select the appropriate setting for any given set of conditions. I pointed out to our chief that the Cessna manual clearly stated for strong crosswinds "to use the minimum flap setting necessary for the field length."

Anyway, this and other rebellious acts caused the removal of my student. Don't ever correct the chief in public; they hate that. I learned from this to always say the right thing around the chief. I also learned to teach students two ways to fly: one was to please the chief on stage checks; the other was to take them and the airplane through the full envelope of legal limitations complete with any survival techniques I could pass on. The worst thing you can do to your students is to teach them only what is required for the checkride. I learned from this the politics of instruction and how to please capricious superiors.

My students would joke about performing for the chief in our farcical but required stage checks. They just went up on nice days for their check flights and used full flaps to keep the chief happy. During the regular lessons, we explored the full capability of the airplane using all flap settings, including no flaps. We found that the Cessna 152 and the late model 150 with the larger tail and rudder can handle direct crosswinds over 20 knots. All you have to do is come in at cruise speed on final and not use any flaps. I tried to tell the chief that landing in a crosswind is safer and easier without flaps.

The necessity to win the argument, however, meant he would hear none of my logic. This is when I learned to listen to chief instructors and then do exactly as I pleased. Is this a good idea? Should you challenge your superiors? Aren't they always more knowledgeable? Won't this send students a mixed message? Well, only if they can't think for themselves. I found that aviation is just like the rest of the world. People can rise to levels of authority solely because of who they know. Any qualifications they may have for the job is mere decoration.

Okay, I was still a new instructor, so even though the manual said use the minimum setting, I still had the voice of the chief, and therefore doubt, in my head. I told one of my advanced students the flap story and of my doubts. We agreed to test the theory of full flaps. On a really blustery day, we took a 150 around the pattern, once with full flaps and once without any. You know what's coming. If I remember correctly, the full flap attempt ended in a go-around. Without those barn doors hanging out for the wind to grab, of course a 150 is infinitely easier to control. I knew it ahead of time but still had doubts which were planted by the chief pilot.

This is where I learned to trust my own research, investigation, and experience. This is where I began to challenge everything I had ever been taught about flying and became committed to finding better ways to fly and teach. This is where I lost all respect for titles and only had respect for wisdom. If I can learn from a gifted superior who will pass on the benefit of his wisdom and experience, that is great. If wisdom comes from a potential student on a demo ride, so be it. Sometimes it is the people who haven't been restricted by being taught what not to think who come up with the most brilliant revelations.

You may have none of these experiences as you begin your teaching stint. I offer them for you to draw on because chances are you will have no idea what to expect and these general areas of difficulty are pretty common. You will definitely find that actual instructing has nothing to do with what you had to learn to please the FAA. There is no excuse for this and the FAA should be ashamed. If they can't come up with a curriculum and standards that accurately reflect working conditions for flight instructors in the real world, then they have no business certifying instructors at all.

When you do emerge from the checkride and start working, you should be filled with blissful optimism. I was. Fortunately, my enthusiasm for instructing never wavered for over a year. This is very unusual. Consider any honeymoon over six months to be a blessing; unless you are of course a lifer. I have tried to steer you clear of some of the pitfalls from my early instructing. You will find different ones of your own, I'm sure.

My early days of instructing were an absolute blast. Demo ride to checkride, whatever the student wanted to see or prepare for was fine with me. I would sit through stall after stall, steep turn after steep turn, and bounce after bounce as students tried to land. What a life! I was actually earning money and sitting all day flying around in airplanes. It was fun. It was challenging. Sure, there were disappointments. The weather would be bad and I would try desperately to find something useful to teach given the limited conditions. Students would cancel at the last minute, and I would immediately call up someone else to fill the void in my schedule. I flew seven days a week and hung out at the airport from 8:00 a.m. until 10:00 p.m. every day. I had boundless energy and unbridled enthusiasm. It was one of the best summer's of my life.

I wish the same for you.

15 • STUDENT PILOTS

Student pilots are trying to kill you; that is their function in life. Your function is to survive student pilots so you can one day actually fly planes again yourself. In the mean time, you have to sit in the right seat, watch the calamities unfold, and when necessary grab the controls so that you can live to fly another day. Most students aren't trying to do you in. It's just that there is something about an airplane that causes people to lose complete control of their faculties and blindly stare danger in the face while passively awaiting divine intervention — the Law of Jesus. If you let them, student pilots will fly right into the ground or into another airplane; anything solid within range.

What is amazing is the total lack of concern or awareness on their faces and in their voices, as if they really have no idea what is happening. In the students' minds, this airplane thing isn't real. If it isn't real, then they can't really be there. If they aren't there, then they can't hit anything. What they see out the window is a video game. When you land, it is as if your student is waking from a floating dream.

Call this, if you will, the ultimate defense mechanism to handle the strange, new world of flight. They aren't responsible for anything, so they don't even have to mentally be there — but you do. You are responsible for your student, the airplane, everyone with whom you share the sky and the airports, everyone and everything you fly over, and your future career. You know damn well this is not a game. It is very real and if it is not done properly every time, you may not walk away. This is why you make the big money . . .

When you take students up for their first couple of hours, you might make the mistake of trying to teach them a lot of flight stuff. Keep the information content light because students, as part of their acceptance of the reality of flight, have to get over the "gollygee factor." This phenomenon manifests itself in many forms. Initially it is where students have to get over the shock and accept that they are off the ground and they made it happen with their own hands and feet. Once that has been internalized (on average three to four hours), they will be able to look out the window and pretty much see what is out there. Now you are able to begin flight training.

Since denial is a huge part of students' repertoires, anything that is unsafe, different from what they want to happen, or anything unexpected will cause them to instantly retreat back into the dream world. That is how they can stare at an oncoming collision hazard and calmly wait to be saved. Some students never get over the gollygees and have a hard time concentrating and applying their rote procedures.

I recently graduated a private pilot who could not concentrate unless verbally beaten by an instructor. This student had had many instructors. Holding on to a sincere desire to pilot an aircraft, he knew he had a concentration problem; however, it was just so much fun to be up in an airplane gawking out the window that nothing else mattered. Besides, he knew the instructor wouldn't let anything happen. When this student's mind is engaged, you see a good pilot so my hope for success was reassured. Finally after huge delays, he got the private. Now he can sit up in the air all day and look out the window; not a bad way to spend your time.

You should be warned that there are a small number of individuals who sincerely have no regard for their own lives and certainly no regard for yours. One of the real fears of instructors (including me) is being overpowered on the

controls by students or having them do something dangerous before you have a chance to react. Someone who is
convinced of their superior ability to fly or who is panicked
has no interest in you being in control. We don't screen
people before we take them up, not that we really could, so
every first flight with someone new is an experiment.

The bosses don't care before you go; they just want the
business. Put it this way: They don't generally use their
greater experience to weed out potential problems for you
before you go fly with someone new. They do care if there is
a safety problem or risk from any pilot, but you can't report
that until after the flight.

You never know what will happen when you meet a
new person. People lie. I had a student pilot with about 50
hours come to me. He had soloed and wanted to continue on
to the private. The possibility of crashing an airplane I
learned well into the session was of no concern; in fact, I
think he was intrigued by the idea. I had been instructing
long enough to have my mental red lights beaming so I
could feel a problem coming on here. Since this person was
somewhat smaller than me, I felt I could maintain control
of the aircraft and handle any situation that came up. We
all make errors in judgment; this was my turn.

A cursory check of the logbook revealed that many different instructors had only lasted for a couple of flights before
dumping this student. This is a dead giveaway that no instructor wants to share an airplane with this person. The
overriding problem that removed any possibility of effective
flight instruction was his disgusting attitude. He had no respect for me, anything I had to say, the airplane, the club, the
people on the ground below us, or any regulations.

The initial sweet disposition I thought had walked
into the school turned chameleon as soon as the engine
started up. It went progressively downhill from there. After
a couple of spin recoveries that should have been normal
stalls, I attempted some touch and go's. Whether this is the
eternal optimist in me or the desire to give someone with
little flight time the benefit of the doubt, I don't know. We

should have gone back right then. However, those inevitable questions about my judgment of this person caused me to gather more evidence. After grabbing the aircraft just before the nosewheel was sheared off for the second time, I flew the aircraft back.

"So what is *your* problem?" came the accusation.

"You," I tersely replied. "I do not trust you. You are dangerous and you have no business in the air until you change your attitude."

Can you as a new instructor be that direct? Although harsh in retrospect, it sure seemed appropriate at the moment. What a student wants to hear from you may have no similarity to what he has to hear.

Back at the club I got rid of this student, after collecting my money of course, and blackballed him from the club by warning my manager and fellow instructors. I then learned that he had been in just two weeks previously. Another instructor had been subjected to exactly the same harrows I had just seen. Somehow the story hadn't gotten to me. I was informed by this student when he walked in that it was his first visit to the club and he was just looking for a little instruction to finish up. What a whopper!

How many instructors at other clubs have the same story to tell? How many airplanes and instructors have been put at risk? I would love to have mug shots of the dangerous pilots who make the rounds, like they have at the Post Office, but that is not the way for aviation to operate, fun though it is to contemplate.

Speaking of being overpowered, have you ever done a demo ride for a body-builder? Anytime you see someone with obviously more physical strength, a small voice in the back of your head tells you that this person could very well wrestle the airplane away from you. My experience with such folks tells me that the larger the person, the greater is his fear of flight, or most other things for that matter. What is it about puffing up the old muscles that makes some people timid? Are they just hiding behind the bulk but still hoping to prove themselves? I have taken some large people for some very smooth and very short demo rides.

The problem with student pilots is that they don't really see you as a human being. You are a flight instructor. Student pilots would not trust their lives to mere mortals. They are mortals themselves, so you, the instructor, must be superior. You are raised to deity status and mounted on a pedestal. You lose your humanity because as far as they are concerned God really is their copilot. The worst part of this is that you lose the capacity to make mistakes in their eyes. It will be a challenge to retain your humanity, but it will be worth it.

When you goof a demonstration, laugh about it. Show your students that mistakes are fine as long as you do something about them. Also explain that everyone needs practice at something. You went into this business just trying to do a job, but it is very easy to get caught up in your students' own propaganda that you have to be perfect for them — which you will not. Your God status takes somewhat more time to die down in your students than do the gollygees.

You may have students who occupy powerful positions in the real world. They will be used to making tough decisions. They are probably used to driving fast and complex machinery down the highway. Even though they are used to decisions, judgment, and speed, all of this gets lost when they hop into an airplane. So much for the transfer of knowledge.

Flying small aircraft is the universal humbler. We all start as equals in an airplane. It will seem a bit strange at first, but even the richest and most powerful person starts off as just another student pilot. Unless they have attitude problems, they will want to be treated as such. No matter who the students are, they will try to put you on a pedestal — don't fall.

Student pilots come in all shapes and sizes, colors, ages, backgrounds, education levels, and both genders. None of this has anything to do with an individual's ability to fly. If you have a prejudice, your students will show up your ignorance and defy your expectations every time. One of the touchier subjects is the interaction of men and women in the flight training environment. Women are still the minority in commercial aviation, which is strange because the airlines have made great leaps to promote women in the business. Sometimes too great an effort. There is a perception that women get a much easier time as they move through training and into the airlines; that current affirmative action practices allow women to get better jobs, advance faster, and slide through checkrides more easily than men. If true, this is unfair to everyone, especially the passengers who deserve the best people up front.

Whether true or not, the perception still causes women to be looked on with contempt by some male instructors who believe that all their best training efforts will go to teach someone who will get a job before they will. Men who feel this way should change students, or maybe even professions. Some women feel that they deserve special treatment because they are women and want to use affirmative action to promote themselves as rapidly as possible. They shouldn't be in aviation either. These women can be very difficult to teach, especially for men who are normally demanding yet bend over backwards to be fair with all their students.

Most women, though, in my experience just want to work their way up through the ratings under the same conditions and in the same time as anybody else. The best way to handle all students is to always look upon them as pilots first, judge them all to the same standards, and make everyone do the same work. Bring out the best in all your pilots. When the perception holds that all pilots are judged solely by their merit, aviation will be better off.

Why women have not flocked to aviation is a mystery to me. I could be wrong, but compared to other fields with glass ceilings, aviation seems to have surprisingly little if any impediments to advancement to the airline flight deck other than the obvious glut of pilots that affects all of us. The treatment of women in commercial aviation, whether equal or even better, must be a secret because there are still relatively few women in the business. I wonder then if they must be limiting themselves by choosing other professions?

I have yet to see a male instructor consciously and overtly discriminate against women students. Well, there was this ex-Marine instructor, but he terrorized everyone. It would be interesting to read an article from a woman who instructs to see how her experience and perception compares to mine, also to see what light she could shed on the working conditions of women in this business. I can't speak for those women, so that story will have to remain untold for now.

There are some problems the guys can work on to help flight training for everyone. For those men with older ideas (not necessarily older men), who believe that women (for whatever reason) can't fly as well as men, they have a tendency to make the training much tougher on women than it should be simply to prove that when any woman fails, they were right in their beliefs all along. That won't do. This is the fastest way to get a bad reputation as an instructor. The good news is that this probably happens much less frequently than it did say 20 or 30 years ago.

The standards are tough, but they are tough for everyone. We all have to fly to those standards. You won't be accused of picking on someone when you require the same legal standards. The worst thing the guys can do is baby their women students. You can be just as tough, give the same challenges, and expect the same standards of flight. You will not be respected by your female students if you do not let them reach their highest potential as a pilot.

Probably the most volatile situation happens when either the student becomes attracted to the instructor, the instructor gets a crush on the student, or they both find each other attractive. This is where all learning comes to a crashing and immediate halt. Here is one topic that definitely gets swept under the rug except for some really juicy gossip. The normal ethical rules clearly state that any person in control (in this case the instructor) shall always maintain a professional relationship with students under his supervision. That looks great on paper; however, in the real world of human emotions anything can happen.

As an instructor, you are just doing a job, so you see people pretty much as they are. You can detach yourself from most attractions and still get the job done just like people do in every other walk of life everyday. It is natural to want to work with people you get along with, so in certain cases the next emotional leap is bound to happen periodically. How you handle it is the real mark of a professional.

When you find that your emotions are interfering with your judgment and your ability to teach, then it is your responsibility to change students. Relationships and teaching in the close confines of a small aircraft don't mix. You can't do them both. When the student switches instructors or leaves aviation, you can go back to being regular people again. What you will feel as an instructor is the dread of weakness. Here you have been trying to live up to this lofty calling, believing all the FAA propaganda and your students who have put you on the pedestal, and you go and do something as human as getting a crush on a student. It happens; deal with it and move on. At least you realize you are still human.

I suspect this happens a lot more than people want to talk about. As long as the situation is handled properly, it's nobody else's business anyway. The instructor is still the one in control, so any emotional attachments should be more easily severed when the instructor is the one attached. However, for the student who gets a crush on the instructor, that is a totally different matter. I don't think you can go through instructing without some students at some point letting their emotions get in the way of their training.

Think about how instructing works. You are teaching people in close quarters who have put their lives in your hands and complete faith in you to teach them how to fly. How easy it is to cross the line from admiration and trust to emotional involvement. You will have done nothing to encourage anything but the best flying you can get from your students. However, simply by virtue of being a good pilot and instructor you will have very attractive qualities. That is why you tell people at parties that you teach flying — it is a great way to meet people.

Where this problem is particularly evident is with women instructors. Because there are so few women in aviation, it seems to happen to them more often. Then again, I have seen women talk themselves into flying some pretty special airplanes owned by old guys who liked having them around, so it all balances out. Flying is still looked on by many people as man stuff, so the women who successfully make it must be extra special. If they are really good looking as well, then this added factor magnifies the effect and the number of potential problems that students can cause.

In a system that removes all responsibility from the student, it also removes the emotional responsibility. For any instructor (even though you have nothing to do with a student crush), you still have to deal with the problem. Yet another occupational hazard. There is nothing that can be done to prevent this problem for we are all still human beings with emotions. It is just a fact of life that we all have to deal with periodically. What it does make is an incredibly awkward situation. How can you go to your boss and complain that one of your students has a crush on you? You certainly couldn't say anything if you found one of your students attractive. This is one of those situations where no one can win.

Some of the best people you can ever work with are high school kids. These young people are so used to learning that hopping into a teaching environment in an airplane is no big adjustment. The only obstacle is getting them to retain the cumulation of knowledge that regular schooling has taught them they can forget, but everyone has that problem. It is easy to see how people can make a career out of teaching at the high school level. High school kids believe what you say without the argument that is customary to older folks who have grown cynical through being lied to and cheated.

There are age restrictions to certificate eligibility so the best time to begin serious training is within a few months of one's 17th birthday, that way there will be no delays. They can solo at 16, but then they have to wait a year to be eligible for the certificate. That may be fine for people who don't want to wait and are willing to spread out their training. I spaced my training out for years having started my flight lessons at 13. There is a certain innocence and purity of trust that high school kids bring to you. Your responsibility towards these people greatly increases because the nature of flight training also teaches them about possibilities, challenges, accomplishment, failure, frustration, self-esteem, confidence, and judgment — all the things that get magnified in a younger, more impressionable person.

Flying an airplane is a lot to ask of kids whose peers are happy to get keys to a car. However, they are choosing to fly so they have to fly to the same standards as everybody else. They will take less time on average to get their ratings because they believe what you say, have better learning and retention skills than folks who have long since left school, and best of all, they don't automatically object to doing their homework.

Two of my best students started flying at the age of 17. One of them was this quiet, shy guy. He could do almost anything without undue effort. You know the type: Eagle Scout, good grades, natural athlete, things like that. He is about as close to a natural pilot as I have ever come across. He also had surprising confidence and was not phased by flying in weather at IFR minimums, with dustings of ice on the airplane, or the one time when we experienced a temporary communications loss in IMC conditions. Whatever the adversity, this pilot could handle it. He has his commercial-instrument certificate and has been accepted into the military flight training program. Don't you just hate people like that, where things come so easy at a young age? He should be in combat jets while most of the rest of us will be trucking around in airliners. The consolation is that flying has made him more confident and more assertive, and this formerly shy kid is now just as obnoxious as all my older students.

My other student was completely different. Off the airport she fits the image of someone you would likely find in a museum studying Florentine art, unless you know she is a pilot. We received the best service from the male controllers when we flew because she had the sweetest radio voice on the air. She is another natural pilot with perhaps the best instincts and sense for controlling the airplane I have ever seen. She had an exceptional attitude, welcoming any challenge with the words, "No problem, it builds character." If I could work with high school kids exclusively, it would not be a problem. If they were all like these two, it would be a piece of cake.

People need to know as they progress through flight training that they are coming along as they should. Since they only see their own progress, they have nothing with which to compare themselves. Some folks are particularly used to regular results and constant positive feedback. Take your basic tycoon surrounded by "yes people" for example. For some people, they think an airplane would allow them to be more productive by saving on travel time. This is a perfectly legitimate reason to fly, although such empty utilitarianism is lost on puritans like those of us who live to fly. There is all the difference in the world between the person who flies for the sheer joy of it all, and the type who flies as a mere function of some higher purpose.

The hard-driving types want results and they want them now! They have a schedule to keep and they need the certificate yesterday. An airplane is just a faster car with wings, right? These are the folks most likely to convince you that they can learn to fly in half the normal time because they couldn't have gotten to where they are by being average. They will tell you how capable and wonderful they are, in case you hadn't noticed. These people prefer to make their own rules based on expediency rather than follow the FARs like the rest of us, so you have to watch them very carefully. You have to think of your future career all the time and don't let any student sabotage it. They will also tell you when they are ready to solo, when to go cross-country, and when they are ready for the certificate; as if they had any experience in such matters. There is nothing like an insecure tyrant who has built a corporate empire in which to hide who now wants you to teach him how to fly to make your job more exciting.

Your salvation is in Part 61 and the *Practical Test Standards*. These folks won't take your word for anything; it has to come in writing. The best way to inform business pilots what they are in for is to give them a business proposition. Spell out a reasonable time estimate, the accompanying cost to them, and the commitment necessary to meet their schedule. You have to clearly spell out their responsibilities. Above all else, do not even imply a contract. These folks are used to negotiating and will gladly try to talk you out of free instruction if they can convince you that you didn't live up to *their* responsibility. When they fail to live up to their part because of some business thing that always comes up, void the estimates and spell out new hours and costs.

Lawyers, doctors, and other people who are used to playing God have the toughest time taking criticism from an instructor playing God and also accepting the fact that they can't do everything in an airplane perfectly the first time. Consequently, they feel the need to control any part of the process they can, which means telling you when they feel ready to take the next step, even though they have no rationale on which to base their decisions. You are under no legal pressure to ever sign off a person for any privilege according to the FAA. However, the economic pressure from the person in the left seat (who pays you the checks that cover your rent and who can complain to the boss who holds your job) can exert severe pressure on you that is countered only by the greater pressure you would feel from the FAA should you, through dereliction of duty, endorse pilots well before they are ready for any privilege. People like this have a reluctance to believe the word of just one young instructor. Sometimes it takes two instructors or maybe a flight with the chief to make a point. This is why we have stage checks. We tell the students it is for a review of their progress when many times it is simply self-defense.

"My instructor never told me that!" Sure — pass the buck, blame the instructor. Why not? It's easy. Tell me, have you ever overheard a student on a phase check admit that they hadn't worked hard enough to obtain the proper level of understanding for any given topic? I would be dreaming to think that is possible when it is so easy to say that the instructor is negligent. This catch-all phrase is the one statement that can bail out the students for any perceived failure on their part when they don't want to take the responsibility for not knowing. Even if you had just gone over the same topic the day before, your students can disavow any knowledge of the information sought in the question being asked and blame you for their forgetfulness. I've seen it done.

I will be convinced that a student takes responsibility for his training when he comes up to me and says something like, "Excuse me, it says here in Part 61 that you are supposed to teach me to be proficient in navigation using the magnetic compass and we haven't gone over that." Well, it hasn't happened yet. The problem for the instructor (as far as teaching specific things that students later say you never told them) is that with a lot of students on your schedule it is hard for you to keep track of exactly who was told what. Student pilots greatly extend their training and your frustration because they are unwilling to take the initiative to inform you of any deficiency in their knowledge. You have to find this out all by yourself without any help from them. You can give them the books, tell them to read everything, raise any question with you, dig into their training, and reach beyond the lessons so that they will understand the big picture of flight; but it never happens. What does happen is that your students will put so much blind faith in you that they will repeat whatever you have said, even if it is directly contradictory to the facts.

We have talked about how myths get passed down the

instructor/student chain. What is even more interesting, though fortunately not as common, is when an instructor makes a mistake or through a slip of the tongue says something contrary to the material he is trying to present. The student will repeat the mistake back as if it were the gospel truth even when the student suspects or in some cases knows from their readings that the material is in question — all because of the blind faith that if the instructor said it, it must be true.

There is a feeling out there that such a thing as a bad student does not exist; there are only bad instructors. The FAA perpetuates this view in the preparation, testing, and supervision of flight instructors through their career. This whole idea is complete garbage. The vast majority of flight instructors I have seen are dedicated, hard-working individuals. The vast majority of students are not because the FAA system removes any reason for them to take responsibility for their training.

In a sense then, the opposite of the current wisdom is true. There are no really good students because they aren't allowed to be good students because the burden for training is on the instructor. Students sense this and quickly abdicate any responsibility of their own. Maybe many instructors like this control? Maybe they want to supervise absolutely everything a student does, thus reinforcing dependence on the instructor? Maybe some instructors are afraid of what the students might do if they don't watch every move because the instructor is absolutely responsible?

This is why students can continually show up after the lesson should have begun, without a briefing, and with the aircraft still sitting untouched on the ramp. The instructor is in a position where he can't trust the student because the student can do no wrong. Therefore, even if a student got a briefing and did a walk-around check of the airplane, the instructor would still have to call to verify the briefing by getting another one, and still have to check the airplane to make sure the student didn't forget anything — like oil for example. This is all wrong.

It is wrong for the students because they never get used to the responsibility of pilot in command, and it is wrong for the instructors because of the undue burden over things for which they have no control. For the most part, student pilots are adults. They can vote, buy alcohol, operate vehicles, sign contracts, own businesses, and maintain normal and legal standards of behavior or face the consequences. All of this is negated when they begin flight training as they are looked upon by the FAA as helpless sheep. You still have to teach these people to accept the authority of pilot in command even though throughout their training their decisions will always have to be your decisions, their judgment will be your judgment, and their actions will be what you tell them, because if anything happens only you are responsible.

Have you had enough of this responsibility theme? Would you like to see a change in the system to relieve you of this unreasonable burden? The question to be asked is how can students be forced to share the responsibility for their training? My recommendation would be a clearly defined legal split right down the middle, where both the instructor and the student are equally responsible for their own actions and for insuring the proper completion of training. The best way to accomplish this is to write it directly into the FARs. In Part 61 for all aeronautical knowledge and flight proficiency requirements, there is a line which states something approximating that:

"A student must have demonstrated satisfactory knowledge and have demonstrated flight proficiency to an acceptable level of performance as judged by the instructor who endorses the student for any privilege."

Inserted in the text should be the following line regarding aeronautical knowledge.

"Student pilots shall insure by securing proper instruction that they understand and can apply the knowledge and material contained in all subject areas listed in this part before exercising any privilege."

With regard to flight proficiency, put this in the rules:

"Student pilots shall insure that they have secured the necessary instruction and practice to be proficient in all maneuvers and procedures listed for any privilege, and that they have secured the written recommendations of two authorized flight instructors before exercising solo and solo cross-country privileges."

With this fundamental change in the rules, the students will be just as responsible for securing proper training and for making sure they understand the material as the instructor is responsible for giving proper training and teaching the material. No student could say that an instructor forgot a subject area because he would be just as responsible to insure that it had been covered.

The regurgitation of facts so common to oral exams would not suffice anymore as the student would have to insure that he had the level of understanding necessary to apply the information. He would have to ask meaningful questions until he knew the information. This would build responsibility in all students. Ignorance of the law is never a legal defense except in our current system of flight instruction. With your help, we can change all that.

The next major change in the FARs would be the idea of "co-endorsements." For every instructor endorsement, there will be a student endorsement certifying that they have "read, understood, and accepted the privileges and limitations of that endorsement."

Take the initial solo endorsement, for example. The students would certify in writing that they have received training in all the areas listed in Part 61, they understand the privileges and limitations of the solo endorsement, and they assume the responsibility for properly operating an aircraft in solo flight according to the flight rules of Part 91. After 90 days, along with the instructor endorsement of competency, the students will endorse their logbooks to again certify that they accept the responsibility for operation of aircraft in solo flight according to the FARs, and that they will secure a subsequent 90-day endorsement from their instructor if necessary. Cross-country endorsements would work the same way.

Instructors and flight schools should have copies of those student and instructor endorsements on file so there will be no confusion as to each student's responsibility. This dual certification should forever change the student/ instructor relationship such that students become equal partners; with equal liability for their aeronautical knowledge and flight proficiency; equal responsibility for learning, knowing, and following the FARs; and an equal stake in the successful outcome of their training.

Until such time as the FAA deems it desirable to have students give up their slovenly ways, your problems will continue. One typical tactic of students who have better things to do than their assigned work is to make excuses. Nobody tries to get away with more lame excuses than student pilots. There are times when you have to fight against your own students to make them work hard so they can get the rating they want. I always preview the next lesson with the student at the end of every session. They know ahead of time all the topics and procedures to be covered and where to read about them in depth. Well, you can lead a horse to water, but you can't force a student to read, study, and prepare. What a mixed metaphor, but you get the idea.

I remember an English teacher back in elementary school who, when you didn't complete a homework assignment, would make you stand up before the class and explain why you had failed to accomplish your task. This teacher taught accountability to ten-year-olds, and it worked very well. Why is it then that student pilots can take their training and instructor for granted and show up for lessons completely unprepared, wanting only to be spoonfed? After the inevitable excuse, you will probably get a complaint about how their training is taking too long. You can't make adults stand up before your flight school (although it's not a bad idea), so instead you have to keep charging them more money as you explain to them how they have delayed their own training. Rather than intelligently discussing a topic and questioning the student to insure understanding, you end up lecturing to a nodding head, with no confirmation that anything is getting through. Without understanding, there is no retention, and they end up denying that you ever covered the topic when questioned by another instructor or examiner.

What goes on in the mind of a student who shows up for a cross-country lesson without a navigation log at the time you should be departing? How did he expect to fly without a plan? Why is it when you want to introduce landings to a student, to demonstrate the technique, and then have him try the various parts of the approach, the student hasn't a clue what you are talking about because he refused to read about landings? Why are you greeted by a blank stare when you ask your student before his first night flight about differences from standard daytime operating procedures? It all comes back to responsibility.

You are a babysitter to student pilots. That is why I started referring to all my students as "babies." As an instructor, you have done everything humanly possible to raise your standards of flying and teaching to the highest level of professionalism you can. Your students, on the other hand, especially those who desire a career in aviation and should know better, treat the whole training process frivolously and deny you by their actions any respect or courtesy for your ability, knowledge, or level of achievement. I don't refer to the top few percent of students who are a dream to teach; I am talking about the vast majority on the bottom, the ones you have to teach.

Most students, if you let them, will cancel at the last minute or not show up at all if anything comes up that interests them even the slightest bit more than your lesson. The FAA tells you that you are a professional, but your students will treat their training as a lark. Barbecues and golf games take a big toll on flight instructors. Movies and restaurants really cut into the study time the night before a lesson. Concert season wipes out all your teen-age students until the bands stop touring as much.

Students are almost always late for lessons. Half your time on the ground will be spent waiting for students. Most instructors undercharge for their ground time with students. You will find that standard practice for new instructors is to charge only for the Hobbs time. This is unfair to you. The competition for business is such that you can't start charging for your time from the beginning of the scheduled lesson whether the student is there or not and stop charging for your time when the student finally walks out the door at the end of the lesson (even though this would be the fair thing for you) because you would lose all your students. Now if every instructor started doing that . . .

What I do is charge for "working time," which is any time I am engaged in any form of airplane training activity with a student, be it in the air or on the ground. You will find that the Hobbs time only covers about half the time you will spend with each student. New instructors are reluctant to charge for ground time because they think their own time isn't worth as much outside an airplane. This practice effectively gives you a 50-percent pay cut from what are already starvation wages. They get to make excuses for not living up to their end of the bargain and you get paid for only half your work — some deal. Student pilots can be so cheap. Anyway, here are some of my favorite weak excuses:

"There was too much traffic on the roads."

Okay, leave earlier, go a different way, or schedule your lessons at a different time.

"I overslept."

Get a better alarm clock and get more sleep next time.

"There were clouds in the sky so I just stayed home."

Did you ever think we might do ground work? I never cancel a lesson. These are the times to visit the tower or do an FSS tour. There is always something to learn.

"I didn't understand how to . . . (fill in the blank)."

This is where the student shows up without his nav-log because he forgot how to do the cross-country you had just showed him the day before and he said he understood. Why didn't he call you in time to help him? I have a standing policy that any student can call me with any question during civilized hours.

"I have a headache."

I don't want to have sex with you, just teach you how to fly. This excuse got even better.

"I had two cups of coffee this morning (it was now 6:00 p.m.) instead of one, so I can't fly."

This person was actually very upset about the upcoming checkride. I can understand that — I just wish he could have said so directly.

One thing you do have to analyze is any potential human factor that may be behind a series of lame excuses. This might be a good time to reread the fundamentals of instructing now that you are a working instructor. The real problem of excuses will continue until the day that our students treat flying as seriously as we do.

Would you like some stress? How about the first time you have to solo a student. These days a phase check before the solo is pretty much the norm, so I wonder how a new instructor used to just take all the responsibility upon himself to let his first student solo back several years ago. After some experience though, you will know when students are ready to leave the nest and the second opinion will become a formality.

When you have to step out of the aircraft for the first time, you will ask yourself many questions. Have I done everything possible? Are they really ready right now on this day? Do they know it's coming and are too psyched up? Do I have a pen to endorse the old logbook? Will they do something really stupid and get us all in trouble?

We try to downplay this solo stuff but the rest of the world won't let us. Plenty has been written about the first solo. However, I have never read anything about the first time you solo a student. We know soloing is one of the most memorable times you have in aviation. Soloing your first student comes a close second. Chances are you will be almost as excited as they are, once you see that they lived.

There is always something special happening whenever a student solos. The airplane has a glow to it. The sky looks somehow different. Something magical is taking place. We have precious few rights of passages left these days, and none that I know of compares to soloing an aircraft. This is a gift from aviation that people can carry the rest of their lives even if they never fly again. Your greatest moments with students are spent during the lesson when they solo. There you are on the ground watching helplessly, yet you know your voice is in their head. My students tell me they hear my voice years after the checkride. What a curse.

Solo flights are always made on perfect or near-perfect days to insure the safest and most enjoyable conditions for the student. Come on, when else would you solo anybody? Sometimes I duck into the tower just in case a student wants to chat with a familiar voice during the flight, although no one has felt the need as yet. Many times, though, I just pull up a chair by the taxiway, tell them to go have some fun, and then watch them do their thing.

I remember one time a student who had many family members in the area was going to solo. One proud parent had gathered the troops to our club and awaited a signal from me that their son was taxiing out. There were about 15 people including instructors, staff at the school, family members, and fuelers all gathered in lawn chairs and seats watching this 16-year-old kid solo on a picture perfect afternoon as the sun was just starting to sneak down in the sky. What a party.

When a student solos, something is gained and something is lost. The students have accomplished something wonderful, but they will never be as dependent on you again. They have the capability to operate within 25 miles of the airport, but they can get into far more trouble than before. You now have far more responsibility for them, but far less actual control. This is all part of the process.

Just to prove to you that almost anything is possible if you set your mind to it, here is an example of a student who was ready to solo. The phase check was complete and we were just waiting for the next lesson. One very proud father had secretly arranged with me to be present at this historic event for their family, complete with video camera. Now, it was summer time on the West Coast and our usual stratus layer had moved in. Since actual experience in clouds is valuable to a pilot of any level, it is fairly routine around here to depart IFR and hold the lesson in the VFR conditions available close by inland.

When we broke out, the stratus revealed a much deeper and wider area than had been reported, and our destination was now under a solid layer. Never missing a chance to instruct, we held a little tutorial on shooting the ILS we would need to arrive at the field where Dad was secretly waiting. We broke out under a 3000-foot overcast with 30 miles of visibility. With the IFR stuff out of the way, it was now simply a matter of practicing a few touch and go's to get warmed up. After three patterns, I stepped out, did all the official signature stuff, and watched with great pleasure as my student soloed, complete with expert commentary from the man with the video camera.

How many student pilots do you know who get to try an IFR departure, in actual, followed by flying their first ILS, in actual, and then complete their very first solo flight — all in one lesson. Only with a very special pilot could such a task even be contemplated. What a flight to remember!

Watching a student solo is always a thrill. Sure you are concerned as you watch them go, but they are always in sight. Besides, how long can it take to bop around the pattern three times? You won't have to stew for long. Whenever you send a student for a solo cross-country, your feeling is entirely different. The real test of a student and your training is whether he can navigate around the countryside and get back without your help or without any major incident occurring. This is where you shall discover worry.

Student pilots have no idea what can happen to

them as they bravely but naively venture forth in their tiny craft — but you do. There is a cumulative effect throughout your instructing career where the longer you teach, the more you see, the more you learn, the more you realize what can happen to your students, the more you will worry. This is why people should not instruct for too long; you worry too much. Ask veteran instructors and they will tell you that the worry never really goes away; they just learn to deal with it and put it aside. They would never sleep otherwise.

Part of the job then is leaving all your concerns at the airport so that you may lead a normal life when away. What happens, though, when students are overdue? Where are they? What are they doing? First you worry, then you get angry. You get angry because they upset you by making you worry. When they finally show up, you are initially relieved and then you really get angry. All that worry for nothing.

Much of your worry can be alleviated ahead of time with a few tricks. One of these is to insist on simple navigation logs. Another is to have them arrive at the airport at least one hour before they are scheduled to take the airplane, two hours if it is their first solo cross-country. Get your students to fly the whole trip in their minds the night before so they can organize their thoughts, plan their actions, and see if they forgot anything. Just like when you were learning, your students have no idea beyond the actual flight time how long it really takes to fly a cross-country.

They waste unbelievable amounts of time with such things as perfecting the navigation plan to the point where it is unusable, overdoing the preflight, the obligatory hour at the destination just to talk about their adventure and the planning and travel time to get home. Students are unrealistically optimistic about scheduling the aircraft for adequate time to make a trip and about deciding if they have to call the school to inform you they will be late. They usually realize they can't make it back on time when the airplane is due on the ground in 15 minutes and they still have an hour or more of flight time to go. Most students don't think about you or anybody else who may want the airplane after them, for they have enough to think about with their own operation, so that is all they think about.

As long as they are not worried, there is no reason for them to believe that anyone else would be worried. They are off on this big adventure. Who needs the instructor anymore? If they are a little late getting back, so what? They are late for their lessons all the time. See how the pattern works. When students start taking responsibility, they will learn the value of courtesy and earn your respect rather than cause you worry and incur your wrath. When your student schedules a cross-country, just double the time they originally scheduled and that should take care of the whole issue.

I had one student fail to return on time. He had scheduled an afternoon cross-country flight that should have brought him home well before the sun would have started to fall low in the sky. This student was new to me. His regular instructor had moved on from our club. Anyway, all the cross-country training had been completed with the appropriate endorsements entered in the log and medical, so all that was left for me this day was to review the planning and send him on his way.

Never a fan of afternoon cross-countries when a morning flight will give students the whole day to have their fun, I was just a tad leery of this flight. Must be those red lights going off again. The planning was quite good, so I could find no reason for him not to make the flight. Off he went . . .

It was well into the night before I stopped calling the various approach controls, flight service stations, and towers along the route to locate my new student. There was no record of the departure back home with the destination tower. The resulting airport search located the airplane on the field with no sign of the pilot anywhere. Once I knew the airplane was safely on the ground, I went home.

The next day the student flew back home as if nothing unusual had happened. All of us at the flight school, especially the manager, were keenly interested in his explanation. It turns out that the flight up was uneventful. He was then met by a friend at the destination. Off they both went for the afternoon. Soon after leaving the airport, this student decided it was too late to come back and try to fly home. He chose to spend the night on the town instead, stay over, and fly home the next day — and he never told *anybody*!

We had words that day. We had so many words that the student had to swear nothing like that would ever happen again or he would be banned from our club for life. This is why I say there is way too much responsibility placed on instructors. Students just do things of which you have no previous warning or control. Who could have foreseen this one? Most of your students will fly conscientiously, legally, and properly. However, you will always wind up with some that cause you to worry.

Student pilots lie. This is not premeditated mind you. It is because they want something, or they want to please you, or both. Say, for example, you are getting a student ready for the private checkride and the old question of night time comes up. Let's say also that this is a new student for you because the instructor who taught him most of the way through just took off for a better job and dumped all his students. This happens all the time; welcome to aviation. Anyway, you are casually going over the requirements for the checkride with the student.

"So, how much night time do you have?" you ask.

The student just looks back with a blank, quizzical look. Students do that a lot.

"You need three hours and ten landings to go for the checkride. Do you have that?"

"Oh yes!" comes the now sure reply.

This is rather strange considering he didn't know what you were talking about a second ago, and he certainly hasn't looked in his logbook to verify anything. You know how this game works because you were a student yourself once.

"Are you sure?" You see in his eyes the hint of uncertainty despite the assurance of his voice.

"Oh yes!" comes the answer again with all the authority the student can muster.

Here is how this game works against you. You ask if the student has the required night time to cover your butt because you have to sign him off. He tells you that of course he has the time (whether he has it or not) because he wants to get the certificate, and something for which he has no responsibility (like this night business) is not going to stand in his way. Besides he wants to look good for the new instructor. When you ask him if he is sure, you are accusing him of lying. He already told you once so when you ask again, of course he says he is sure, whether he is or not. What you have done is set a trap for yourself.

If the student has the night time, you look stupid for not trusting him. You will find this out when you are so curious that despite what you have been told, you have to check his log anyway. Now, should you find that he does not have the time (which is usually the case), as far as you are concerned, you just have to go night flying. However, as far as the student is concerned, you have just made a liar out of him and he will resent you for it.

How instructors handle this issue will tell you how long they have been instructing. The brand new instructor will simply ask the students if they have the night requirement. Not wanting to offend the students, the instructor will trust them and simply sign them off for the checkride. The examiner will halt the checkride should the night time be deficient. The examiner resents the instructor for wasting his time. The instructor resents the student for making him look bad. The student resents the instructor for not checking to make sure, despite what he said about meeting the requirement because ultimately he isn't responsible and he knows it.

The mid-time instructor who has had this happen once or twice will ask the student about the night time, not believe the student, end up checking the log anyway, and got the resentment of the student before the checkride. This is the scenario we went through just above.

Now the old-time instructors know students will say anything, so they don't even bother to ask, and simply grab the logbook and add up the hours and landings for themselves.

The fascinating thing about working with student pilots is that what you think you are teaching and what they are actually learning may not have a whole lot in common. The reason is modeling. Every facet of your personality serves as a model to your students. You are the only pilot they carefully observe engaged in real pilot stuff. Every attitude, offhand remark, reaction to any given situation, solution to a problem, in fact every action you take, both good and bad, is internalized and in many cases mirrored. Most times this is a good thing because, as a professional instructor, you try to set a good example by demonstrating the qualities that you think should be mimicked by your students.

It is the stuff outside your carefully planned exemplary behavior that also gets picked up by your students which will surprise you. What about those times when you cut corners? Every time you require something of a student that you are not willing to do yourself you send a message. A blatant example would be passing within 1000 feet horizontal to a cloud right after beating VFR cloud clearance requirements during your preflight briefing.

Many times, though, the message is far more subtle. How about something as innocent as descending for your airport while flying in the yellow arc in slightly bumpy air because your next lesson is coming up. When your students ask why they can't do that, you might explain that they can when they know what they are doing. At what point did you decide you knew what you were doing? Probably when you found yourself in the yellow arc inadvertently, got away with it, and continued the practice in progressively rougher air. Your students could arbitrarily decide they know what they are doing the same way you did; possibly on their first solo cross-country, for example.

You will be startled to find out just how much of your

personality is reflected in your students when they start throwing your lines back at you. Every instructor has sayings and phrases that are common in our speech patterns. Part of being an instructor is developing a boatload of one-liners appropriate to any situation that arises as we teach. Oh, when entire lessons consist of endless one-liners, it is time for a new profession. Anyway, you will find that your students start repeating those lines back before you do. They know your speech patterns and mannerisms so well that they can do "you" before you can be you. This is a scary thought because if they are that receptive, what else have they picked up?

Your particularly impressionable students will become virtual clones. Your fears will become their fears. Your joys will be their joys. Your strengths and weaknesses will be transferred to them. Whatever you are, they will become. This is not healthy for anybody. Better your students be similar to you, but not try to be you.

You will notice that students who already have tendencies in your direction will gravitate towards you. You will enhance these qualities as you train them. Flight schools, especially the larger academies that assign students at random, miss the chance to let students match up with the instructor of their choice. The dynamics of which student eventually ends up with which instructor is interesting. If you hang around a flight school where you know the individual students and instructors but didn't know who was flying with whom, just by their personalities and temperaments you could probably match many of them up correctly. In a flying club or school where students have total choice, this effect is really dramatic. This is where working with all the different types of instructor can make your life very interesting.

I remember one bureaucratic instructor attracting bureaucratic students. They all work rigidly within the system but also know how to get around it when necessary. One instructor is the scout leader of their club. Those students are honest, thrifty, do good deeds, and can find their way very well when flying through the aviation wilderness. I know an instructor who makes up his own rules and procedures for teaching that always borders the edge of legality with minimum safety margins. Those students are the terrors of the skies, or so they think, and are proud to live on the edge. I know bookworm instructors who turn out phenomenally knowledgeable pilots, many of whom go on to become instructors with similar qualities. For every instructor, there is a student with a compatible personality that needs you. How do your own students match up?

We need a new strategy to help our student pilots as the amount of information they must master continues to pile up on their shoulders while the minimums for the certificate remain completely unrealistic. Twenty years ago, there wasn't half the complexity to our national airspace system compared to today. Students are having an increasingly difficult time mastering the new information and then ripping out brilliant statements before an examiner. Your basic examiner has taken years to absorb all the subtleties of the FARs and the system of airspace. They feel free to test pilot candidates as if the student had the same luxury. Your career-minded students want the ratings as quickly as possible, which squeezes them with a time constraint while they still face the same mountain of material.

How can we help them? Fortunately, the human brain has the most amazing capacity to act like a computer to store and retrieve massive amounts of information. You just have to give the brain a hand, so to speak. If you can clearly focus on one simple idea surrounding a complex subject, recall of the entire topic can be achieved. You access the entire subject by putting a catchy mental tag on to a larger subject that is too detailed, complex, and hard to grasp and organize into one coherent statement without help. All you have to remember then is the key that unlocks the individual treasure chest of knowledge.

The way to remember all the keys or tags is to play a word association game. This game is what allows for the placement and recall of information so readily. This is kind of an expansion of the acronym idea. The difference is that with acronyms there isn't the same association capability. Consequently, you become overloaded with acronyms with no way to remember when and how to use them all. There is no such limitation with the word association method; therefore, the ability to store and retrieve information is limitless. In the process of accumulating your own flight information using this method, you will store all topics in one vast and efficient mental filing system.

As with all such ideas I have presented to you, this has been tested and proven through experimentation on myself first and then with my students. Here is how it works. Take "special use" airspace as a typical category of information that confuses student pilots. I don't even use the term special use for it doesn't mean anything to me, whereas a key like "military airspace" makes much more sense. Anyway, you give your students the topic "prohibited area," and they repeat back the key "national security." From that word key, they will be able to retrieve their mental files on prohibited areas and expand on their knowledge before an examiner. If they remember nothing else due to a mental block, at least by mentioning the key out loud sometimes the examiner will know the student is operating in the ballpark.

The key is not designed to be spoken, however, as some keys are silly even though they are effective. It is only designed to open the floodgates of knowledge on one particular topic. For example: My key to "restricted areas" is to have the student use the key "where they blow things up." This really sounds crazy, but I bet you remember it. Try this with a student. You name the class of military airspace on the left and have your student repeat the key on the right. They should be able to classify all the types of military airspace in about five minutes. Once they know the keys, it is up to them to study and store the detailed information associated with the topic. Here are my keys to special use airspace:

Prohibited Area — National Security
Restricted Area — Where They Blow Things Up
Warning Area — Where They Blow Things Up Off The Coast
Military Operations Area — Dog-Fighting Jets
Alert Area — Big Jets Training
Controlled Firing Area — Guns With Binoculars

Having now read through this yourself a few times, is there any class of special use airspace which you cannot instantly recall, and for which you do not have a permanent quick mental reference? See how it works. No one would understand if you came out with a phrase like "guns with binoculars" to explain controlled firing areas. However, that cue can immediately allow you to retrieve the information that tells you this is where there are hazards to nonparticipating aircraft, like guns going off, where they are watching for such aircraft with radar, participating aircraft and spotters with binoculars, and they will stop shooting when they see you.

You can make your own mental file for every topic you have to teach and you have to learn for future ratings and certificates. You will be astounded at the amount of information that you can have available when you have your brain organized with such a mental computer system. Your students will be surprised themselves as well.

Students make so much of checkrides because the system gears their whole training towards them, so it is no wonder they come to near panic as they approach their first one. What can we do? We can change our language for one thing. One examiner explains that a pilot can never fail a checkride — he only fails to complete it. Most tests in our society (including FAA written exams) allow you to fail a certain percentage while still passing the test. The flight test, however, requires that you pass everything; until you do the test is merely "incomplete." It may take more than one meeting then to pass every task to the satisfaction of the examiner.

This should take some of the fear out of checkrides

because it sounds much less harsh than being told you *busted* the flight test. Busted in our current slang is synonymous with being either broke or arrested. Our world is rapidly becoming very sensitive to language, the so-called "politically correct" speech. I'm not saying we have to baby our students with nice words. However, the air surrounding checkrides can be so charged that students can easily convince themselves they are failures before they even get started, which often results in a self-fulfilling prophecy. What we can do with our attitude is remove the myth of checkride terror and student incompetence while still keeping our language focused on the serious reality.

I hope we never have to resort to politically correct speech in instructing, for the removal of our well-developed sarcasm would take away all the fun. Should we ever have to change our methods and subdue our wit, I have taken the liberty of preparing a small glossary of new terminology that you would find handy.

"Gee, that landing really sucked," would become "I see we are earthwardly challenged today."

"Do you have any idea where we are?" could now be phrased "Do you have a navigational disadvantage?"

"What do you mean you haven't gotten a briefing yet?" should be changed to "Haven't you chosen to verify the conditions of our shared atmospheric environment?"

"You will never make the airport from this altitude," should be rephrased as "We have a conflict in our independent decisions of appropriate pathways."

"Of course we are going to practice stalls today. I know you don't like them, but I can't have you spinning into the ground because you never learned to fly properly." This rather lengthy explanation requires a most comprehensive modification. "I hope you respect my preference for certain maneuvers, as I believe they shall ultimately enhance your total self."

Student pilots bring with them the best and the worst that aviation has to offer. This can be packaged all in one person. They will be your greatest frustration, keenest

challenge, and your most satisfying reward. I believe we
have to change our system so that students are held as
accountable as the instructor for their own training and
actions. Right from the start, instructors must make it
clear to students what is expected from them and what
they have to do for their ratings. The economic influence of
attracting customers often makes this impossible as many
in the business feel the reality of flight training will turn off
potential students if they knew the whole truth.

Chances are, though, these folks would leave anyway,
so what is the loss compared to the goodwill gained by
giving determined students the truth up front? No matter
what you do, you will probably lose half or more of your
student pilots before they ever get to the checkride. Get
used to it. For those who make it, they will be pilots in
spirit forever and they will never forget you.

Your students will never fly as well nor know as much
as they do around checkride time. From the moment they
leave you, a gradual slide begins into degraded skills, forgotten knowledge, and the cutting of progressively rounder
corners. Keep up with your former students for they will
always need help and improvement. They are also your
best source of instrument students.

Recurrent training is woefully inadequate for the
private pilot. The biennial is not enough. All private pilots
should have an annual review where the FAR clearly states
that "The instructor shall raise the pilot to the level of
aeronautical knowledge and flight proficiency required by
the *Standards* to pass a checkride for all certificates and
ratings held by that pilot." This should be the standard for
all flight reviews.

Most of the people you will teach will be student pilots.
Your students will teach you about flying by placing you in
predicaments that you would never have thought of on your
own. We are learning about flying all the time we are
pilots. The name student doesn't apply just to those going
for the private, as we are all student pilots.

16 • INSTRUMENT STUDENTS

Try to get as many instrument students as you can. Instrument skills are the fastest to leave you, so by going for the most instrument students you can get, you can at least observe the IFR system in operation. Your career is the most important priority for you, so with lots of instrument instruction, particularly in actual weather, you will be in a better position to get a job when you have the hours. No employer cares how many stall and slow flying hours you have acquired. They want to know if you have the experience and ability to move the people and goods in everything short of dangerous weather. Teaching student pilots will not help you in this regard.

Besides, student pilots get very predictable and very boring, very quickly if that is all you are doing. Instrument training breaks up the monotony. Your best source of instrument students are your primary students. They should also be your best instrument students because they are familiar with how you teach and what is expected of them. Cultivate your market for instrument training with every primary student, every checkout, every biennial flight review — in fact, any place you can.

It is called "suggestive selling." This is the same technique that gets you to buy a dessert at a restaurant that you had no desire for until your server made it sound like a great idea. Try some enthusiastic selling to your students. Spell out the advantages of instrument training. As always, you have to be honest. Clearly state all the requirements and what the training will entail. If the student can justify

it in his own mind, then you have an instrument candidate.
Your sales burden is removed when you deal with professional career students. Any pilot is commercially useless
for all but the most specialized operations without an instrument rating, so they have to get it. This is where
teaching at a big academy has its advantage because you
are just handed instrument students.

It is really not up to us as instructors to question the
motives of folks who just want an instrument rating for
their private use. They can swear up and down that they
will maintain their skills, but you know that isn't as likely
as a commercial pilot who has to use them all the time.
What you do with these folks is keep in touch so that after
the rating when they need recurrency training, they will
think of you.

Pilots can justify an instrument rating for a variety of
reasons. Some people get rated simply for one big family
trip which would be easier and safer with an instrument
rating even if they never fly through a cloud. Some people
go flying VFR for most of the year and brush up on the IFR
skills for certain times when the weather warrants it. I
have had students whose sole purpose for the rating was to
be able to bust through the thin coastal stratus common to
my area. This is a great reason to get rated if that is what
the pilot wants to do.

The key as always is to be honest and to match the
training to the need. People with no intent to be commercial pilots don't have to be trained with that same drive for
efficiency of operation and the "get there come hell or high
water" attitude. Every pilot must have complete and thorough
training because you have no control over where he will
actually go once rated. However, the individual emphasis you
place on certain areas can be tailored to the student. This
means work on your part because you may have to modify the
standard program your school wants you to teach.

The generic training idea that one instrument rating
fits all pilots is absurd. Most programs offer no training in
small airport IFR. You wouldn't neglect IFR procedures at

uncontrolled fields for a pilot who comes to you from a small field, would you? Pilots will show up for the rating with a variety of desires, motives, and uses for the rating. Your job is to match their needs to the best training and make them competent in operating in the IFR system.

The biggest problem you face with new instrument students is that their general flying is so sloppy that the bulk of your training will be spent just getting them to be able to maintain the basics — heading, altitude, and airspeed. You will work on this all the way up to the checkride as this problem never goes away. What would be nice if you had the luxury would be to fly with potential instrument students just to evaluate their basic skills before taking them on as students. You could then analyze their weaknesses and send them on their way until they had practiced their basic skills until they were sharp enough that their instrument training wouldn't be a waste of time. Dream on.

In the real world, you can't take the chance that another instructor or school is going to tell your student that "Of course, you're ready to begin training," despite your advice. You can't do the right thing in this business even if you want to. What happens then is that you take what you can get and do the best job you can because you need the hours and the money. That is life.

One myth that gives me great chuckles is the idea that instrument training will make students better VFR pilots. Wrong! The only experience that is of any value or significance is direct experience. Nothing is transferable as we have proven ad nauseam. You cannot stare exclusively at the instruments and hope to be a better horizon pilot. Since proper VFR flying results only from the integration of the VFR scan of the instruments with outside visual references, breaking that link through the exclusive use of the instruments removes the integration. This completely destroys any possibility of becoming a more capable VFR pilot.

There is another problem. In order for you to believe that instrument training will make you a better VFR pilot, you have to accept the assumption that flight by reference to instruments is what the instrument rating is all about. This belief is what limits students to the mindless repetition of procedures, causing them to miss the 90 percent of information they need to operate safely in the IFR system under IFR rules. These faulty core beliefs are the cornerstone of a training system that is incapable of turning out good instrument pilots at the time they are rated.

Those flawed assumptions then lead to the perpetuation of absurd myths like instrument training will make better VFR pilots. Any serious analysis will reveal this claim to be a direct contradiction. Tell me, does staring exclusively out the window to become a great VFR pilot make you competent on instruments? The only thing that instrument training will do for sloppy VFR pilots is to turn them into sloppy and therefore dangerous instrument pilots — period.

Instrument pilots come to you with a very different attitude than student pilots. These folks have already made an investment in their training. They know how the system works, what a checkride is all about, and what a training program entails. So aside from the specifics of the training, they pretty much know what they are getting into. They won't have the degree of faults present in your typical student pilot. They will have all the same faults of course, just to a lesser degree. Your instrument students won't be quite as late to their lessons, cancel them quite as often, leave as much studying until the last minute, give as much of the responsibility of their training to you, or expect to be spoonfed the same quantity of information. Instrument students are more serious about their training.

Simulators bore me. I don't know why, but watching a student twiddling at a simulator has to be the most boring activity in aviation. The smell of electronic machinery puts me immediately to sleep. What possible good are we doing by starting off a student with a simulator? There your student sits at a fake airplane under the delusion that this is great training. Okay, so you don't believe me; fine. How many students have you ever taken right from a simulator, put them into the clouds, and witnessed a great flying performance? They can't do it of course, because simulators and clouds — are different.

Ask your students some time what they think they are getting from simulator training. If they respond with something like "It improves my scan," that kind of canned answer could only have come to them from someone else, so it doesn't count. What it shows is that your students have no reason or purpose behind their training other than the usual — you told them to. We don't ask students often enough what they are getting out of any method of training. This might explain why they learn not to think their opinions are valid.

I don't know what my problem is with simulators because virtually every expert raves about them. Then again, those are the same people who brought us the rectangular course, which teaches students how to fly parallel to a runway. You can't use simulators to justify saving the student money because the cheap ones are so removed from real flight that they are useless, and the expensive ones cost more than the rental of an airplane.

My real problem with simulators is that you can talk any student into making those needles look exactly as you want to see them without the student understanding anything. Students believe that they are receiving wonderful training from simulators because all they have heard from sources they respect is that simulators are wonderful. The perception becomes the reality even though there is nothing on which to base the perception. When an instructor raves about simulators, he gives the impression that something fake is as good as something real. This is a lie.

I do recommend simulators, but only to practice procedures that are already fully understood, where the limitation of such a machine is clearly stated, and the student knows it has nothing to do with operating in the active IFR environment. For those who value computerized TV screen simulators like the "Elite," how many aircraft can you fly using a "mouse?"

Now, just because I consider these devices useless does not mean that my students do. My personal belief comes because my simulator training actually retarded my understanding of the IFR system, so my rationale is based on personal experience. I found, though, that my students picked up my prejudice without ever having tried simulators for themselves. No instructor should have that much influence over anybody and no student should let them. Since my own students were so quick to go against the trend and condemn simulators without any evidence to back their claims, it is all to easy to see how the majority of students can be convinced that simulators are valuable simply on the basis of that is what everyone else is telling them.

I began to insist that all my students at least try a lesson or two with an instructor who loved teaching on simulators. That way they could make up their own minds. I lost a couple of students doing this, but after reflection it was clear that they would not have flourished under my style of instruction. My regular human dynamos always came back because they wanted to train in the clouds, not in office buildings. These students, having at least tried simulators, never felt cheated out of a possible aid to their learning. Some would still use them for practice, or even take some lessons on them, which was fine since they had a good perspective on simulator limitations. As long as you don't get jealous and consider students your exclusive turf, experiences like this with other instructors can only help them. They will respect you more for recognizing your own limitations as an instructor.

Once your students can make the distinction between simulator training in an office and real training in the clouds, the simulator will be useful to them. Since most formal programs force you to start teaching in the simulator, this distinction will never be made in the minds of your students and you will have to start their instruction all over again when you get them to the airplane. What is the point?

What does simulator training do for you and your career? In a word — nothing. The employer does not exist who will go through your logbook and say, "I see noted here way in the back you have 500 hours watching people in simulators — that's impressive!" Simulators are in buildings and buildings don't fly. You can't log simulator watching time. Why even call yourself a flight instructor when you teach on simulators? You are a ground instructor.

For career instructors who can afford to spend all their time on the ground because they have no further ambition, they should do all the simulator training. For those of you who desire bigger, better, and faster aircraft that you can actually fly (not just watch other people fly), every hour spent on the ground on a simulator that could have been spent in the air is an hour wasted. For those who think this attitude is selfish, so be it.

I insist that all my students take a simulator lesson and they are free to do so whenever they choose. My students dump them as a primary source of training of their own free will with very few exceptions. You might say that students don't have your insight into what is best for them; therefore, they don't know what they are doing and you have to tell them. Isn't that the attitude, though, which allows students to abdicate any responsibility for their training and place it all on the instructor? Think of it this way: The more time students spend on a simulator, the more they are cheated out of real life experiences in the air.

Are you going to have what it takes to teach students in the clouds? Will you be able to keep your composure while bouncing in driving rain? Can you take your students into weather below minimums to show them what a real missed approach looks like? The majority of you are great instructors and will be able to do all that. Most instructors pride themselves on offering the best training.

For those instructors and examiners who do not want to fly in the real stuff for whatever reason, please stay away from instrument students because you are only cheating them out of knowledge and experience that could save their lives. How can you condone your former students experimenting in the clouds for the first time long after you have supposedly trained them? An interesting and possibly revealing study could be made by analyzing the amount of actual time logged compared to simulator and hood time among recently IFR-rated pilots involved in weather related accidents.

Back to the gritty weather instructors. You will never go into the clouds with students without feeling just a twinge of anxiety and excitement. This is because you never know exactly what they will do, especially if it is their first time. It is one thing to be an airplane god out in the clear blue; it is quite another to have to play god while sitting on your hands trying to talk a student brand new to instrument conditions through his first encounter with the real thing. There are no lackadaisical instructors in the clouds. This may be the best thing you can do for your students, but what about your own adrenaline and blood pressure?

You are not God, so don't be bothered when you feel trepidatious about hurling your craft into the clouds with a student. I was very fortunate to have previously flown with almost all the folks I took into the clouds in some other capacity, so I pretty much knew what to expect from them. It is critical to know the temperament of the person with his hands and feet on the controls as you leave VFR behind. Personality clashes on the ground are a pain in the ass. Clashes in the air can be dangerous. Clashes in the clouds are suicidal. Make sure you can work with the person for whom you intend to give all this real world experience.

For the most part, you will find teaching in the clouds an incredibly rewarding experience and a booster of your own confidence. Your students will respect what you are trying to do for them. However, if you have any doubt about your ability to fly in these conditions, your students will pick it up immediately. They will magnify your doubt such that your anxiety could become their panic. I was very fortunate that both my primary instrument and CFII instructor loved the weather. After many hours in the soup I loved it too.

Sometimes one flight can spark brilliance. Sometimes one flight can tie it all together. Sometimes you jam so much into one flight that everything else pales by comparison. I had just such a flight right before taking my CFII test. On a cool, rainy, spring afternoon two instructors went out to battle the elements. A warm front was lazily moving through; although warm in this case was a relative term. Layers of rain-soaked clouds were stacked around and above us, up through the atmosphere, absolutely drenching untold thousands of square miles in all directions. The wind howled steadily about 50 knots from the south at 3000 feet and increased as we climbed. It was dark and wet as we flew through that mess in a Cessna 172. What a great airplane. Oh sure, we bumped around, but there was no serious turbulence.

This was my first exposure to real weather. I had flown through nothing but fluff up until then. We were in solid IMC conditions except for those rare instances when we popped out over a runway. I was surprised at how little any of this bothered me. This is where I realized that I could teach students how to fly in the clouds. Think of the occasion when you realized you could do it; we all have them.

I also learned that you must do whatever it takes to navigate. If that means 40-degree wind correction angles, then that is what you fly. That may seem out of the norm for correction, but this was not a normal day. We actually got a hold from ATC. Why I don't know as there were precious few other aircraft on the radio. This was the first of only two holdings ever assigned to me.

Anyway, we couldn't pass the opportunity to try our skills. We told ATC to take their time, we were in no hurry for the approach. As you might expect, trying to hold in 50-knot or greater winds makes for a less than ideal pattern. We had no DME, so the only person who knew where we went was the person at ATC. With the wind blowing pretty much down the outbound course, we elected to start with a 30-second outbound leg on the entry. When it took seven minutes to get back to the fix, we dumped the whole idea of an outbound leg and on the first pattern just turned full circle to the inbound. This got our inbound leg down to about five minutes. We were learning. We got the approach after one more circuit.

The wind rapidly deteriorated as we descended, making it impossible to maintain any kind of consistent correction angle. Whatever was required to stay on course, that's what we did. I found out why approach lights are so valuable, even in the daytime. I discovered how far out you can see runway end identifier lights even through the fog and rain. I found that on a wet, slippery day it is nice to have a long runway, and that there is no reason not to use it all if you want. I realized how valuable that person on the other end of the radio is and how critical a radio failure can be. I wondered how I would handle a power loss or total engine failure? Would we break out in time to make some kind of a landing?

I learned that ATC with radar is your best friend. We had instant position reports and groundspeeds should we need to know them. I learned there is no shame in asking for help. That is how we found out that our groundspeed on the way back was a scant 60 knots, which explained why it took so long for the intersections to appear on the VORs. I realized that no simulator could give my students any of this wisdom and vowed to limit their use as much as possible.

Like the caveman who first comes out to see the light, in aviation you must venture into the cave where it is cold, dark, and wet if you are to pull together into one coherent picture all the disjoint bits and pieces of insight our current training system throws out. Once you have gone into the cave, only then can you look back with the crystallized vision that forces you to pause and whisper: "Oh — now I understand." Only by going into the dark will you ever see the light. I learned also that there are times when all the rules go out the window and you have only your guts, your brain, your imagination, resourcefulness, and your survival instinct to get you home. When you have taken this journey, you will be able to walk your students through it.

When we teach students about things that are dangerous, through constant repetition of the warnings, we can build up such a fear in their minds that when the students actually encounter the offending condition they are in a state of near panic and beyond the capacity to learn from the experience. The best example of this is icing. We really get the message out on icing. Icing is bad news. Well, it is! However, a couple of crystals on the airplane aren't going to immediately remove you from the sky. Try telling that to a student who has just been through one of those juicy ground schools filled with ominous sounding "experts."

Icing presents you with a dilemma. You can't purposely go into known icing conditions because our little trainers are prohibited from it. Yet sooner or later, you will encounter a minor icing situation. Icing is so uncooperative. It shows up at the oddest times. It occurs when all the correct conditions come together to create icing, and no one knows where and when that will be. Students should know this. They should know that you are not free from ice just because it is not forecast or you are in cloud where the temperature is above the freezing level.

For safety, of course, you would never want to be flirting with the freezing levels unless you have lots of warm air below you, just in case. But there have been many times when I have had dustings of ice and found the occasion perfect for teaching students who had never seen ice before other than in horrible pictures where the aircraft looked like it had been swallowed by a glacier. At least now they knew what the stuff looked like in its beginning stages when you can do something about it. I remember one student who epitomized the "panic because they were told to panic" school of learning.

"Oh no! There is *ice* on the wing!"

You know you are truly jaded and burning out when you calmly lean over to the student and whisper: "So tell me — is it clear, is it rime, or is it conglomerate?"

That was tacky beyond the call of duty, but at least I got a laugh out of the student, and we could put this ice thing in perspective. We had some trace ice on the leading edges and windshield. After getting over the initial shock of accumulating ice, and then seeing what it looked like, the next realization was that the aircraft would continue flying even with some ice on it. The student got a valuable look at ice so he could recognize it the next time. However, as soon as this student realized that the aircraft could keep flying, his reaction took a complete pendulum swing.

This is where I fault such overdone warnings to students without any perspective to the warning because this student was initially convinced we were about to die, followed by the realization that he can't feel any difference with the airplane carrying some ice, so what is the big deal? Now the warnings look stupid to the student. This causes the most dangerous reaction of all — complacency. Students have been filled with dread over something that to them when they encounter it doesn't seem to have any effect. What happened to all that great ground school training?

Here is where perspective is critical. I had to impress this student with the seriousness of our condition however harmless it looked at present. I told him to think of a dusting of ice as a warning of a potentially disastrous situation. This is the perfect time to try all those ways you have memorized for removing and avoiding ice that fit so neatly into dots on written tests. What are your students going to do about real ice when they get it? Will they deal with it or watch it accumulate? Are you observing their reaction when they see ice on the airplane to see if they need prompting?

Students often know what to do and still refuse to do it because of the need to fly normally even in unusual circumstances. With known warmer air below, what we have in this instance is a learning situation that is very real, easily recognizable, and with a safe escape route. You have to make the most of every serendipitous teaching opportunity without ever jeopardizing safety in the process. In this case, our encounter was purely by accident as we had taken the usual precautions to avoid it. It is good for students to see real ice so you may help them put it in perspective.

We talk about structural ice a whole lot in our weather chats, which is funny in retrospect because the worst encounter I ever had with ice had nothing to do with ice on the wings and everything to do with ice that I think was gradually encasing our carburetor air filter. Induction icing receives way too little attention in training; especially considering the lack of engine alternate air in most trainers. Consequently, even as an instructor, the whole experience was new to me.

What do you do when you have a student on a training flight in a 172 when the RPMs drop from 2400 to 1800 in a little over a minute and you are flying in solid cloud? Just what I need now I thought, engine problems in this muck. Well, it wasn't life-threatening. The ground below was flat and at sea level, the ceiling was at 4000 feet and we were at 6000 feet. We had our out. You should always have an out, especially in a single.

My challenge was to solve this as if we were in solid cloud without an out. Never missing a chance to teach, my voice found its way over the intercom and headset to a very quiet student. "Okay, now what?" Why do students no matter what their level of flying experience always give back that same blank look? What would they do if the trusty instructor wasn't on board?

Well, the trusty instructor had neither seen nor read about any situation quite like this. I was improvising. There had to be a way to solve this without ducking lower, and I wanted to find it and learn from the experience. That's the thing about instructing — you have to have or find the answer. Your student counts on it. They will of course be useless at any time like this, so you are on your own.

Okay, back to basics. Check the engine instruments. Hmmm, everything in the green. Great, now I have even more questions. Flow check time. Hmmm, everything is on and set properly. The engine was not running rough at all; it was very smooth. However, it was now down to 1600 RPM, and we were just starting to have trouble holding altitude. I called ATC and advised them of our predicament and that we may need a lower altitude on very short notice. No problem, as we were informed the sky was ours.

I thought of carb ice as a possibility, but that had never been a problem when operating at cruise speed. Looking out the window, this slushy stuff that wasn't quite ice yet was splashing on the airplane. The temperature was above the freezing level, so icing didn't immediately cross my mind. Some would stick for a bit, but most of it just blew off. The little bit that did stick was starting to freeze onto the metal surfaces of the airplane.

As we started to lose altitude, all my light bulbs started flashing on. Applying full power and full carb heat brought a tiny but immediate rise in power. Then we waited. We waited for many precarious minutes. Nothing moves slower than a tachometer when you really would like full power. I will never know exactly where the airflow was being choked-off. My best guess is that the slushy muck we were flying through was getting caught in the air filter and then freezing to it.

This weather phenomenon was limited to this one, big cloud that was darker than the others we were flying through. I maintained full power and carb heat until the weather had changed to where there was no chance of any induction clogging. As an instrument student, I was never taught about the possibility of induction ice when none of the rest of the airplane is getting any noticeable accumulation. I had never heard of any such thing, which is why it happened to me.

You will most likely encounter a situation in training for which you have not been prepared. It doesn't matter why it happens, just know that it does. As an instructor, you have to find the answer to bring you and your student home. From this experience, I learned to apply full throttle and carb heat first, and then try to figure out the problem. I also learned that the carb heat *is* the alternate air. When the normal rules don't apply and the standard training doesn't cover your problem, make up your own solution — whatever it takes.

Vertigo is your greatest danger as an instrument instructor. Not because you will get it. Instructors are not allowed to get vertigo — but your students are. I told you of the student who tried to fly me into the hangars near the decision height because for some unexplained reason he chose that exact moment to trust his feelings rather than the instruments. It happens all too frequently, so it is something you have to learn to handle. Part of being an instrument instructor is knowing when a problem exists before the student does, and before it gets you both in trouble. Call it "mindreading through instruments" for lack of a better name.

There we were (I just had to start out one story this way) being vectored onto the localizer for an ILS. Imagine a night of broken to overcast fluffy clouds. Imagine a night where the lights of the city blend perfectly into the stars of the night sky. Flying with some visibility can be a hell of a lot harder than flying with no visibility. At least with nothing to look at but the instruments, you can't get as distracted. This was a beautiful night to be bopping in and out of the clouds. What a sight. I even became sidetracked by the view.

I was with one of my best students in a Cessna 310 approaching the ILS final approach course. These things always happen so fast. We got the final vector from ATC. My student started turning — and just kept right on turning! We were at 45 degrees and heading for a spiral in less time than it has taken to read this sentence. The airplane had to be wrestled back to level because despite my loud warnings and directions, my student was perfectly happy in this attitude. He had no perception of the illusion and consequently no recognition of the impending steep spiral, which is why he kept trying to put us back in it. What can you say but "my airplane!"

On this particular night, the conditions were perfect for a horizon of lights and stars somewhere about 30 degrees of bank. You didn't feel the effect of the illusion until you started in a turn. I love those instruments. They keep me coming back for more and more flights. Vertigo can strike students in any attitude, at any airspeed, in any configuration — just like stalls. It will probably strike you too, but you have to develop a way for you to get rid of it instantly. There is something about being responsible for someone else that brings out that kind of inner strength. Now if we could just teach that to students . . .

Instrument students give new meaning to the concept of being "behind the airplane." They can be so far behind in fact, you would think they were still on the ground. Well, what is happening? Students get a complete mind-block and become vegetables during complex procedures like holds and approaches. The fault lies in the organization, function, and design of the human brain. When the brain evolved eons ago, humans didn't have to juggle as many functions as they do in an airplane when flying an approach on instruments. The technology has changed vastly over time, but not the brain.

Of the thousands of little things students are taught by instructors to try to accomplish, to remember, to say, to listen to, and to communicate while flying, none of these procedures, methods, or acronyms are organized by any priorities. Consequently, students give every single one of them equal weight, because they have no capacity to be selective — because no one ever tells them they can. This effect is magnified if more than one instructor with more than one laundry list of tasks to accomplish is teaching the student. Students have no power to question anything on their own because they take no responsibility for their training even at this level, again forcing you into spoonfeeding. Nothing ever changes. They want to please you by being able to accomplish absolutely everything you give them to do.

The brain has a limited capacity for the number of conscious ongoing processes it can handle. Therefore, when the student tries to do absolutely everything all at once because they have no sense of priority or selectivity, they become completely overloaded and end up not being able to do anything. Then they give up. That is how they can be fine on limited tasks like basic attitude flying, but are not able to make the transition to the next level of difficulty, so they fall apart during every single approach they try.

This root cause of all their problems is never understood by instructors because it is not part of your instructor training. You are only taught how to treat the technical mistakes, never the human limitations. Now you understand the real core problem in instrument training. Your job is to teach them how to select, organize, and prioritize all the individual instrument tasks and activities. The secret of instrument training is to give students the ability to do more things than their brain can normally do. Like everything else, it is just another skill.

The way to begin this process is to teach your students how to take as many activities as possible and make them instinctive. This leaves the space in the brain for the few activities that require ongoing conscious thought. Instrument pilots on approach should only be concerned with headings, altitudes, airspeeds, timing, tracking, communicating, and most important, *landing*; the rest has to be automatic. We give students exact pitch and power settings so they never have to think about how to configure the aircraft for either stabilized or nonstabilized approaches. We give them the "T's" so they never have to worry about forgetting an action step. We tell them to set up the approach long before it ever gets started so that a minimum of twisting is necessary once the approach is initiated.

What we don't do is give all these different things priorities so the student knows what to do, what not to do, when to do them, and when not to do them. Many times in trying to make it simple for the student, we actually make it more difficult and draw out their training. You don't notice this effect anymore because you have already molded yourself to accept these procedures. Once again, we see the flaw of molding people to the procedures rather than the other way around.

Take setting up for the approach, for example. Many schools have long acronyms, exact orders to follow, and specific methods of setting up for the approach. The student tries not only to remember how to get everything done, but in exactly the right order at exactly the right time. This is a totally unnecessary burden on them. Here is why.

Let's say you have one of those laundry list procedures like: "ATIS, altitude, airspeed, approach, alternates, marker beacons, comm 1, comm 2, nav 1, nav 2, ADF, (we hate the ADF!) transponder, minimums, directional gyro, altimeter, missed approach, and oh boy, watch the ball, gee I hope I said them all." Your typical students (because their only priority is looking good for you) will concentrate on all of these things at once. They will say them when they think you think these things should be said, and they will do them when they think you think these things should be done. Your students, by dwelling on all three tasks of remembering, saying, and doing your laundry list, all at the same time, has effectively tripled their own workload.

Somewhere mid-laundry list, the heading and altitude will wander because they are trying so hard to do things in your arbitrary order. You will point this out and they will correct the problem. They will then lose their place in your laundry list and because they are too embarrassed to tell you this, will try and pick it up on their own. As they become more frustrated at forgetting their place, trying to catch up where they were will dominate progressively more of their thinking and energy right up to the exclusion of everything else (like flying the airplane). You will again remind them that they are heading off into unknown space and the approach will fall apart. Does this sound familiar?

Think about what you have taught them. Laundry list procedures teach that memorizing the order of procedures is more critical to the instrument pilot than knowing what they are doing to fly the approach. Their goal is to become the best memorizer of procedures. They have no purpose for those procedures. They have no strategy for the approach. They have no reason for anything they are doing other than to make the needles look like you want them to and to impress you with their memorization powers. Why?

Working backwards, the goal of any approach is to make a landing. To make a landing, you have to see the ground. To see the ground without hitting anything, you have to fly a known path clear of obstacles. That is why we have approaches. Try this: Tell your students simply to "set the approach." As a guideline, tell them to start from the top and work their way down.

You have then removed any burden to please you by doing something in your order; they can use their own order. You have forced them to think about what they actually need for the approach, which makes them find a purpose for every navigational instrument, which teaches them how to come up with their own strategy to fly the approach. The regimentation of standard instrument training convinces students that if they don't exactly follow the method you lay out for them, worse than any worry they have of crashing, they suffer from a fear that you will think they look like a bad pilot.

You can fly an instrument approach any way you want as long as you stay on the approach course, don't bust the minimum altitudes, go outside the safe area, or fly a speed off the chart. Students have no idea of the freedoms available in instrument flying because your list becomes more important than the objective. Students do, however, need a framework from which to work and base their strategy.

The basis and building block of all structural training is our most valued and treasured T's. Although somewhat of a laundry list themselves, they are general enough to justify the exception. The reason is because the T's are simple, flexible, and work for any phase of any instrument procedure. They can prompt any missing step. They give your students a logical order to execute their tasks. This way you both know what to expect as you are on the same general wavelength.

What the students have to learn for themselves is when and how to make the best use of them so that the T's have purpose and meaning and don't remain just another laundry list. Purpose is inspired because the T's make you think. Let's run through them:

Turning will be pretty specific without a lot of options.

No room for creativity here as turning is controlled by the procedure and ATC. The exception is when and how you execute reversal turns.

Timing, however, has a strategy all of its own. Students are usually taught to time every single leg of an approach using digital timers requiring constant resetting, where they often push the wrong button ruining the whole procedure. One option discussed earlier is to have the digital timer set way in advance for use only on the final segment, where you have a second timer such as a stopwatch which you start when crossing the initial approach fix, then two to three minutes later you turn outbound, let the stopwatch run another minute and turn back in, all of this without ever touching the stopwatch again after passing the initial approach fix, just let it run. Your students may develop better strategies that work for them.

Twisting can have many varieties of how and when to set and twist the navs. This again puts the responsibility on the students such that they are learning to develop the strategy rather than the usual situation where you make all the decisions for the approach using your laundry list. Now you can work on critiquing the various methods your student tries and guide them to the one most efficient for them.

The *Throttle* comes under the realm of instinct, in that all pitch and power settings have to be automatic, where the only thing to consciously think about are the minor adjustments in power necessary for the individual situation.

Talking is pretty straightforward as well. There are places to call in, minimal requests they can make from ATC, and things to repeat back. Students simply have to learn to work with ATC. With proficiency they will learn how to negotiate with ATC.

We teach the T's so the student does not have to think about every step, only how to carry it out. We try with the T's to take as much of the burden off them as we can for that is the way to success in instrument flight. When the student takes the responsibility for testing and developing his own strategy, he flies to please himself. Students also learn how to be the instrument pilot in command. This drastically reduces their burden to please you and enables them to concentrate more on their training.

Whereas instructors have a problem completely overteaching procedures, students have a problem refusing to take our advice, even when it is correct. You can take your instrument students and give them all of the pitch, power, and airspeeds they will need in instrument flying, then give them the T's, then tell them to go sit on a mountain at dawn and commit these simple things to memory for they will constantly come into play in every phase of their training, and yet they refuse to put the most minuscule effort into making their lives easier. Even after several lessons where you show them how handy it would be if they knew their stuff, they still refuse to memorize and use their pitch and power settings or T's, preferring instead to play improvise catch-up after the fact. Why?

I used to think my students were just lazy. Then I found not only that all my students were lacking in effort, all my fellow instructors had the same problem. Do the T's and the pitch and power settings just get lost in the shuffle of procedures? Do instrument students think purely in the present and expect to be spoonfed the next step so they don't have to plan ahead? Could it be that students are so overloaded with the need for approval that they can't possibly exercise the initiative to invoke the T's or use the pitch and power settings, without the acknowledgement from the instructor before every single procedure that it is the right thing, at the right time, to do?

Maybe it is all of these things. So much of a student's standard training involves unnecessary memorization that he may not believe us anymore with this stuff. That is why I always try to keep the memory work down to the bare minimum. When such work is required, I fully explain specifically why they have to do it. Students, because they are new to instrument flying, cannot see the real benefit of the T's and pitch and power settings until after they have learned how to implement them. It is only then that they realize how much time they wasted by not committing them to memory earlier.

I have tried having students post their T's on the bathroom mirror, the desk at work, and on the car dashboard; any place where they can be imprinted on the brain. I have tried making students repeat them during basic flight maneuvers, only to have them neglected on the approach.

"Why, why, why don't you memorize your pitch and power settings and your T's?" I ask.

They never have a good answer, just that same old blank stare. Well, it's their money and time. I think about innovative approaches to learning, like rounding up all the instrument students at a school and marching them into any large field near the airport. Dressed in orange robes and carrying small cymbals and drums, the students would chant the T's until they were so ingrained that students would be saying them in their sleep. The instructors, of course, would get small cups to pick up whatever small change could be cajoled and wrung from sympathetic passers-by. Sure, we would look like Hare Krishna's; how else would you get a bunch of people to chant in a field?

Laziness, although a convenient reason for a lack of memorization, is far too simplistic an answer. For all the work the FAA has done to investigate the learning process, one area yet unexplored is the deliberate resistance to learning, even when the student has seen ample evidence that such learning would be beneficial. Why do people go out of their way not to memorize the information? Could it be as simple as because we tell them to? Is it as simple as teen-age style rebellion? You may demonstrate an approach using the top-to-bottom setup, the pitch and power settings, and the T's, and still your student will refuse to use them, preferring instead to deal only in the present and try to do every step in the procedure as it happens. We as instructors need to find the way to break through this resistance so that the training time may be greatly shortened.

Much of our instrument training problems start back in private training. I teach all my private students to use consistent pitch and power settings from day one. That way by instrument training, they are instinctive. Even if pilots stop at the private, the inherent consistency of fixed settings greatly shortens their training.

Too many students are taught without specifics and develop the habit of improvising everything and relearning procedures as if they were brand new to each lesson. Once this habit is ingrained, it will naturally follow into instrument training. Students need the framework to develop their own strategies and to give each procedure purpose. It is just as bad to provide no framework as it is to mandate every single nitpicky segment of the procedures. Another good study for the FAA would be: How can instructors set good learning habits right from the beginning that would follow students throughout their entire training?

One thing for which you have to be very careful as an instrument instructor is forgetting just how hard it is to both control the aircraft and fly the procedures. When all you have to do is watch others fly, you are relieved of the burden of aircraft control, so you are free to devote more of your concentration to the specific procedure. This is why autopilot use should be integrated into any program where the student's goal is single-pilot IFR. Anyway, when free of the tasks of aircraft control, you get to sit back and observe how the whole IFR system fits together. This is great for you.

However, the further you get from actually flying the airplane, the less tolerant you become of aircraft control errors. You will see any problems immediately, while the students are diverted with their own concerns. One thing you should do is periodically demonstrate an approach in actual conditions. This way you would get some desperately needed practice, and your student would get to sit back and see how the system works. We don't do this because students insist on paying to repeat their mistakes without the benefit of a model, and because students are trained for the checkride, not how to use the rating.

Instructors do get out of practice because they can't

afford the luxury of paying for an aircraft to get their practice. The FAA should require as part of their operating certificate, that all clubs and schools make an aircraft available to each instructor for one hour every month free of charge in order to maintain flight skill. Instructors could double up and get a plane for two hours to maximize their practice and share teaching techniques. I'm dreaming I know, but something has to be done to insure that the people who teach the flying — can still do the flying.

Currently, an instructor can maintain IFR currency simply by watching students fly through the clouds. Sure, you are pilot in command, but how sharp are you as time goes by? It is a good thing that we can maintain currency this way because instructors can't afford to rent airplanes. The more you instruct, the more you know, the less proficient a pilot you become — interesting phenomenon.

Try reaching beyond your curriculum when you teach. If you stick rigidly to most instrument training programs, flying becomes a set of monotonous, complex procedures with no objective, that seem really funny to the students when looked at all on their own. We have to integrate all these funny procedures into a coherent and understandable system by giving every phase of instrument flying a purpose.

All IFR flight plans start their life at a Flight Service Station so that is where training can also start. Take your students to an FSS for a detailed tour. Arrange it ahead of time with the supervisors. Go into depth about how the weather information is acquired and interpreted. How much of your briefing comes from a computer and how much is subject to the judgment of the specialist?

What happens to a flight plan once it is filed? How does it become a clearance? Why do they issue different clearances from what you have filed? Why is a certain amount of time needed between filing and departure? How is your flight plan fit into the traffic flow? Who figures the engine start times and how does flow control work?

Why does every FSS briefer request pilot reports when they know we have to remain on ATC frequencies to maintain positive control and separation? If the code for pilot reports is UA (which stands for "unsolicited advisory"), why are we solicited for them all the time? Why do the reports we give to ATC never get back to the FSS?

What are the benefits of "flight watch" for the IFR pilot? How are the surface analysis and constant pressure charts used by the FSS folks? Your FSS people should be able to both raise and answer questions that your students will never even think to ask. You may have questions yourself.

Your students will see you asking questions. Hmmm, they think, the person who knows everything is asking questions. Hmmm, maybe you never know everything. Maybe this is the most important lesson of all.

The next place to visit with your students is the nearest, busiest tower. How do they separate IFR and VFR traffic? Who has priority in various scenarios, the IFR Cessna or the Gulfstream jet? What are the concerns of tower people when working IFR traffic? What kind of weather shows up on ATC radar? How much real help can you expect from the ground folks when you are operating in an area of thunderstorms?

How does the tower decide which approaches are going to be used? Why is it we ask the tower for IFR release into the system, yet they clear us for takeoff (like in VFR)? If the tower came back and told you "you are released," would that clear you for takeoff? You see all the things we take for granted. Instrument students should know the concerns of controllers by seeing the system from their point of view. That way they can learn how to make the whole system more efficient by working better with the controllers.

Students should then visit approach controls or en route centers. Here they could see the traffic flow in their training environment. My local approach facility has a great model made of coat hangar wire that conspicuously details all the flow patterns in the TCA and around the various ARSAs (Classes B and C). After seeing this, my students have a total picture of our area in their minds. They understand the clearances much better because now those clearances have a purpose.

The questions for the approach folks might be for example, how do they separate and regulate the flow of IFR traffic? How do they resolve conflicts? At what point do they decide to discontinue the service to VFR pilots? How do they handle an aircraft with which they have lost communications? How does canceling IFR affect your service and protection?

How do they set up the letters of agreement that regulate the tower en route clearances? Why are they called tower en route clearances when they are handled exclusively by approach controls? What are they required to do when you request a pop-up clearance? What are the pilot and controller options and responsibilities? How do they decide on vectors and airspeeds? What do they expect from you when they clear you for a visual approach?

From the center you might want to know how they set up gate holds? Where do center weather advisories come from? How are various sectors derived and combined?

Under the current system of instrument training, you are taught how to fill your students' heads with enough mush to completely block out the purpose of IFR training, which is how to operate safely under Instrument Flight Rules.

There are so many questions that never get asked when you stick only to those same, old procedures. We spend huge amounts of time perfecting holding patterns, for example. Your students have to know how to do them because some time during their career they might actually get one assigned to them. This puts holds on about the same level as an emergency procedure. How precise do holds really have to be when all ATC wants from you is to park? Only you know how well your student is flying the hold because only you can see the pattern on the ground. They have no idea how good it is, so what are you really teaching them?

We know procedures like perfect holds are only designed to please examiners in VFR conditions. The effect of wasting huge blocks of time overdoing these procedures is that it detracts from learning so much else. Much of what you teach your students, therefore, is how to take the same things for granted that you do. This is why it is critical for you to break the straightjacket of your syllabus and expose your students to questions that would never occur to them because most pilots haven't even been taught how to ask challenging questions.

Most instrument students go through training never having a clue where they are, what they are doing, or why they are doing it. That's how it was in my training. One of the many reasons is the hood. Instructors use the hood from runway to runway to maximize that simulated instrument time. Consequently, students never get to see what an approach looks like. Worst of all, they never get the visual backup to confirm in their minds that what the needles are telling them is what is actually happening. This is the video game mentality at work again. When the hood is used down to minimums all the time, the student suddenly appears over the runway without any way of intrinsically understanding how he got there.

In the mind of the student, he is taught that when he makes the needles do what the instructor wants, he is doing the right thing. Students miss, however, the verification that the needles actually are making the airplane do the right thing. This is all taken on blind faith, which inevitably leads to spoonfeeding. There is no link between the tracks printed on the approach plate and the objective of finding a runway. What you learn as a student through all this is that you have to teach yourself to fly, because nothing the instructor says makes any sense until you figure it out on your own first. Then you can go back to see what they were getting at.

When I was a student, I found that if I made a mental map of the world beyond the hood by imagining the ground under me, I could make sense of the approaches and pass the checkride. It worked. I passed the checkride only because I went to very familiar ground where I had memorized not just the approaches, but the whole terrain. Even with the hood, I could "see" in my mind exactly where I was. The question then becomes how many newly rated pilots who learn this way are not safe to fly on instruments anywhere beyond familiar places?

What imagination should never become is a substitute for understanding the system and the procedures. However, by using imagination as a training tool, it allows students to understand the logic and purpose behind the procedures, so they can then go on to fly in unfamiliar territory. After teaching my first few instrument students by the usual vapid techniques, I thought back to how my imagination had allowed me to figure out this instrument business.

This lead to an experiment. Never be afraid to experiment with your students. Most of our problems as instructors result from trying to beat success out of students from proven failures in methodology; the old "this didn't work the last dozen times so let's try it again" school of instructing. My experimentation has lead to some amazing breakthroughs. Every technique in this book has been tested through experimentation.

Anyway, I had this student who came to me complaining he was completely lost on approaches. He was perpetually lost, period; however, approaches were in a class by themselves. After witnessing one debacle, I tossed the damn hood to the back of the aircraft. Something drastic and unorthodox was required here. I then had him watch me fly the same approach visually. The next time around (still without the hood), I had him fly the approach, taking special note of how particular landmarks matched up with the different phases of the approach.

Then I put him under the hood while I again flew the approach and had him call out the landmarks, telling me where we were as we went through the procedure. You could see the fireworks going off as revelation exploded in his mind because he had finally made the connection between the approach procedure and the world beyond the aircraft. Once the student realized how the whole approach is related to ground obstructions, traffic flow, and putting the aircraft in the best position to see the runway to make a landing, everything became clear. What a difference! Now of course, I make every student do it.

Try this technique with your students. The power of the mind and the magic of imagination can be your greatest aid in helping the student see the logic behind IFR. After this realization, all the separate pieces of the puzzle should fall into place. You, of course, visually see approaches all the time, so you tend to forget how it was in training when you had no idea what you were doing.

The hood has to be used selectively. Once students have the visual reinforcement to see and understand what an approach looks like and what it is for, their imaginations can fill in the gaps for the rest of their training and the needles will take care of themselves. It is at this point where instrument training really begins. Please remember that your main objective as an instructor is to redress the imbalance between overloading the procedures on the one hand, without teaching how to operate safely under IFR on the other. The goal of instrument training is that students can actually use the rating when they get it.

17 • COMMERCIAL / MULTI / CHECKOUTS / BFRs

——————————————————————————————✈

Teaching commercial students is probably going to give you some of the most fun in your instructing career. Your students pretty much know how the training game is played, so they shouldn't drive you nuts with the same insidious problems that get under your skin with private students. Pilots go for the commercial rating so that they can earn money and get someone else to pay for their flying. To get to this lofty position, they are willing to put up with learning the ludicrous maneuvers required for the certificate. They have resigned themselves to the inevitable and can sit back and have some fun as they train.

The nice part of commercial training is spending your time in a high-performance single for an extended period. Oh, I know there are some large academies that want you to train for the commercial in a low-performer. What a bore. This is such a waste as some flying club down the ramp more than likely has a high-performance aircraft available for the same price as a basic trainer would cost from the inflated prices of the academy. Commercial pilots should be fully competent and experienced in high performance aircraft. You have to decide if where you work is also where your students will get the best training. There are virtually no low-performance aircraft on the job, so why allow any training in them for the rating?

Think about all the maneuvers the FAA mandates that you teach. There are chandelles and lazy-8's so that you may "dance" about the sky. There are steep spirals to bring you back to earth. You have 8's on pylons to prepare you for air races. I always like 8's on pylons. I remember the simple joys one day of whipping an A-36 Bonanza around a couple of pylons while flying between 600 feet and 1000 feet above the ground. Not only is it legal, it is required. The commercial: the VFR rating in an IFR world — how appropriate.

Once you get over the giggles of actually making money wasting a student's time teaching him maneuvers that are of no use, for a rating that gives him a privilege he can't use either, you have to take this opportunity with the student to give him at least something of value. The best thing you can do is to change his thinking from the amateur private pilot to the efficient, money-making commercial pilot. Money-making for the boss that is; commercial pilots don't make any money.

Since you have no training in this yourself, it could be difficult. If you have no idea how commercial pilots operate, then go to a commercial chief pilot and find out what are the common problems with their newly hired pilots. Armed with this knowledge, you can turn out commercial pilots who might actually be of some use as commercial pilots one day. The whole strategy of commercial aviation is to get the job done. Move the payload and get the pay. This translates into a completely different training environment for the student, but it will help him in the long run. Make every operation the student engages in as efficient and utilitarian as possible.

Private pilots take forever to get off the ground. This is fine for them. They don't have a schedule save for what they impose on themselves, so they can take all day. A commercial pilot, however, always has a boss. Whether it is the manager, the customer, or the passenger, the commercial pilot always works for someone else. That responsibility comes with the duty to be competent and efficient. A good commercial pilot with a decent checklist can accomplish more operations, far safer, in much less time than your basic private pilot. When you have a commercial candidate who can get far more done without physically moving any faster, you will have engaged in genuine commercial training. Of course, nothing like this is ever mentioned in the *Standards*.

Commercial pilots in almost all cases have to immediately go on to become instructors so that they can get the hours to become real commercial pilots. Therefore, the first responsibility they will have in their aircraft will be student pilots. They certainly won't be carrying passengers or cargo anytime soon. You have to help them gain the confidence to live up to their approaching responsibility. We don't want any weenie-babies up there operating for hire, so your commercial candidates should practice in the most adverse weather they can safely handle. Then you should move them up to weather and situations they cannot handle, just to show them their limitations.

Of course, you had better be able to handle the situation when your student gives up. My experience tells me that even commercial students are willing to give up much too easily. They are surprised to find that with some practice they can deal with crosswinds they never before dreamed that they could negotiate. Should you be up on a particularly blustery day, just perform all of your normal maneuvers. The added turbulence makes things that much more interesting. Besides, it is a really good indicator of your students' competency if they reduce to maneuvering speed without any prompting. Most won't however; they know they should but they still won't unless you tell them it's okay.

Many students prefer to stay on the ground if they think it may be bumpy, so they never get to see a situation where it is necessary to slow down, nor do they learn how slowing aids the aircraft in its ability to navigate rough conditions. Anyone engaged in aviation in a regular business can't avoid running into nasty stuff some day. It is the responsibility of the commercial pilot to do whatever is necessary to insure the safety of the flight. The best thing that can be done for commercial candidates is to give them the experience ahead of time to deal with rough or adverse weather before they have that responsibility.

When your students are comfortable with rough

weather, they will be able to keep passengers and their own students comfortable — well, somewhat. At least they won't transfer unnecessary fear. The public expects nothing but rock solid confidence from their commercial pilots. I still have no idea what the FAA expects.

Whereas the private students know they can't fly, the commercial students may actually think that they can. Something happens in all those intervening hours after the private flight test to pilots who aren't in the stranglehold of an ongoing, highly structured flight program. They gradually convince themselves that their own improvised short cuts and style differences are revolutions in flight technique. Uh-huh.

Some pilots love their autopilot. The classic case is the lawyer/doctor type who has more money and airplane than ability, so he uses the autopilot from rotation to the flare. They have forgotten how to fly. It is amazing to see what people come to believe and what they practice the further they get from regular instruction. You would think the biennial flight review (BFR) would catch this stuff. However, there are just enough good old types who are satisfied with bopping once around the pattern and calling it a BFR.

The private student thinks you are God; the commercial student requires more convincing. All this is leading to the fact that you will get a lot more arguments from commercial students as you reteach them how to fly. Just like your instrument students, your commercial students will suffer from problems in their basic flying that you must overcome before getting into the advanced maneuvers.

Pilots who were formerly enrolled in accelerated flight programs under Part 141 have problems of their own. They will worship their rote procedures knowing full well that the training wasn't adequate, which is why they left, but will fall back on those same methods because that is all they know. Flexibility and imagination will be alien concepts that will require much work to develop.

If you are working at an academy, you will have become aware rather quickly of that school's institutionalized myths and mistakes, such that you will be prepared for commercial students as they work their way through the program. Of course, if you have a job where you trained, you will do nothing but perpetuate those myths and mistakes because you can't even recognize them. The big problem with these schools are some of the designated examiners, who are virtual employees, who put private students through with marginal skills because they know full well that the student is never going to get to use their certificate. That student is going right back to dual lessons for the commercial before the ink is even dry. Pilots can't get into too much trouble when they are babysat by the academy.

Students who learn to fly out in the clubs and FBOs receive far more scrutiny from examiners on checkrides because they may fly for a lifetime on that one flight check. Consequently, you should have an easier time training students who have recently gone through regular Part 61 programs. Wherever your students come from, plan on lots of remedial instruction.

We teach private pilots how to use the national airspace system and the FARs so they don't get into trouble or take chances. Actually we teach them this stuff so they don't get us in trouble. Don't believe me? I'll bet if you have a student you recently graduated come back to you reporting that he just clipped the TCA (Class B), your first thought is that you hope the FAA does not want to talk to you . . . See what I mean. Anyway, that is for private pilots.

When we teach commercial pilots the system, we should teach them how to use it to their best advantage to save the boss money. This is where to explore options like the departure scenario: It is marginal VFR; do you go out IFR or take a chance on special VFR? The thing to do next is to take your students out flying on a special VFR departure just to show them why they shouldn't make this a habit. How about VFR through a TCA? Commercial candidates should be well-versed in negotiating the best course of action from ATC. Get them to request a clearance into the TCA and make sure they can do it. It is not enough for commercial pilots to simply know the system; they have to know how to work the system.

Commercial students should be pushed to the maximum to prove that their own self-imposed limitations are well below their potential capabilities. Your job is to take them beyond where they thought they could go. Pilots always restrict their field of options and possibilities to the comfortable and familiar world of what they have been previously shown. That is except for the dare-devil types who fancy themselves test pilots. These pilots have other problems.

Generally, I think pilots get into far more trouble for what they don't know, don't do, are unwilling to do, or have not thought to do because it is something new to them rather than from any course of action they take that turns out to be a mistake. When pilots are paid to be responsible, they may have to reach beyond what is known to reach the solution which to them is previously unknown. This is the realm of the commercial pilot. They have to have the maximum capability to insure the maximum safety.

When you take a commercial student out for short-field work, find the shortest, narrowest field with the biggest trees around it from which you can safely operate. Many pilots never work around real trees, preferring instead to imagine the FAA pygmy fir which only grows to 50 feet. These are the same pilots who also use simulators. When you have to work on crosswind landings and takeoffs, wait for a real doosey of a wind. Every commercial pilot should be able to land, roll down the runway, and take off again, all on one main wheel. They should also be well-versed in slipping into a massive wind while making near vertical approaches.

For the ultimate challenge, there is a little thing I call "the overload test." Sometime during a lesson, you have to pull the landing gear circuit breaker without the student seeing it. Even if they do see it, just wait half an hour and they will forget. Next you simulate an engine failure. While the student is getting organized, ask a whole bunch of distracting questions — things a passenger might ask the pilot when faced with a real engine failure like "Are we going to die?"

If the student is on the ball, he will glide to the airport you hid under the right wing while simultaneously trying to follow the emergency checklist. The gear warning horn will be blaring because you will be below the safety manifold setting, but your student won't hear it. "Gear, gear, gear," it keeps repeating; however, the warning doesn't register. Your student gets used to the noise after a while, considering it just an annoyance. Because of this, when the time comes to lower the gear, the horn will have no effect in telling the pilot that the gear is not down. Most students proudly line up on their key position, do their final simulated shut down checks, lower the gear handle, and completely miss the fact that the circuit breaker is out and the gear hasn't moved. This is so much fun.

It is also one of the most memorable and valuable lessons you can give your students on the hazards of distractions and overload. Anyway, there you are on downwind coming abeam your touchdown spot.

"Everything okay for landing?" you ask casually.

"Uh, unlatch the doors and guard your face," is the usual reply.

Well, you gave them a chance. As soon as you turn base while still at a safe altitude, ask them if they intend to belly a perfectly good airplane in on a perfectly good runway. While they are recovering from the shock of feeling so incredibly stupid, get set for a go-around. Don't try to drop the gear at the last minute and land or you might create a real emergency while trying to simulate one.

There is a certain amount of risk to this lesson in that you both may forget the gear — another reason for the planned go-around. This can happen because you can get just as used to the sound of the warning horn as your student. After a while, it may not register with you either. I never attached any significance to the horn when this was pulled on me because I was so busy. So when the time came, I lowered the gear and just assumed it would come down. None of my students have heard the horn after becoming overloaded and they never checked the gear either — but we all remember the lesson.

Ever since my instructor pulled this ritual harassment on me, whenever there was the chance of feeling or becoming overloaded, I have gone right to the established procedures and checklists to maintain my focus. It has worked out fine so far. Many times the most valuable lessons cannot be taught without some risk. You have to decide if it is worth it. We treat commercial students like glorified private students. The *Test Standards* give you little other choice. What you have to do is turn out good commercial pilots in spite of the constraints and standards of the certificate.

Would you like to be busy? Would you like to be incredibly busy? Try teaching single engine instrument procedures in a 310, including circling approaches, in a pattern crowded with single engine trainers. That — is busy! It is also one hell of a lot of fun.

Multi students come with the biggest egos in flight training because they will soon be able to fly "real" airplanes. Twins are for the gods; singles are for weenies. Twins spoil people because they never want to go back to singles, kind of like the Escort driver who rents a Ferrari for the day. It's fun while it lasts, until the costs start mounting up. Twin students suffer a short but bad case of the gollygees.

The problem with a twin is that you have an airplane that is bigger, faster, heavier, and can ruin your whole day in an instant if something goes wrong. Be especially careful your first couple of flights while your student is gawking out the window, marveling at seeing engines on the wings, and feeling the power available from two throttles. The worst thing you can do with a multi student is give in to the pressure to start chopping engines before the student has any grasp of normal operations.

Your student will pressure you in the same way as the new instrument student who wants to leap right into approaches. Unless you are a freelance instructor or have great freedom in your instruction, most multi programs will force you to give the barest introduction to normal procedures while you dwell predominantly on the emergency stuff. Consequently, your students will have precious little knowledge or experience in standard operations, which will be their weakest area. The basic twin rating is, therefore, only an emergency rating.

Never underestimate the power of ego in the new multi student. There is something about a twin that changes pilots. It is hard to say which is more fun: flying them or telling your peers you are flying them. These aircraft give the prestige that every student seeks. This can lead to some incredible overconfidence. As a wise instructor once said to me, "It's just an airplane." Meaning, all the physical laws still apply. You have to show your students that twins are just airplanes and they are just multi students. Maybe they will be more receptive to learning when they aren't tripping over their own glory.

One thing you can do is try to give the widest amount of experience in the short time that you have your students. Insist on a couple of night lessons. We tend to teach at night only when it is required. However, flying twins at night, especially if the aircraft has extensive light and other complex systems, will be different. How many instructors have ever had a student practice a single engine approach at night?

Twins are different than singles for obvious reasons, but they are still airplanes. Part of wide experience involves planning and flying longer distances at higher altitudes. I learned long after I was multi rated how to fly out of the practice area by taking various trips with students who wanted to learn how to maximize their twins when flying cross-country. You will have to take the initiative if your students are going to learn from hands-on experience about things such as fuel management, extended cross-country procedures, night flying, and the in-depth understanding of multi-engined aircraft that gets lost because students can't wait to lose an engine.

The transition up to multi-engined airplanes is going to be a big jump no matter what the aircraft. I sought out a school with a 310 because I wanted experience in that class of aircraft. The students who came to us usually had previous multi experience, or they had a fair amount of heavy high-performance single time, so the jump to the 310 was comparable to the low-performance pilot moving up to say a Duchess or Seminole. Whatever the twin, there is going to be a transition or adaption period to the new environment. What you should do is try to keep things as familiar to the student as you can during this time.

For example, if you fly wider patterns, even though the aircraft is faster than the students are used to, the actual time spent flying each leg can be the same as a slower aircraft in a tighter pattern. This way they will have the same amount of time to accomplish their tasks. Take the time to get the normal procedures firmly planted before the excitement really begins.

Multi students are long on demands and short on cash. These folks have already invested heavily in their future career and they know how the game is played. They can't bargain too much because they still need the training. The problem is the exorbitant (or as some would say extortion) rate of twin instruction. Commercial operators use twins because they are profitable, and because some operations require twin-engined aircraft. The advantages of speed, rate of climb, and load capacity are completely lost on instruction because you don't go anywhere or move anything. Therefore, the full cost of operation falls on the student.

Your job then is to shorten the training time as much as possible. Since normal operations are not a significant part of the checkride, that is what will usually be left out as you try to save the student money. Because of the expense, multi students tend to shop around. The exceptions are those students locked into a Part 141 program. Unfortunately, they are the ones who end up paying the most money for the least aircraft.

The acquisition of multi time can be an instructor's

prime preoccupation. The lucky folks who teach in big schools will be handed lots of multi students; however, the regimentation of the program limits the potential experience. The rest of you will probably get less time, but it will be in a wider variety of aircraft and situations.

In our area a small industry has spawned to serve the market-wise multi student. Many instructors have collaborated to buy twins, usually Apaches, and advertise training at rock bottom prices. The student gets a good price and the instructor gets multi time. I worked at a school that offered a 310 for a price comparable to many Duchesses in the area, so we attracted a fair amount of business ourselves. It's all a racket.

Twin students are such a short-time proposition because the ratings aren't that involved and because the students know how to save time by doing the preparatory work on their own. You have to attract enough students to give you the time required by commercial operators and the airlines. One great place to work to build multi time (and a difficult racket to compete against) are the weekend wonder programs with examiners on staff that guarantee multi certificates at "going out of business" prices.

I have some questions about these programs. If you cram all your training into one weekend, how much of that can you retain? Sure you have a rating, but are you really a confident multi pilot? What have you earned if the rating is guaranteed? Can you teach students with a multi-instructor's rating you acquired in a couple of days?

Whatever racket you participate in to get multi students, someone else will be competing with a racket of his own. I had one potential student even have to gall to tell me that I should train him, but he would refuse to pay me anything because I needed the multi time more than he did. What a jerk! I felt ill. There are instructors who give away time just to get those multi hours. They are that critical, and there is no way for you to compete against that because all instructors have the same pressure to get the time.

What you might think of doing is cutting some deals of

your own. Trade a big trip for some training. I know instructors (including me) who have gotten to take very long trips with some special students. You have to be resourceful because multi time is precious and difficult to come by.

Just as it takes confidence to take students into the clouds, it takes confidence to practice single engine procedures with students. Once again, you never know exactly what they will do, so there is a risk. One thing I noticed after all my multi training was that I had never flown a twin by myself. I was shocked to find that many schools do not even let their instructors take a twin out solo. As if practicing with a student is somehow safer. You have to find a way, if you can, to go out solo in a twin before you take on students. This is the best thing you can do for your twin confidence.

My policy is that if I can't do it, I can't expect anyone else to either. That meant treating myself by taking out the old 310 for a solo practice flight. If typical, like in my case, you may only be able to afford such a luxury just once. It will however be worth it.

Talk about concentration. I had to try single engine procedures and landings just to see if I could do them on my own. How many gear checks can you do when things get busy? Try a solo multi flight and you will find out. It won't be as stressful as working with a student because you are in control of everything and you get to fly to your own standards. At least you try to anyway.

There is no doubt in my mind that multi training is dangerous. Single engine procedures are about as close to actual emergencies as you want to get on a regular basis. The dichotomy of twin training is that what works in your favor to enhance safety goes against the quality of training for the student. The type of airplane also makes a big difference as everything is a tradeoff.

The 310 can give you wonderful experience in a commercially viable airplane. However, the single engine stuff can get pretty exciting on a hot day or in a crowded traffic pattern. It has a big rudder with lots of travel so Vmc seems low for an airplane of its size.

The popular training twins, the Duchess and Seminole, are less powerful and less likely to get away from you, but no one flies them commercially. They are, however, a great place to start twin training.

The Apache has probably the worst single engine climb performance, but that big wing which generates lots of lift because of a lower loading than most twins, makes it perhaps the safest for single engine landings. One way to land an Apache on one engine is to fly abeam your touchdown spot, idle the good engine, and just glide on in. That way you have no asymmetrical thrust.

The Twin Comanche will give you the best performance for the price, but takeoffs and landings can be tricky. It can also be squirmy on one engine.

You will have observations like these for the twins you fly. With every twin there will be constant tradeoffs of expense, safety, performance, and applicability to commercial operations. For each airplane you fly, you also have to decide how far you can let students go before you take over.

Multi training is dangerous, but there is no other way to prepare students for real emergencies. You will develop your fastest reflexes when you teach in twins. If you have to take the airplane, you will have to take it *right now*! The obvious thing to watch is getting low and slow, especially on one engine. You can get into a real pickle here, and it gets worse the higher the wing loading, and therefore the greater the reliance on power to manage the aircraft.

Takeoffs are trouble also. We make elaborate plans for the first takeoff because we have all been impressed with the dangers in twins. However after that, every subsequent takeoff is given the same consideration as if you were flying your basic single.

Multi instructors usually neglect such emergency planning whenever they get into touch and go's. Your student certainly won't have an emergency plan, so it is up to you to always have the scenario worked out in your head to handle any emergency any time.

The faster the twin, the larger the speed gap with

other traffic, the more hectic will be your pattern work. The training twins fit pretty well into most airport situations because they operate comparably to a high-performance single. Faster twins will require advanced planning and juggling on your part. Another advantage of wider patterns is that you are just out of the majority traffic flow. Should you become distracted even for an instant with something inside the aircraft while flying around other traffic, your student may try to kill you with a mid-air, so it helps to have most traffic out of your way.

Don't expect any special consideration from a tower as you practice the single engine stuff. They have an airport to run, so unless you have a real emergency you are no more special than any other aircraft. This distorts emergency training because if you really lose an engine, you get to clear the airport. In normal lesson practice however, you have to work with that Piper Cub on the long final.

Oh, examiners really like to have students tell them that they would immediately declare an emergency when operating on one engine and would request the equipment standing by. This is what you should do anyway; it's just that examiners also like to hear it.

The last common pitfall is the crosswind landing. Multi students somehow expect something different from an aircraft that has all the same flight controls, operates under the same laws of nature, and flies from all the same airports where they have spent all their training. They don't know what that new thing is, and you can't read their minds to know that they don't know, so they end up waiting for some divine inspiration while not doing any crosswind correction at all. These students are surprised to find that the old sideslip method works just fine. Multi students are very leery of propeller strikes and are therefore reluctant to try to land on one wheel. However, once you give them permission to do so, they feel really cool wheeling in on one in front of their peers at the flight school.

Why students have to be able to demonstrate Vmc situations to examiners is beyond me. They should of course be aware of them so they will know what they feel like so they are able to recognize, avoid, and if necessary, recover from them. My question is why should new multi pilots have to know how best to get into them? The peculiarity of practicing Vmc is that many manufacturers and flight schools go to all the trouble of carefully warning you of the dangers of practicing Vmc at density altitudes where the reduced power available lowers the Vmc speed near, at or below the stall speed, and then mandates some arbitrary altitude like 5000 feet for all your Vmc demonstrations. This altitude under the right conditions could easily be in the realm of where the Vmc and stall speed meet. Recovery from Vmc is relatively easy; however, you may not recover if it is accompanied by a stall.

Your school may be requiring you to put yourself in danger when what they ought to do is let you pick the altitude where you can safely avoid the ground but still have an adequate safety margin between the Vmc and stall speeds. For example, on hot days 3500 feet might serve all your purposes much better than would altitudes in excess of 5000 feet. When operating at higher altitudes, you might be advised to limit the rudder travel to see the effect of Vmc earlier. Unfortunately, this gets students in the habit of not using full rudder, which could be disastrous if the situation arises when they actually need it.

There are very few situations in all of flight training when full control input is ever used. Students are reluctant to do anything for which they have not been given express permission by the instructor in the airplane, even when they know from ground training it is the right thing to do; so anything less than the actual emergency procedure is a waste of time. Therefore, requiring high altitudes and blocking the rudders defeats the whole value of Vmc demonstrations. Different twins have different Vmc characteristics, so you have to be intimately aware of when the aircraft in which you are training people is going to break from the heading.

The other problem with arbitrary altitudes is that it fails to take into account the unpredictable student who may blow the recovery, which is exactly what you do not want to have happen anywhere near the stall speed. I had a student who not only forgot to reduce the power during a Vmc recovery, but also applied full rudder into the idling engine. This is where I discovered what real Vmc looks like. Recovery from this unusual attitude in a 310 requires about 1500 feet of full opposite rudder, idle throttles, and a window filled with the earth below. Thankfully, Cessna put big slabs of rudder on their 310's, and we were low enough and fast enough to avoid any stall tendency.

I wouldn't want to hop into an airplane with the intention of trying any kind of instrument work in a twin until the students had memorized all the pitch and power stuff, all the airspeeds (both multi and emergency), for both stabilized and nonstabilized approaches. There is just too much happening, too fast, for them to try to learn as they go, and you will end up flying the approaches yourself. Fun though it may be, this isn't the goal of instructing.

I had a multi-instrument student show up for some training and a checkout. I followed the usual practice of giving him the aircraft manual and our training manual. Then I told him to go rehearse for a week in the airplane on the ground and next weekend we would fly.

Students seldom disappoint your lowest expectations, and this one was no exception. All the information I had supplied was given a quick, cursory glance the morning of our flight lesson. Someday we may live in a world where instructors make enough money to send students home to do their work so they don't waste either their time or yours. However, in this world I had a student with money who wanted to fly the twin, so despite my earlier qualification, of course we flew.

I, however, knew what was coming next. It was a cool morning with scattered showers in overcast over the entire state. The ceiling was ragged at 2000 feet with rain moving through all the time. I love this stuff. It is funny how students react with hesitation to flying in the rain where visibilities can easily be five to ten miles, but have no qualms about flying in three to five miles visibility in haze and fog. We took off and quickly streaked into the overcast.

This student was amazed at just how fast things could fall apart in a twin. Sometimes a lesson which teaches nothing but the dangers of unpreparedness can be most valuable, expensive though it is. After the initial daze wore off, he gradually began taking over more and more of the flying. I had to fly though every so often so he could catch up. This was purely a normal flight with normal procedures. I don't do simulated emergency stuff in the clouds. What is clearly evident is that just the normal IFR operations can frazzle the cocky student who thought he was above studying.

This flight was a great experience for the student because it became a flight of discovery. This is what I call a lesson where students aren't so much expected to know or be able to do anything, but where they see how the system works and what is possible from their aircraft. I later found out that this was his first experience in the clouds with a twin and his first operations from wet, slippery runways.

I really believe in discovery flights. Students need to be shown for what they will be capable with practice. Flying in the sunshine under a hood won't do that. Landing on windswept runways, even if the students need help, will teach them more than simply talking about hydroplaning in a warm and dry flight school. Along the way though in flights of discovery, you may have to correct for localizers you are blowing past, altitudes that you may have missed, and runways that you approach at some rather peculiar angles. Think of it as just another day at the office.

One of your riskier endeavors in instructing is the checkout. You have to deal with a pilot who has one singular objective: to get his hands on your airplanes as quickly and cheaply as you will let him. Some pilots, even though they have never flown the model of airplane, insist on one bump around the pattern and then the keys. In my experience, very few pilots are qualified to fly an unfamiliar aircraft in just one lesson. Two lessons, though, work out quite well.

The first lesson is for seeing what everything looks like and the second is for reinforcement. This works best when there is some time for reflection between the flights. It is the second lesson that gives students the ability to retain so much more information than they could possibly hope from just one flight. Although many pilots piss and moan when you suggest that two flights might be necessary, their complaints are replaced with grudging acceptance of the wisdom of two flights, and they are eventually grateful for the suggestion when they see how much better they fly the second time out. There is much more to operating an airplane safely than just flying the airplane, yet this seldom gets investigated in your basic checkout, which amounts to little more than "stalls, steep and slow, touch and go, here are the keys, look out below."

Do you investigate what a pilot intends to do with your airplane? If he desires to take a big trip over huge variations in terrain, do you gear the checkout accordingly? Checking a pilot's ability to fly slowly tells you nothing about his ability to navigate. Besides, pilots don't rent airplanes in order to fly them slowly. Most checkouts become a rote formality; however, you are still responsible for whatever that pilot does when he flies off into the wild blue with your signature.

For pilots planning on big trips, I like to have them plan me a cross-country for discussion. I want to know how well they know the charts, what they can tell me about maximizing aircraft performance, how best to take care of the aircraft, and how well they understand weight and balance. Checkouts are notorious for being given in lightly loaded aircraft to pilots who show up later with the family and baggage.

Are your pilots night current and do they intend to fly at night? Part of any checkout is checking their experience. If this is a deficiency, then just make one of their two flights at night.

Some checkouts (like mountain ones) are mandated by insurance requirements, so there is not a lot of flexibility and discretion that you may exercise. However, a general checkout is solely up to your judgment. Try not to get locked into a rigid pattern — you may miss something important. If you are unfamiliar with the pilot, there are subtle things for which you should look. Carefully observe a pilot's attitude for it reveals things he may be trying to hide from you.

I have a theory that the pilot's ability to handle the airplane safely is directly related to his level of cooperation. This means that the brutes who bully you into the shortest checkout are the most likely to get you in trouble. You should always check a pilot's certificates.

Some time ago I made it a policy to always have any instrument-rated pilot shoot an approach as part of the checkout, whether he intended to fly IFR or not. How do you know what he will do once he leaves town? Anyway, it has revealed some startling things compared to what pilots have told me of their qualifications and abilities. How much can you learn from a pilot if all you check are his presolo maneuvers? If your school has one of those pat checkout forms, modify it. Most forms like that always include things like soft-field work that the school or club prohibits anyway; so what is the point? Tailor all your checkouts to the pilots and their intentions.

How do you feel about passengers riding along during checkouts? Many times pilots will show up with friends who want to come along for the ride. These folks may never have been up in an airplane before. Any surprise passenger is going to put undue pressure on you to keep the flight straight and level. You won't check the things you should because you are worried about the passengers; yet it is the safety of those passengers which demands a thorough checkout. I learned to warn all passengers who were coming along for what they thought was some scenic tour flight.

"This is not a joyride; it is a training flight. We will be performing maneuvers not necessary for normal flight at the maximum capability of this airplane. If you want to come along, you are welcome; but I won't give you any special consideration, nor will I change my lesson one bit because you are here."

That generally gets their attention and sends them to the flight school for a long, leisurely cup of coffee. The ones who stay for the flight have a great time because they accept what they are getting into. Strangely, the pilot getting the checkout is more relieved to have the passengers they brought with them sit on the ground. They don't want to look bad in front of a future passenger. Looking bad in front of an instructor is normal. The question comes up as to why the passenger was invited in the first place? I think people make offers they hope won't get taken up, like when you offer to help because it is the right thing to do, hoping to be told that no help is needed, but you really want people to know you asked.

I remember a Centurion checkout where the pilot showed up with two colleagues in business suits. I warned them, but they insisted on coming along anyway. All you can do is warn people because the person paying the bill may want to show off, with an instructor safely along just in case. Anyway, it was a steaming hot day and the deadweight behind us was complaining constantly. This was really unfair to my student who was doing his best to fly an aircraft that is quite a handful your first time. After I had heard enough bellyaching, I popped the side window flooding their faces with fresh air.

"Is that better now?" I shouted.

Well, no, not really for it made them very nervous, but it did shut them up. The next flight my student came alone and we had a most relaxing and productive flight.

There are certain characteristics about pilots that are predictable. Airline pilots getting checked out in general aviation craft will want to know all the numbers before the flight. They actually use all the checklists regularly. They can hold altitudes very well and are very close on the airspeeds except that they get uncomfortable flying very slowly if they are not used to it. Depending on the equipment they use on the job, you will find them flaring anywhere from 50 to 100 feet off the deck. They also lead turns way too much. Keep in touch with these folks because they are full of great information and they may help your career.

You may get to fly with jet fighter jocks. These folks haven't a clue what normal maneuvering is all about. They are great fun to fly with because nothing bothers them. Much of their flying is visual, so the instrument skills could be rusty. I remember checking out an A-6 pilot in a Cutlass. The biggest problem he had was with the mixture and prop control. In his jet, one lever fits all. We ended up discussing the finer points of Cutlass flying during extended 60-degree banks at various pitch angles to the horizon.

If you want to learn from the gods themselves, see if you can check out an old master pilot sometime. I remember early in my instructing one of my students brought his grandfather for a checkout. I think he had earned one of the first ATP certificates given out. This person hadn't the slightest idea what the speeds were for the aircraft, and it didn't make the slightest difference. He flew on pure gut instinct better than anyone I have seen who was intimately familiar with the type. What a learning experience. Who was checking out whom? What a great attitude as well. He obviously didn't need my opinion of his flying yet insisted on the same checkout as everybody else.

All your checkouts will be different. Here are some hints that might give you insight into someone with whom you are about to share an aircraft. People with a lot of checkouts and only a few flights per school are making the rounds for some reason that they will do their best to hide from you, so be careful. People with cooperative and friendly attitudes and lots of experience in the area will be a lark. People who insist on telling you how wonderful they are at flying and how quickly they can prove it are usually the most dangerous. Be especially careful of bifocal wearers because sometimes they look through the wrong lens and lose all perception of when the ground will meet the airplane. Watch out for sloppy preflights because they are often followed by sloppy flying. Pilots who do not ask you questions think they know it all. People who claim poverty and therefore must have the minimum checkout are trying to get away with something that could put you in jeopardy. How could they rent the airplane after the checkout if they were that poor? You will see all this and more yourself; however, forewarned is forearmed.

What is the biennial flight review? That is where you sign over two years of liability to pilots you may never see again, but are nonetheless responsible for because you certified their competence. I asked my first chief instructor how to do one of these things because I had never given a BFR, nor had the need to get one for that matter because I acquired new ratings that precluded the requirement for a BFR.

"Get that pilot up to the level of his highest certificate where he could pass the checkride again," was the response from the chief.

That has been my standard ever since. As I have said earlier, those words should be written right into the FARs.

The biennial is whatever you want it to be because you get to decide when the pilots are competent to exercise the privileges of their certificates. The people who come to you for BFRs will use this flexibility against you. They will bring as much economic pressure as they can to keep their own costs down. They do this by forcing you to keep the review of their skills to a minimum in return for getting their business. If this is their attitude towards flying, you will probably find some interesting gaps in their skill and knowledge. What you find is that the folks who want the shortest review will need the most work. They have nothing to lose by persuading you to give them a quick signature. However, your liability with the FAA and any lawsuit from their families should something happen remains the same, so there is no excuse for anything less than a thorough BFR from you. The whole idea of the checkride standard is actually a protection for instructors, so that you would have that stringent written standard to back you up when pressured by a customer.

The other problem with pilots who seek the cheap and easy BFR is that the conscientious instructor will lose business to the cheap and easy instructor. The system actually rewards instructors who engage in leniency and slipshod reviews with more business. You have heard the old ploy of "You should go to so-and-so — they're easy." The pilots who go for the thorough flight review usually don't need it because they are always keeping up with their recurrency training. So here we have a review system that rewards the shoddy instructor and lets the bad pilot slip away with the minimum review.

The other problem with BFRs is that you are certifying pilots for every operation for which they are allowed to engage whether you investigate it or not. This is too much responsibility. If you have any doubt about a pilot, or don't care for his attitude, simply decline to give him a flight review. One bad review can ruin your whole future in this business.

One day this bozo stormed into my flight school demanding a flight review. He held commercial-multi-engine-instrument and helicopter ratings — or so he said. The next demand was that he would only fly in a Cessna 150 because he didn't want to waste his money. I had to laugh. You can't take responsibility for a pilot who won't let you check him out, yet that is what he wanted. He was loud about it, too. Everyone in my flight school heard me get accused of price gouging when I refused to even consider a flight in anything less than a high-performance retractable, and even then I would limit my BFR endorsement to single engine airplanes. I later found out you can't do that as BFRs are unconditional. I often wonder where such people go when they stomp out with such angry, red faces? The best recourse for any problem is to just say no. If you ever sign off a BFR that you later regret, you have signed yourself up for two years of worry.

I used to have this big, elaborate checklist full of rules and procedures that I made up for my BFRs. It covered lots of stuff but was way too rigid. I was reluctant to give it up because it had taken so much work. It is funny how we are resistant to change, preferring instead to hang on to familiar yet obsolete things. Anyway, I soon found it more advantageous to tailor the BFR to the students based on their qualifications and the type of flying they do. Now what I do is just pull the *Practical Test Standards* for their highest rating and off we go.

Many pilots use the BFR as a way to get back into flying after a long absence. Such a flight review becomes more like a small pilot course, possibly covering weeks of training. What a deal. Sometimes you get a pilot who has been out for ten years or longer. The world has changed completely since he last flew so he will have to learn virtually from scratch. What about the new airspace? That will be a real surprise for someone who has been sailing or doing something equally boring for the last decade.

Anytime you have the chance to work with pilots for longer than one day, you have the opportunity to do them some good by taking them through anything they have been reluctant to try. Widen their horizons by flying in whatever is unfamiliar to them; in congested airspace or empty airspace, big terminals or small fields, flying at night or in marginal conditions. See what they need and boost their weak areas. Pilots have avoided some particular things for years just because no instructor took the time to show them how. The best examples of this are pilots who bypass ARSAs and TCAs (Classes C and B). You have a wonderful opportunity to teach on the BFR, not just confirm what they already know.

There are a lot of pilots with instrument ratings who are afraid to go into blustery clouds because they trained with the damn hood and simulator. Make an effort to give them the experience they should have received during their training. Maybe you will have a flight that opens up a whole new world to them. Maybe they will think of you when they need more training.

Should your students feel you are taking advantage of them because the training is beyond their overly optimistic goals, then send them up with another instructor for evaluation. One conflict that arises on BFRs is when the students think you are wasting their time and you think they aren't safe to fly on their own yet. One of my first BFRs was like that. The pilot was a nice person but he had an incredibly arrogant streak. His idea of good enough was definitely not good enough. I suggested another instructor as we were starting to degenerate into a personality conflict. I could not sign him off because he was not safe, but I knew I was not the instructor to help him because of the arguments he had with my review and training. I thought he was just dangerous in an airplane.

My fellow instructor did not and signed him off after a couple of flights. This student resented me because he thought I was gouging him. However, I watched with horror at the precarious landings that were near crashes in a crosswind, and the abuse of power in taxiing that put undue stress on the aircraft. I heard the delays in ATC communications resulting from his very imprecise language. The worst part was his general contempt for the system and his feeling of superiority. Nothing serious has happened yet, and it has been years since then. Was I right to be so critical? I don't know. I think so. He did not and promptly gave all future business to my more lenient colleague. All you can do in this business is call them as you see them.

The BFR is a time bomb with a very long delay because you may be the only instructor certifying a pilot's competency for two years. Well, a lot can happen in two years. You will always be faced with continuing economic pressure, student pressure, and your legal responsibility, all of which are in direct conflict. Who said life was fair? With a student going for a certificate, at least you have the *Standards*, a second opinion, and an examiner with which to share the responsibility and the blame. You are on your own with a BFR. Most instructors will be lucky and nothing serious will happen to the folks you sign off. Should any incident occur, the last instructor to fly with that pilot could be questioned. Try not to let that instructor be you. Kind of like Russian Roulette, isn't it?

18 • BURNOUT

———————————————————————————————————➤

"Weenie-Baby-Nosewipes!" That's how I referred to my students. I hated their sniveling little faces. I despised their constant drivel of whiny, whimpering, petty excuses for every personal failing. I dreaded my lessons as they became torture sessions inundated with the same, stupid, tired old mistakes that were repeated over, and over, and over again. I had no patience yet I had to have patience. I was a professional. I had no enthusiasm save that which could be superficially mustered. The brutal honesty of my plight was so carefully buried just below a facade, where a rush of fury waited for any student provocation to lash out at the continual incompetence. This is the face of burnout, and it deserves a full and complete investigation by the FAA.

The sole motivating force to continue teaching becomes the excruciating fear that this is my only marketable skill. Since starving is not an option, my torment and misery would remain my food and rent money. The guilt associated with such hostility derives from the realization that getting paid to sit in airplanes all day is a privilege, which only complicates the emotional turmoil. The untouchable glimmer of hope that someday I would be rescued from all this and once again permitted the sheer joy of piloting my own craft became my personal salvation. Depression, always lurking in the shadows, became an ever more familiar companion as the economy waned, airlines folded, and movement out of the pilot positions I so desperately desired slowed to a trickle. How long would I be forced to endure student pilots? How long would I have to watch others relish the intense passion of flight while I had to be content to vicariously and passively observe from the right seat?

Despite all this drama, I want you to know that without burnout this book never would have been written. Burnout permeates every chapter. Burnout creates energy. Burnout demands change. Burnout is the dark side of instructing that gets pushed under the carpet, ever simmering, waiting to be exposed by some cheap, tabloid talk show. Burnout can, and probably will, affect all instructors at some point in their careers. The classic case will be the full-time instructor who has taught more than a year and accumulated more than 1000 hours of dual given and has had enough.

What do you expect from a system which mandates that commercial pilot aspirants sit for endless hours observing maneuvers irrelevant to their future careers for years at a time? Instruction is the FAA's version of conscription and burnout is the result. No one will ever convince me that spending two or more years of your life talking other people through procedures you can recite in your sleep does anything to create a qualified commercial pilot. After that much time teaching, the typical instructor requires remedial practice just to retrieve his former flight skills.

Does the FAA offer a workshop for recovering instructors? Is there a 12-step program for curing burnout and the animosity toward the harmless student pilot? This is a serious condition that deserves some real attention. Most working people when they burn out get to take a vacation. Flight instructors can't do that. Yet another tragic consequence of a servant's income which allows no time off if one is to keep his head above water. The only full-time instructors I ever saw getting a break were married to a second income. I instructed seven days a week for two years and still lost money. Each successive rating saps several months of income.

We force pilots into an activity raging with internal hatred, yet critical for economic survival and professional advancement. Why? Because flight instructors are cheap sources of labor, and all flight schools have a vested interest in keeping it that way, that's why. Since flight instructor burnout is neither recognized by the FAA nor any health insurance company that I know, there is no help or treatment for your condition, save for a better job. Besides, how many instructors can afford health insurance? You have to accept the fact that there is no cure; you just have to live with it until you can find a way out of instructing.

For those instructors who love the profession and have no idea what I am talking about, I wish you all the best. I hope this never happens to you. This chapter is dedicated to the vast majority of instructors who, if they had their choice, never would have stepped into an airplane with a student.

This condition is inevitable because there is no way to avoid it. With the current requirements for employment, cost of recreational flight, galley slave wages, and temperament of student pilots, it simply cannot be prevented — it can only be dealt with. Recognition is the first step in your recovery. Burnout is such a subtle and beguiling process. Someone else will inform you of its progression before you recognize it yourself. It's like a hearing loss; you can't perceive what you have lost until someone points out what you didn't hear. Burnout is like that. You can't know how much patience you have lost until a student tells you what you used to be like. Burnout in its final form will lead inescapably to a total and complete breakdown of your ability to instruct. By pointing out the early warning signs, I hope you recognize its onset and deal with it in time.

This subject really isn't explored much by instructors in conversation. It is embarrassing to reveal to your colleagues the contempt you feel for the people who are paying for your future career. Professional instructors just aren't supposed to feel those things. I recognized my own burnout long before I ever exposed those feelings to another instructor. Denial, therefore, is one of the early giveaways. You don't really hate your students, do you? It's not really their fault they are so stupid and incompetent, is it? Funny how

they seem so much more incompetent the longer you instruct. Have the students changed? No — you have! The
longer you teach, the greater will be your denial.

What is new to the student you have seen countless
times. The brilliant discovery to them is not even worth your
attention. The more you see, the more jaded you become. The
near mid-air that just took ten years off the life of your
student is just a walk in the park for you. You will joke about
how a collision would just cut into your income by delaying
your next lesson. Denial of something potentially fatal is
not normal — unless you have burned out. I used to tell my
students that I wasn't working hard enough if I didn't come
close to at least one airplane per week. Gallows humor is
the result of denial. You will joke about anything.

You will develop a cruel streak that cuts without mercy.
Listen to what you say as if you had the ears of a student
pilot. What do you think of yourself? Are you teaching, or
simply lashing out? You can have contempt for your students
and still not blame them for their flaws; an accommodating
emotional twist. Since you still need an outlet for your hostile
emotions, sarcasm becomes preferable to screaming. You
should hear it from the receiving end where it can be painful.
Students hate paying good money to be insulted.

You might ask how as a burned-out instructor could I
ever get into an airplane with a student? Part of the trap of
burnout is that there is no choice. Every day all across the
country, there are burned-out instructors hauling themselves into airplanes. In my case, I was completely honest
with my students about the fact that I could not wait for
another job and how I hated seeing all of those ridiculous
maneuvers they had to do. My students took up joking with
me as well and we all had fun. I dealt with my burnout by
seeking out dynamic students who could handle an aggres-
sive learning situation. Timid souls would be eaten alive, so
I left them to the newer instructors. My folks in return got
someone who knew the pitfalls, knew the common mistakes, could lead them directly to the source of their errors
without any bullshit, and greatly reduced the training time

it would have taken a neophyte instructor to fumble through the futile conventional methodology. In return, they had an instructor with less patience, less tolerance of error, and no use for weenie excuses. Everything in life is a tradeoff. This was the price we all had to pay.

Boredom is another dead giveaway that you are headed for burnout. Instructing becomes a formula after a while because everyone has to learn the same, stupid things, and then makes the same, stupid mistakes. I made all of those stupid mistakes myself when I was a student. How else can you recognize them as an instructor? However, any instructor with a good case of burnout can easily forget how they used to fly way back then.

I used to joke that it would be handy to carry a box with buttons on it to play back the most commonly expressed lines used in instructing. Things like "right rudder, use your checklist, the tower just called you," stuff like that. Maybe it could even include a red panic button that yelled "my airplane" at twice the volume. Boring, boring, boring; lessons get so predictably, tediously, painfully, repetitively boring. So what are you going to do?

You might try spicing up the lesson a bit. Take a few more chances. Relax your standards ever so slightly. You may take your students out in weather that is a little more challenging than usual. You may try to operate larger airplanes from ever smaller runways. You might squeeze the performance and density altitude margins of safety to nothing. You might take a simulated emergency situation too close to a real one.

There is all the difference in the world between teaching students the maximum capability of their airplanes and reaching down into their souls to inspire and motivate students to attain the highest level of proficiency and skill, and purposely putting yourself in challenging situations simply to relieve your boredom. You have to analyze your own motivation for therein lies the critical difference. One is to be commended, the other is despicable. That line only exists in the mind; a tricky place to operate. When you can't honestly determine in your own mind how much you are teaching and how much is for your personal amusement — that is burnout.

How long are your lessons? How much time can you stand to be with a student? Many instructors will start shortening the old flight lesson as they become ever more burned out. Anything to get away from students. Wouldn't it be nice if students would just pay us not to fly with them, we could save them a lot of consternation.

Take a look at your log. In the beginning when you started teaching, you should see relatively short lessons. This is because you were new to instructing and found that teaching is as fatiguing for you as learning is for the student. As you gain experience and learn the basic formula for everything you have to teach, you can stand longer time in the air. Your lessons will be longer so you can make more money. When there comes a point that your sanity becomes more important than your income, meager as it is, your lessons will start to get shorter again. See if there is any correlation in your logbook between how you feel about teaching and the length of your lessons.

Patience erodes as burnout steals your ability to function. Not just your tolerance of mistakes and the constant repetition of procedures necessary for students to improve their skills, but every single peripheral interconnected piece of aviation that touches you will crawl under your skin and take up residence. Auto traffic will move slower both to and from the airport. The weather will more frequently get in your way. Imagine the arrogance of thinking that nature is getting in your way. Well, how can you make a living without the proper weather, right?

The boss will seem to provide you with more distractions, obstacles, paperwork, and aggravations on the job. Do you question policy changes without giving any consideration to the merits simply because they upset your routine? More burnout. Aircraft maintenance will appear to get worse. Whereas you may have leapt from aircraft to aircraft to find one in airworthy condition when you started this job, you will now be filled with complaints and disgust when any airplane is less than pristine.

There was a time when you would do anything to make a lesson happen. When students called to cancel at the last minute, you would simply reschedule them at the earliest opportunity. You would call any amount of students, juggling lessons to keep constantly busy. Now you couldn't be bothered, preferring instead to slap your students with a no-show fee. Patience. You will have no patience for any of this.

Any instructional problem that your students may have is amplified as you weigh in with all this burden on your shoulders. Burnout is an internal war that rages within an instructor. You will still function quite well for your students even though you seethe inside, for the mechanical formulas of instructing provide adequate cover. Every once in a while, you will lose control and your evil twin will emerge, try as you will to remain in control. You could function for a great length of time, while your patience like an ever-tighter string tears strand by strand to its core. However, all you will do is function, for your heart will not be in your work. When you run out of patience and lose control, you must leave instructing.

As patience withers, the hate builds; hate for your students and all their immeasurable faults that you have no interest in fixing; hate for a system that forces you to instruct against your will; hate for all employers who require such outrageous hours just to scrape out a living; hate for the income you didn't get but so richly deserve; hate for the respect and courtesy that is due any other true professional in our society that you never see from anyone in this business. The only thing that will exceed your hatred for teaching will be the dread of not teaching, for the only thing worse than being off the ground with a student is not being off the ground. Flying is life. Flying is breathing. Instructing is strangulation. That is our system. That is burnout.

Okay, bucko, what the hell are you going to do about this? I wrote a book. What is your plan? Positive action is always the best cure. Pilots are action people who have to make things happen. You never would have gotten this far if a little challenge destroyed your spirit. That is why the frustration of being stuck with burnout, with no foreseeable way out, is the cruelest punishment of all.

If you are burned out, do everything you can to get a real pilot job as soon as possible. The truly burned-out instructor could appear desperate to any future employer. This is neither healthy nor profitable, so get on the stick before you fall too far. You might try some diversions. Cut back on your instructing and take a part-time job completely away from aviation. Keep a few of your best students, however, to maintain proficiency.

The best thing you can do is rediscover why you got into this business in the first place. You did it to fly airplanes; the one thing an instructor is not permitted, with rare exception, to do. That is why the system can't possibly work. Everything is backwards. It keeps you from the one thing you want most and trained so hard to get and then teases you by dangling the controls in front of your face for thousands of hours while you sit and watch people who can't do it properly.

What do you consider yourself, a pilot or an instructor? If your answer is a pilot, then you will most surely become burned out some day. You have to start flying and stop watching other people fly. Find any way you can to get back in the left seat and get those controls in your hands. Go up by yourself, or with friends, and just fly for fun. You will be amazed at how quickly you will want to hand over the controls to whomever is up front with you and start teaching them.

Burnout! You can't get away from it — instructing is pure reflex by now. If you can salvage some measure of flying with your instructing, it might make your life bearable until you get a real commercial job.

Flight instructors are the greatest victims of our current system of training. That is why you read so few criticisms of instructors in this book. Unless, of course, they are lazy, corrupt, have endorsements for sale, are brown-nosing, sleeping their way to the top, dangerous, or any of that ilk. This just isn't the case very often. You would think that the FAA would take better care of their instructors. We are the backbone of the industry, or so they say. No one learns to fly an aircraft without an instructor.

The truth is that they don't care about us one bit. The only interest the FAA will ever take in instructors is when they violate the rules. There are no standards put on flight schools for working conditions, hours, wages, benefits, health insurance, or occupational safety. No one watches out for instructors. There is no union or collective bargaining unit to negotiate livable contracts for instructors. The Airline Pilots Association doesn't seem to want us. There is a flight instructor's association out there somewhere, but they haven't changed anything that I can see. The Aircraft Owners and Pilots Association will fight to save airports and aid manufacturers, but doesn't advocate on our behalf to improve our lot. What we have then is a system that uses up and throws away its instructors. That is why we have burnout — no one cares.

Sometime during your burnout, you will be ready to go for your airline transport pilot (ATP) certificate. After 1500 hours of accumulating a variety of experience, the largest block being some 1200 hours of watching, you will be eligible for the ATP. Having forgotten how to fly, it took me as many hours to adequately prepare for the ATP as it did to get my original commercial-multi-instrument certificate. This would never happen if flight instructors could maintain their flight skills at no cost to them. Had I flown the last 1200 hours instead of observed, there would have been no need to practice anything.

The act of instructing actually degrades your ability to fly the longer you engage in the activity. Why therefore do you have to wait 1500 hours until your skills are so weak? Why not lower the ATP eligibility down to 300 hours so you can earn it when you are a better pilot? If the commercial certificate actually meant anything, we could completely abandon the ATP for it would be superfluous. Pilots could then get real jobs when we have all of our enthusiasm, all of our flight skills, and none of the cynicism of burned-out instructors.

Nowadays you have to pay a lot of money to get back what instructing took away. The only redeeming quality about practicing for the ATP is that it gets you to fly an aircraft again. This is great therapy. Just the action of studying and preparing for a checkride invigorates the spirit. When was the last time you faced the possibility of not completing a checkride?

Something insidious creeps up on you as you instruct. You begin to think you know just about everything. Here is how it happens. When you question a student, you are just like examiners in that you will only ask a question for which you already have developed a pat answer. Over time the only questions you hear are your own. Your students get questions from many sources so they are used to inter-rogations from different perspectives.

Since the only questions you hear are the ones you ask, and since you know all the answers to them, you begin to think you know the answer to everything, when in real-ity you only know the answers to your own questions. You will be shocked to find out what you do not know as fellow instructors ask you their own pat questions to help you prepare for the ATP. This will shake you up a bit.

This is the story of the other side of flight training. The side you won't get from any other source except this book. It is my story, and the story of the collective wisdom and experi-ence of everyone I have come in contact with in this business. If I have caused you to think; if I have sparked some debate; if I have forced you to question everything you have been led to believe and teach; if this results in the complete review and overhaul of the way we treat both students and instructors, to make a better system; if we can do that, then I will have done my job. If not, at least you are now armed with a view from the other side. Congratulations to those who have made the full journey from your earliest dreams of flight, to completely burning out, for you will understand.

GLOSSARY

——————————————————————————————————→✈

LIST OF ABBREVIATIONS

-A-

AD — Airworthiness Directive

ADF — Automatic Direction Finder

AIM — *Airman's Information Manual*

AOPA — Aircraft Owners and Pilots Association

ARSA — Airport Radar Service Area (Class C)

ATA — Airport Traffic Area (Class D)

ATC — Air Traffic Control

ATIS — Automatic Terminal Information Service

AWOS — Automated Weather Observing System

-B-

-C-

CDI — Course Deviation Indicator

CFI — Certified Flight Instructor

CFII — Certified Flight Instrument Instructor

C.G. — Center of gravity

CTAF — Common Traffic Advisory Frequence

-D-

DME — Distance measuring equipment

-E-

ETA — Estimated Time of Arrival

-F-

FAA — Federal Aviation Administration
FAF — Final approach fix
FAR — Federal Aviation Regulations
FBO — Fixed Base Operator
FSDO — Flight Standards District Office
FSS — Flight Service Stations

-G-

GMP — See pages 357-358 (Ch. 9) for Author's terms
GMT — Greenwich Mean Time
GUMPS — See pages 355, 357-358 (Ch. 9)

-H-

-I-

IAF — Initial approach fix
IFR — Instrument Flight Rules
ILS — Instrument Landing System
IMC — Instrument Meteorological Conditions

-J-

-K-

-L-

-M-

MAP — Missed approach point
MDA — Minimum descent altitudes
MEA — Minimum en route altitude
MOA — see Military Operation Areas
MVA — Minimum vectoring altitude

-N-

NDB — Non-Directional Beacon
NTSB — National Transportation Safety Board

-O-

-P-

PCA — Positive Control Area (Class A)
PIC — Pilot in command
PTS — Practical Test Standards

-Q-

-R-

RCO — Remote Communications Outlets

-S-

-T-

TAS — True Air Speed
TCA — Terminal Control Area (Class B)
TEC — Tower en route control
The "T's" — acronym for "turn, time, twist, throttle, talk"

-U-

-V-

Va — Maneuvering speed
VFR — Visual Fight Rules
Vmc - Minimum control airspeed
VOR — Very High Frequency Omni Range

-W-

-Z-

Index

ABOUT THE AUTHOR

Greg Penglis first became fascinated with airplanes at the tender age of five while departing from Toronto, Canada, in a Vickers Viscount on his very first airplane ride. What followed was a relentless pursuit, without mercy, on unwitting parents to someday begin flight lessons.

The first lesson in Melbourne, Australia, at age twelve, while designed to intimidate the youngster, only emboldened his spirit to fly. A traumatic move to the United States interrupted events; however, the commencement of regular lessons followed shortly at the age of thirteen. Savings and odd jobs provided for one lesson per month for the next three years. He soloed exactly one week after the magic sixteenth birthday, as hail and snow storms filled the skies on the big day.

His private certificate was acquired at seventeen when he had just started college and had completely run out of money to fly, save for the summers. Timing has never been his specialty. This was back in the late 1970s when Viet Nam combat pilots were taking all the good airline jobs anyway, so a commercial career was out of the question.

Deciding to make his mark as a consumer/environmental advocate, he studied economics and environmental science. He graduated soon after President Ronald Reagan took office, which immediately put him out of the career for which he had prepared. Timing — once again.

Since the cost of becoming a commercial pilot was prohibitive, a career in the military became the next objective. A private pilot fresh out of college with two bachelors degrees made him a desirable candidate. However, at twenty-two, his nearsighted vision eliminated him from consideration.

Eight years of odd jobs, begging and borrowing money were to intervene before he had the resources for commercial pilot training. If not for his father and some very special friends, the cause would never have become a reality. He moved to California for the best training environment — right before that state had its biggest recession . . .

After two full years of flight instruction, working seven days a week, he was qualified to apply for all the airlines that were now going broke and out of business. Timing — is everything.

Most aviation books are written by the lucky superstars. This work is by one pilot, yet written for all pilots, who may have been shut out by the system. The only way to try to reach all the dreamers and pilots who are struggling both privately and commercially is to let them know that they are not at fault; the system is stacked against them. Hence — this book.

THE COMPLETE GUIDE TO FLIGHT INSTRUCTION

For additional copies of *The Complete Guide To Flight Instruction*, telephone TOLL-FREE 1-800-356-9315 or FAX TOLL FREE 1-800-242-0046. Mastercard/Visa accepted.

To order *The Complete Guide To Flight Instruction* directly from the publisher, send your check or money order for $29.95 plus $4.50 shipping and handling ($34.45 postpaid) to:
>Rainbow Books, Inc.
>Order Dept.1-T
>P. O. Box 430
>Highland City, FL 33846-0430.

For QUANTITY PURCHASES, telephone Rainbow Books, Inc., (813) 648-4420 or write to Rainbow Books, Inc., P. O. Box 430, Highland City, FL 33846-0439.